Lecture Notes in Networks and Systems

849

The series "Lecture Notes in Networks and Systems" publishes the latest developments in Networks and Systems—quickly, informally and with high quality. Original research reported in proceedings and post-proceedings represents the core of LNNS.

Volumes published in LNNS embrace all aspects and subfields of, as well as new challenges in, Networks and Systems.

The series contains proceedings and edited volumes in systems and networks, spanning the areas of Cyber-Physical Systems, Autonomous Systems, Sensor Networks, Control Systems, Energy Systems, Automotive Systems, Biological Systems, Vehicular Networking and Connected Vehicles, Aerospace Systems, Automation, Manufacturing, Smart Grids, Nonlinear Systems, Power Systems, Robotics, Social Systems, Economic Systems and other. Of particular value to both the contributors and the readership are the short publication timeframe and the world-wide distribution and exposure which enable both a wide and rapid dissemination of research output.

The series covers the theory, applications, and perspectives on the state of the art and future developments relevant to systems and networks, decision making, control, complex processes and related areas, as embedded in the fields of interdisciplinary and applied sciences, engineering, computer science, physics, economics, social, and life sciences, as well as the paradigms and methodologies behind them.

Indexed by SCOPUS, INSPEC, WTI Frankfurt eG, zbMATH, SCImago.

All books published in the series are submitted for consideration in Web of Science.

For proposals from Asia please contact Aninda Bose (aninda.bose@springer.com).

Erik Puik · David S. Cochran ·
Joseph Timothy Foley · Petra Foith-Förster
Editors

Proceedings of the 15th International Conference on Axiomatic Design 2023

Springer

Editors
Erik Puik
Professor of Smart Manufacturing
Fontys University of Applied Sciences
Eindhoven
Eindhoven, The Netherlands

Joseph Timothy Foley
Reykjavík University
Reykjavík, Iceland

David S. Cochran
Professor of Systems Engineering,
and Director
Purdue University Fort Wayne
Fort Wayne, IN, USA

Petra Foith-Förster
Stuttgart, Germany

ISSN 2367-3370 ISSN 2367-3389 (electronic)
Lecture Notes in Networks and Systems
ISBN 978-3-031-49919-7 ISBN 978-3-031-49920-3 (eBook)
https://doi.org/10.1007/978-3-031-49920-3

This Springer imprint is published by the registered company Springer Nature Switzerland AG
The registered company address is: Gewerbestrasse 11, 6330 Cham, Switzerland

Preface

In modern engineering, the ever-accelerating pace of technological innovation is matched only by the increasingly complex challenges we face. From manufacturing systems and smart robotics to healthcare technologies and aerospace designs, our efforts to innovate are set against a backdrop of intricate, interconnected systems. It is within this context that we gather for the International Conference on Axiomatic Design 2023. The Axiomatic Design community has advanced the field of more and more complex systems through research on methods for systems engineering, particularly on applications of design axioms and associated methodologies. AD has been applied by and for organizations to gain added value but also by universities to teach novice designers to produce better systems. AD has proven to be a logical and rational scientific framework for making the best decisions during the synthesis of a broad range of systems.

The aim of the ICAD is to unite scholars, practitioners, industry experts, and future leaders of the field to share research findings, best practices, and new applications of AD. Our focus is not merely to dissect individual components of systems but to understand how these components interact and coalesce to form integrated wholes. This holistic perspective is fundamental for tackling the multi-dimensional challenges of our near future and beyond, from sustainable development and cybersecurity to automation and data analytics.

We are grateful to our keynote speakers who have provided thought-provoking insights into their respective domains presented in original ways, for example, by comparing the turnover of ASML semiconductor processing systems with selling quite a few bunches of tulips. Special thanks goes to our sponsors Dr. and Mrs. Park for financially supporting the Axiomatic Design Research Foundation, without which this conference would not have been possible. I must also express gratitude to the members of the Program Committee and our dedicated team at Fontys Applied University of Sciences Eindhoven for their hard work in ensuring the success of the event.

Best wishes for an intellectually rewarding experience when reading (parts of) these proceedings. Our aspiration is that they serve as a valuable resource for further research and real-world applications of AD. The perspectives and approaches described herein are also a call to action to pass on the legacy of AD to a wider audience and future generations. And obviously, we encourage readers to engage with these contributions, collaboratively, constructively, and as always critically.

Erik Puik
Conference Chair,
ICAD2023

Contents

Axiomatic Design and Product Design

Conceptual Design and Learning Axiomatic Design

Axiomatic Design and Health

Axiomatic Design and Manufacturing

Design Decomposition for Cyber Resiliency in Cyber-Physical Production Systems

Tanel Aruväli[1](✉), Matteo De Marchi[1], Erwin Rauch[1], and Dominik Matt[1,2]

[1] Free University of Bozen-Bolzano, Piazzetta del Università 1, 39100 Bolzano, Italy
tanel.aruvaeli@unibz.it

[2] Fraunhofer Italia Research, Via A. Volta 13/a, 39100 Bolzano, Italy

Abstract. Digitalization and related networked systems integration and automation have increased the performance of manufacturing. At the same time, the vulnerability of the systems has increased significantly as networks are potential targets for attacks to compromise companies. Therefore, the study focuses on the functional design of cyber resiliency in cyber-physical production systems. To support functionality while emphasizing the resilience of manufacturing systems, Axiomatic Design is used as a design methodology for the concept design of a cyber resiliency module. Based on functional requirements, design parameters were decomposed and design guidelines for preparedness for cyberattacks were provided. The guidelines were applied to a cyber-physical demonstrator that realizes the Industrial Internet of Things with a digital twin. As a result, physical/virtual solutions for the system were found. Such an axiomatic design-based approach allowed for studying solution-neutral functional requirements that resulted in functional cyber resiliency solutions. The provided guidelines have practical value in the planning phase of manufacturing system networks to increase their long-term resiliency. This study fills the gap in the solution-neutral design of cyber resiliency in manufacturing companies.

Keywords: Axiomatic Design · Cybersecurity · Resilience · Sustainable Manufacturing · Industry 4.0

1 Introduction

In Cyber-Physical Production Systems (CPPS), cybersecurity is essentially important as the machinery and its processes are vulnerable due to network integrations. In traditional manufacturing, the link between machinery is a human. In the age of the internet of things, connectivity, remote control, and unidirectional data flow are enabled by virtual networks. Compared with physical access, digital access and intrusion to the shopfloor can be hidden, although the consequences may be even more harmful. In recent years, many companies have been attacked by threat actors and suffered while losing control over their digitally generated processes, workflow, sensitive customer data, or trade secret data. Often cyberattacks are targeted at companies that in addition to performance and credibility loss, must consider environmental impact [1].

E. Puik et al. (Eds.): ICAD 2023, LNNS 849, pp. 3–14, 2024.
https://doi.org/10.1007/978-3-031-49920-3_1

The research aims to derive design guidelines for today's intelligent manufacturing systems by decomposing and decoupling functional requirements (FRs) to derive the most inevitable design parameters (DPs) for cybersecurity purposes. More specifically, to find the concept DPs for CPPS to increase the level of resilience by applying an Axiomatic Design (AD) [2] approach. The work is limited to cybersecurity functions for preparedness for potential cyberattacks. It does not cover the avoidance of cyberattacks.

The paper is organized as follows. Section 2 explains the theoretical background of resilience, cybersecurity, and relevant AD studies. Thereafter, in Sect. 3 the research methodology AD decomposition and decoupling process is presented to derive design guidelines for resilient CPPS on cybersecurity. Section 4 presents the decomposition results used in the cyber-physical demonstrator. Finally, in Sect. 5 the results are further discussed, future perspectives found, and further research recommended.

2 Theoretical Background

2.1 Resilience and Disruptions

According to Gu et al. [3] resilience is the ability of a system to withstand potentially high-impact disruptions, and it is characterized by the capability of the system to mitigate or absorb the impact of disruptions, and quickly recover to normal conditions. In resilience, three main features and phases can be distinguished: absorption, adaptation, and restoration [4]. In the absorption phase, disruptions or the impact of disruptions is eliminated without loss in productivity. In the adaptation phase, the disruption has influenced production performance and adaptive changes are needed to restore productivity. According to the multi-criteria decision-making Analytic Hierarchy Process analysis [5], the Penalty of Change (POC), proposed by Alexopoulos et al. [6], was selected as the most practically usable resilience metric. It divides resilience into two main components: the probability of changes and the cost of changes. The method of POC originates from Chryssolouris and is calculated as follows [7, 8]:

$$\mathrm{POC} = \sum_{i=1}^{D} \mathrm{Pn}(X_i)\mathrm{Pr}(X_i) \tag{1}$$

where D is the number of potential changes, $\mathrm{Pn}(X_i)$ is the penalty (cost) of the i-th potential change and, $\mathrm{Pr}(X_i)$ is the probability of the i-th potential change to occur.

On a shop floor, disruptions can be internal such as product quality flaws or machine failures [9], or external such as pandemics, natural disasters, shortage of materials, cyberattacks, etc. [10, 11].

2.2 Cyber Resiliency

Cyberattacks are up-trending disruption sources. In addition to cyberattacks' probability to occur, also their influence has increased significantly. In the year 2022, the average ransom payment for cyber criminals to decrypt the hijacked data increased by nearly to 1 million $ [12]. Ransomware is just one type of malware. The other three most common types of malwares are viruses, worms, and Trojan horses. Malware's main goal is to get the payload delivered and installed in the victim's system. This enables a variety of network-related remote attacks to be taken.

In addition to overall resilience in manufacturing, CPPS are focusing on cyber resiliency. Cyber resiliency is the ability to anticipate, withstand, recover from, and adapt to adverse conditions, stresses, attacks, or compromises on systems that use or are enabled by cyber resources [13]. For cyber physical systems' cyber resiliency, Haque et al. [14] proposed a metric and related simulation method to automate the resilience assessment process. From a cyber resiliency perspective at the industry level, critical infrastructure-related industries have been in research focus such as the oil and gas industry [15] and power plants [16]. In the manufacturing field, cyber resiliency is mainly studied regarding additive manufacturing. Medwed et al. [17] describe the system to provide self-monitoring for IoT devices to increase their cyber resilience. Rahman et al. [18] developed an index of cyber resilience for the additive manufacturing supply chain, while Durling et al. [19] analyzed the cyber threats to additive manufacturing system security.

2.3 AD for Systems Design in Manufacturing

AD is a methodology used for systems engineering and the design of complex systems. The main pillars of AD are Suh's axioms [2]: (i) maintain the independence of the FRs and (ii) minimize the information content. The central idea of the AD is to concentrate on FRs and remain solution neutral, meaning openness for all possible solutions and technologies, rather than proposing modifications for existing solutions. The main problem (customer need) is translated in a technical language in form of a functional requirement and decomposed into multi-level FRs and corresponding design guidelines as DPs are found. The design matrix connects FR vectors with associated DP vectors (Eq. 2) [20]. Whereby, FRs must be collectively exhaustive with respect to a higher level and mutually exclusive at the same level (having no overlapping nor redundancy). The goal is to reach uncoupled or at least decoupled design matrixes. In the uncoupled matrix, the DPs are independent of each other and provide more freedom. Coupled matrixes must be avoided. Decoupled matrixes are allowed, but the implementation of design guidelines needs to follow a certain sequence in this case. The design matrix can be described as follows:

$$\{FRs\} = [A]\{DPs\} \tag{2}$$

where FRs are functional requirements, DPs are design parameters and A indicates the effect of changes of the DPs on the FRs.

Cochran et al. [21] used AD and a lean approach in manufacturing system design decomposition and provided design guidelines that are suitable for a wide variety of manufacturing systems. Later, the lean approach was extended with a sustainability view [22]. Matt et al. [23] proposed DPs for the design of scalable modular manufacturing systems. In addition, the specific parts of manufacturing systems have been studied more deeply using AD approach. Vickery et al. [24] focused on smart data analytics in manufacturing SMEs. Manufacturing systems design studies in AD approach mainly consider productivity and neglect the importance of long-term resilience. No AD approach for resilience and especially for cybersecurity requirements decomposition in manufacturing was found in the literature.

3 Resilient CPPS Design Decomposition for Cybersecurity

To increase resilience in manufacturing, the AD methodology was used to derive conceptual DPs for CPPS planning. FRs, FRs metrics and corresponding DPs were mapped. Design matrices were used to check DPs independency. POC resilience metric was used as a support for the highest-level DP decomposition. The decomposition was finalized in three upper levels. From the fourth level, only minimizing the cost caused by cyberattacks was investigated in this work.

3.1 First Three Levels Decomposition of Resilient CPPS

As during last years, manufacturing companies have suffered due to the hectic external environment, the long-term performance measure resilience was taken into focus as a customer need. According to customer need, FR0 as the highest-level functional requirement was defined as "Increase the resilience in CPPS" (Fig. 1). The metric POC was selected for measuring the goal as it considers the strong booster - economic impact of disruptions and related changes. The second reason was the practical usability of the metric. DP0 as the highest-level DP was thus defined "Resilient manufacturing system".

Considering the POC components (probability of the potential change to occur and penalty/cost of the potential change), the first level FRs were defined similarly (minimize the need for changes and minimize the cost of change caused by disruptions). From a manufacturing perspective, the cost (time) of change is influenced by preparedness for potential changes and their on-time discovery. Preparation means the ability for rapid and anticipated changes. The probability component is related to minimizing the occurrence of disruptions or even avoidance of them. Therefore, the first level parameters were found avoidance (DP1) and preparedness (DP2) for disruptions and their caused changes. In theory, if bringing one of these two components to zero, the other component could be neglected to observe. In practice, it is not possible to completely control the inputs to the system nor be aware of all possible changes a disruption can cause.

In the second level, both branches were divided between internal and external disruptions as they have different behavior. Internal disruptions are more predictable, and their occurrence is highly influenceable, while the causes of external disruptions are often out of manufacturers' reach. Thus, to avoidance of disruptions occurrence, there is a need for responsible (DP1.1) and quality manufacturing (DP1.2). For preparedness, the most influenceable external (DP2.1) and most influenceable internal disruptions (machine faults) (DP2.2) must be considered. As the range of possible disruptions is not limited, focusing on the most influenceable ones provides the best results.

In the third level, focusing on the specific system modules takes place. From this level, we continue only with FR to minimize the cost caused by cyberattacks.

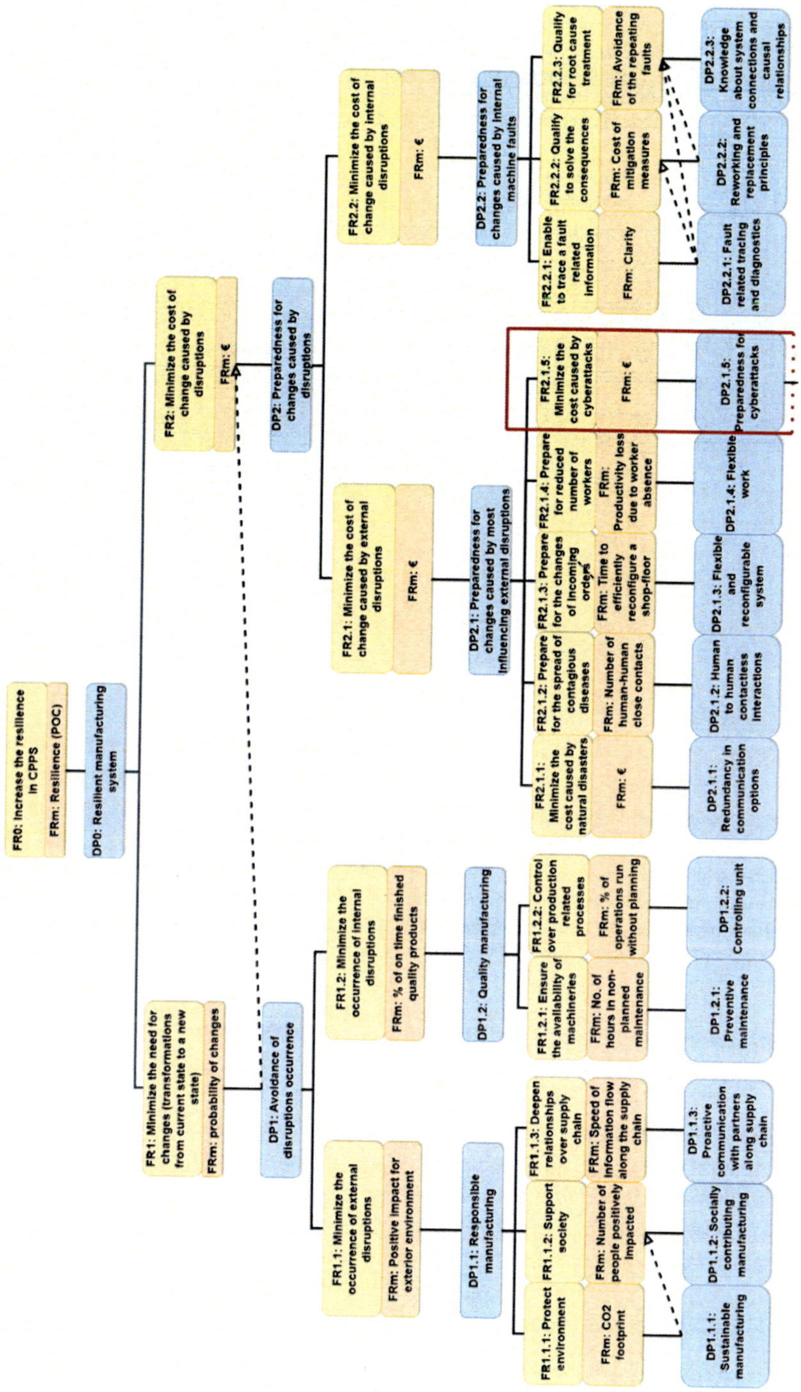

Fig. 1. Main branches of the design decomposition of resilient manufacturing systems.

3.2 Cybersecurity Decomposition

Recently, virtual networks have become one of the most vulnerable systems of the company. Protecting them against external disruptions (attacks) is more complex compared with physical resources. To minimize the cost caused by potential cyberattacks, preparation is essential. Most of the cybersecurity mitigation measures must be executed before the attack to minimize the spatial and temporal reach of the attack. This allows for minimizing the cost of changes for virtual networks and entities in these networks. Cybersecurity branch decomposition provides DPs to execute the preparation measures for virtual networks (Fig. 2).

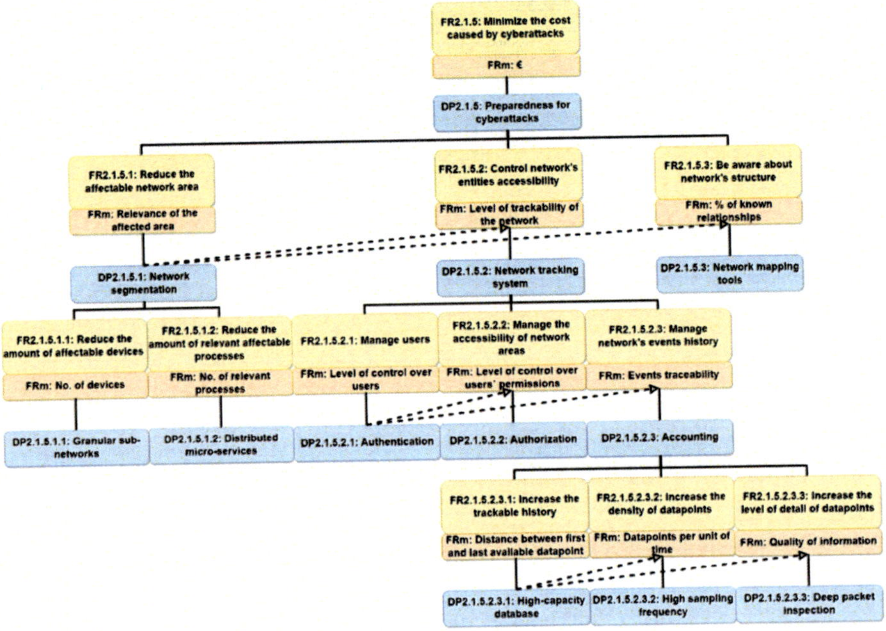

Fig. 2. Decomposition of the cybersecurity branch.

Preparedness for Cyberattacks. Preparedness and mitigation measures for cyberattacks consist of three main components: minimizing the reach of an attack, controlling the accessibility to the network, and maintaining the knowledge of the Full network structure. Network Segmentation (DP2.1.5.1) stands for dividing the network into smaller parts to limit the dimension of consequences of unauthorized access. The network tracking system (DP2.1.5.2) enables tracking suspicious events, related parties, and data packages sent and received. Network mapping tools (DP2.1.5.3) help to remain an awareness of large network structures and their relationships. Network segmentation is the prior activity for network mapping and enabling Full network control as it defines the structure of the network. Therefore, the DPs are partly decoupled (Eq. 3).

$$\begin{Bmatrix} \text{FR2.1.5.1} \\ \text{FR2.1.5.2} \\ \text{FR2.1.5.3} \end{Bmatrix} = \begin{vmatrix} X & 0 & 0 \\ X & X & 0 \\ X & 0 & X \end{vmatrix} \begin{Bmatrix} \text{DP2.1.5.1} \\ \text{DP2.1.5.2} \\ \text{DP2.1.5.3} \end{Bmatrix} \tag{3}$$

Network Segmentation. Segmentation can be realized for network entities (devices and machinery) and processes executed in the network. The network segmentation for its entities forms password-protected granular sub-networks (DP2.1.5.1.1) to reduce affectable devices in case of cyberattacks. It enables continued manufacturing of devices in other segments. Segmentation areas can be compared with physical spaces (shop floors). If one of the spaces is physically attacked, it does not affect the condition of the other spaces. In the same way for processes, distributed micro-services (DP2.1.5.1.2) allow controlling only small particles of the operations. In this way, unauthorized access can only receive limited control over the process. Granular sub-networks and distributed micro-services design matrix is uncoupled (Eq. 4).

$$\begin{Bmatrix} FR2.1.5.1.1 \\ FR2.1.5.1.2 \end{Bmatrix} = \begin{vmatrix} X & 0 \\ 0 & X \end{vmatrix} \begin{Bmatrix} DP2.1.5.1.1 \\ DP2.1.5.1.2 \end{Bmatrix} \tag{4}$$

Network Tracking System. The network tracking system's purpose is to control user rights and monitor network traffic. The users can be managed through the authentication process that controls access to the network. Authentication can be realized by using methods such as username and password combination checks, token cards, and challenges with response questions. Authorization services determine which network resources the user can access and which operations the user is allowed to perform. Accounting stands for monitoring of network traffic. Thus, it tracks who and how the network resources are used. It Records the access time and changes made in the network. The prior process is the user's authentication to enable authorization and accounting, therefore the DPs are partly decoupled (Eq. 5).

$$\begin{Bmatrix} FR2.1.5.2.1 \\ FR2.1.5.2.2 \\ FR2.1.5.2.3 \end{Bmatrix} = \begin{vmatrix} X & O & 0 \\ X & X & 0 \\ X & 0 & X \end{vmatrix} \begin{Bmatrix} DP2.1.5.2.1 \\ DP2.1.5.2.2 \\ DP2.1.5.2.3 \end{Bmatrix} \tag{5}$$

Accounting. From the network traffic monitoring perspective, the characteristics are the length of historical traffic data (DP2.1.5.2.3.1), the density of data points (DP2.1.5.2.3.2), and the completeness of the data that is recorded (DP2.1.5.2.3.3). Historical data of the traffic is beneficial to preserve as a new more advanced type of scanning method may disclose old attacks that were undiscovered. In the first phase, the threat actor establishes access to the system, gathers the data and may search for options for expanding its access area. The culmination of any attacks often arrives in later phases such as encryption of the data to request a ransom. Therefore, a high-capacity database is a prerequisite for high sampling frequency and deep packet inspection, which outcomes in the partly decoupled relationship between sixth-level DPs (Eq. 6). Sampling frequency becomes important if the collected data is not event log based, but real-time monitored. Different network monitoring tools provide packet inspection at various scales. Some tools provide only access time, the accessed user, visitors' IP address, and the type of transferred data. In a network monitoring system, there could be distinguished various data modules such as network traffic, network flows, system logs, endpoint data, threat intelligence feed, security events, etc. Deep packet inspection enables the identification of exact data packets that were transferred and provides access to their content.

$$\begin{Bmatrix} FR2.1.5.2.3.1 \\ FR2.1.5.2.3.2 \\ FR2.1.5.2.3.3 \end{Bmatrix} = \begin{vmatrix} X & 0 & 0 \\ X & X & 0 \\ X & 0 & X \end{vmatrix} \begin{Bmatrix} DP2.1.5.2.3.1 \\ DP2.1.5.2.3.2 \\ DP2.1.5.2.3.3 \end{Bmatrix} \tag{6}$$

4 Application Use Case: Cyberattacks Prevention Solutions for Cyber-Physical Demonstrator

According to AD-based decomposition of minimizing the cost caused by cyberattacks in resilient CPPS, the conceptual DPs were found in Sect. 3. Based on the conceptual DPs the physical and virtual solutions (Table 1) were found for the cyber-physical demonstrator in the learning factory 'Smart Mini Factory' at the Free University of Bozen-Bolzano.

The demonstrator consists of the following physical entities (see Fig. 3): a Montrac transfer line with three shuttles for transportation; a warehouse rack; a Universal Robot UR10 collaborative robotic arm for loading components and products from the warehouse to shuttles and manual workstation; a 3D-printer; a manual workstation with digital assistance system; and an Omron Adept Quattro fixed robot for servicing the 3D-printer. All the entities have IoT functionality which allows them to communicate with each other through the uniform communication system. Input for decision support system is provided by other virtual network entities: enterprise resource planning system, database, analytics, and simulation. The human worker in the manual workstation is in the loop of a production process. Nevertheless, manual workstation servicing processes will be executed automatically (servicing with physical components and providing step-by-step digital work instructions). The transfer line allows to the addition of up to seven workstations, which makes the demonstrator extendable.

4.1 Network Segmentation

Network segmentation's aim is to limit the potentially harmed area in the network if a threat actor should get access to the system. It can be limited by separating connected devices by the creation of multiple access-protected networks. One option to establish it is to use several gateways to physically separate the networks. Virtual segmentation allows using a single gateway that separates the gateway-connected devices into separate networks. For the demonstrator, the Endian 4i Edge X gateway was selected that supports virtual segmentation.

The second option for limiting the access area is limiting the reach of the machine-related processes. It could be implemented by dividing the services that field level entities provide into smaller parts. In this way, a threat actor cannot take over the full macro-services. For instance, the macro-service "Bring the finished products from the work

Table 1. The physical and virtual solutions for minimizing the cost caused by cyberattacks.

Design parameters area	Conceptual design parameter	Physical/virtual solution
Network segmentation	Granular subnetworks	Endian 4i Edge X gateway network segmentation module
	Distributed microservices	Recognized functions of field level entities
Network tracking	Authentication	Endian server Switchboard (multi-factor authentication and authorization)
	Authorization	
	High-capacity database	Relational SQL database
	High sampling frequency	Endian intrusion detection system
	Deep packet inspection	
Network mapping	Network mapping tool	Endian Network Awareness application

station (WS) to the warehouse" can be divided into multiple micro-services such as "Check available bins in the warehouse", "Select and book the bin in the warehouse", "Choose the optimal transportation unit", "Bring the transportation unit to the WS", "Pick the finished products from the WS", "Place the products on the transportation unit", "Choose the optimal path to the warehouse", etc. Micro-services can be realized due to frequent communication between Python script supported IoT devices and decision support system.

4.2 Network Tracking System

Remote access to the network, provided by Endian switchboard (server), is authenticated by username and password. Additionally, device type recognition can be added for authorization. The switchboard also provides permission management based on users and device types. Therefore, different users can access previously defined areas only. It provides access to the network, data aggregation and customizable dashboards for data visualization.

High-capacity database, high sampling frequency, and deep packet inspection provide additional functionality to support network tracking and accounting. A relational SQL database will be used to store network tracking data. Traditional hard drive or solid-state drive hosted databases such as PostgreSQL and SQLite are preferred over "in-cache" database such as Redis.

Zero-trust architecture for remote networks is complemented with intrusion detection system. Intrusion detection system is seen as a sensor, that detects abnormal activities in network. It works based on rules that trigger security alerts. It covers the function deep packet inspection in real-time traffic monitoring and inspection. The data acquisition frequency is based on events occurrence frequency in the network. Therefore, in this case, intrusion detection system also covers high sampling frequency function. The selected solution is Endian intrusion detection system as it connects smoothly with the system. For instance, the system provides transmission control protocol window scaling, support for untagged virtual local area network traffic, bonding mode configuration in the web user interface, and support for dynamic host configuration protocol relay. Alternative software options for deep packet inspection are network protocol analyzer Wireshark and data-network packet analyzer tcpdump.

4.3 Network Mapping

Network mapping is the visual representation of the connectivity between interconnected devices. It facilitates network connectivity management and enables to detection of all connected devices. It provides maintenance for IT infrastructure.

For network mapping, the Endian Network Awareness application with graphical user interface was selected. It provides real-time network bandwidth information with top applications in use on the network, identification of top network activities and flows (for eliminate devices or applications creating bottlenecks and enables to see historical network mapping history. The alternative non-Endian options could be Nmap, Libre NMS and NetworkMaps.

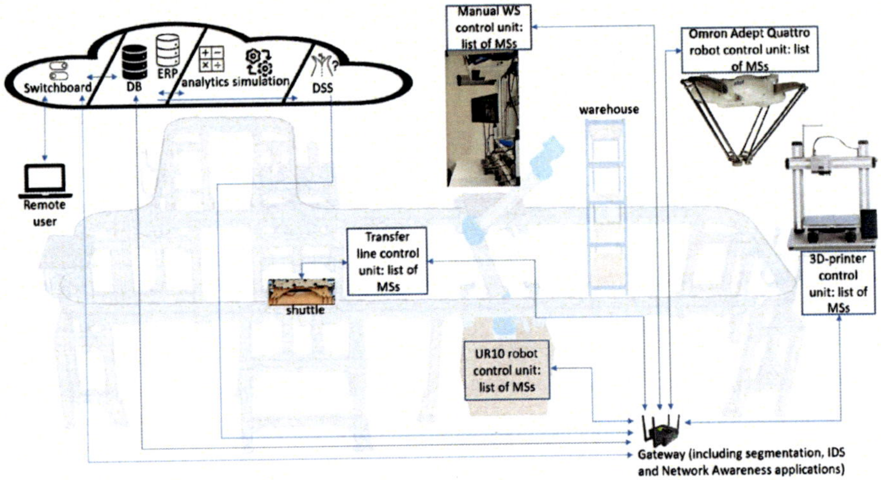

Fig. 3. Application of design guidelines in a cyber-physical demonstrator. DB – database, ERP – enterprise resource planning, DSS – decision support system, WS – workstation, UR – Universal Robot, MSs – microservices, IDS – intrusion detection system.

5 Discussion

AD theory was applied to increase the level of resilience in CPPS. The conceptual DPs of cybersecurity functions for preparedness for potential cyberattacks were derived. The DPs were applied to the Industrial Internet of Things and digital twin supported cyber-physical demonstrator. Based on this, physical and virtual solutions for the demonstrator were found.

The provided concept DPs have practical value not only for CPPS but also for traditional manufacturing systems that use virtual networks in their processes. The derived parameters facilitate in the planning phase of manufacturing system networks to increase their long-term resiliency. This study filled the gap in the solution-neutral design of cyber resiliency in manufacturing companies.

The current research focused on preparedness for disruptions in cyberattacks aspect. The other side of cyber resiliency is minimization and avoidance of the occurrence of the attack which needs further research. Additionally, the other branches (Fig. 1) need further decomposition to derive specific concept DPs from the CPPS resilience perspective.

Acknowledgment. This project has received funding from the Autonomous Province of Bozen/Bolzano, Department Innovation, research, universities and museums (ASSIST4RESILIENCE: Increasing Resilience in Manufacturing - Development of a Digital Twin Based Worker Assistance).

References

1. Mohammed, A.S., Reinecke, P., Burnap, P., Rana, O., Anthi, E.: Cybersecurity challenges in the offshore oil and gas industry: an industrial cyber-physical systems (ICPS) perspective. ACM Trans. Cyber-Phys. Syst. **6**(3), 1–27 (2022)
2. Suh, N.P.: The principles of design. Oxford Univ. Press, New York (1990)
3. Gu, X., Jin, X., Ni, J., Koren, Y.: Manufacturing system design for resilience. Procedia CIRP **36**, 135–140 (2015)
4. Tran, H.T., Balchanos, M., Domerçant, J.C., Mavris, D.N.: A framework for the quantitative assessment of performance-based system resilience. Reliab. Eng. Syst. Saf. **158**, 73–84 (2017)
5. Aruväli, T., De Marchi, M., Rauch, E.: Analysis of quantitative metrics for assessing resilience of human-centered CPPS workstations. Sci. Rep. **13**(1), 2914 (2023)
6. Alexopoulos, K., Anagiannis, I., Nikolakis, N., Chryssolouris, G.: A quantitative approach to resilience in manufacturing systems. Int. J. Prod. Res. **60**(24), 7178–7193 (2022)
7. Chryssolouris, G.: Manufacturing Systems Theory and Practice, 2nd edn. Springer-Verlag, New York (2006)
8. Chryssolouris, G., Lee, M.: An assessment of flexibility in manufacturing systems. Manuf. Rev. **5**(2), 105–116 (1992)
9. Jin, X., Gu, X.: Option-based design for resilient manufacturing systems. IFAC-PapersOnLine **49**, 1602–1607 (2016)
10. Murino, G., Armando, A., Tacchella, A.: Resilience of cyber-physical systems: an experimental appraisal of quantitative measures. In: 2019 11th International Conference on Cyber Conflict (CyCon) vol. 900, pp. 1–19 (2019)
11. Okorie, O., et al.: Manufacturing in the time of covid-19: an assessment of barriers and enablers. IEEE Eng. Manage. Rev. **48**, 167–175 (2020)

12. Palo Alto Networks Blog https://www.paloaltonetworks.com/blog/2022/06/average-ransomware-payment-update/. Accessed 12 Jan 2023
13. Ross, R., Pillitteri, V., Graubart, R., Bodeau, D. & McQuaid, R.: Developing cyber-resilient systems: a systems security engineering approach. In: No. NIST Special Publication (SP) 800–160, vol. 2 (Draft), Natl. Inst. Stan Technol. (2021)
14. Haque, M. A., Shetty, S., Krishnappa, B.: Cyber-physical system resilience. In: Complexity Challenges in Cyber Physical Systems: Using Modeling and Simulation (M&S) to Support Intelligence, Adaptation and Autonomy, p. 12301 (2019)
15. Beteto, A., et al.: Anomaly and cyber fraud detection in pipelines and supply chains for liquid fuels. Environ. Syst. Decisions **42**, 306–324 (2022)
16. Ghiasi, M. et al.: A comprehensive review of cyber-attacks and defense mechanisms for improving security in smart grid energy systems: past, present and future. Electr. Power Syst. Res. 215, (2023)
17. Medwed, M., Nikov, V., Renes, J., Schneider, T., Veshchikov, N.: Cyber resilience for self-monitoring IoT devices. In: 2021 IEEE International Conference on Cyber Security and Resilience (CSR), pp. 160–167 (2021)
18. Rahman, S., Hossain, N.U.I., Govindan, K., Nur, F., Bappy, M.: Assessing cyber resilience of additive manufacturing supply chain leveraging data fusion technique: a model to generate cyber resilience index of a supply chain. CIRP J. Manuf. Sci. Technol. **35**, 911–928 (2021)
19. Durling, M. R. et al.: Model-based security analysis in additive manufacturing systems. In: Proceedings of the 2022 ACM CCS Workshop on Additive Manufacturing (3D Printing) Security, pp. 3–13 (2022)
20. Cosmin, R., Stavarache, Ermolai, V., Irimia, A. & Etcu, M.: Using axiomatic design principles for the development of a device to measure the positioning error of an industrial robot. In: IOP Conference Series: Materials Science and Engineering. vol. 1174(1), p. 012018, IOP Publishing (2021)
21. Cochran, D.S., Arinez, J.F., Duda, J.W., Linck, J.: A decomposition approach for manufacturing system design. J. Manuf. Syst. **20**, 371–389 (2001)
22. Cochran, D. S., Hendricks, S., Barnes, J. & Bi, Z.: Extension of manufacturing system design decomposition to implement manufacturing systems that are sustainable. J Manufact. Sci. Eng. Trans. the ASME **138**(10), (2016)
23. Matt, D.T., Rauch, E.: Design of a network of scalable modular manufacturing systems to support geographically distributed production of mass customized goods. Procedia CIRP **12**, 438–443 (2013)
24. Vickery, A. R., Rauch, E., Rojas, R. A., Brown, C. A.: Smart data analytics in SME manufacturing – an axiomatic design based conceptual framework. In: MATEC Web Conferences, vol. 301, pp. 1–11 (2019)

Smart Mobile Factory Design Decomposition Using Model-Based Systems Engineering

Ali Asghar Bataleblu(✉), Erwin Rauch, Andrea Revolti, and Patrick Dallasega

Industrial Engineering and Automation (IEA), Faculty of Engineering, Free University of Bolzano-Bozen, Bolzano, Italy
Aliasghar.Bataleblu@unibz.it

Abstract. Nowadays, providing an automatic agile process in the design processes relying on Model-Based Systems Engineering (MBSE) to speed up innovation creation as much as possible is a progress key as well as a survival factor in the competitive industrial environment. Therefore, companies should make a cultural shift from traditional document-based information exchange and iterative time-consuming serial design procedures, to communicate the information based on visual modeling in a common language such as SysML, which is easier to follow. In this respect, although the capability of Axiomatic Design (AD) in product work breakdown structure has been proven, from stakeholders' needs to functional requirements and physical solutions, it seems that now is the time to automate and speed up this critical process in the product life cycle practically using developed MBSE tools. That means, when changes occur, updating a model is more straightforward than documents that require manual revisions of tables, glossaries, requirements, etc. To show the application of such a work, this paper proposes the AD of a smart mobile Hyperloop transportation factory through requirements modeling and analysis in the Cameo System Modeler software. As the main goal of the project is the decentralization of producing tube elements, and easily disassembling and building up again along the planned track/construction side, the AD is focused on the mobile factory than the Hyperloop system. Results illustrate how MBSE could alleviate difficulties in dealing with AD problems in real-world complex applications with lots of requirements.

Keywords: Model-Based Systems Engineering · Cameo System Modeler · Axiomatic Design · Smart Mobile Factory · Hyperloop Transport System

1 Introduction

The main idea of Smart Mobile Factories (SMFs) relied on industries that could operate in remote areas with limited logistical capabilities. Using SMFs and operating locally can gain competitive advantages by reducing logistics efforts and costs while improving operational efficiency. As SMFs can install, implement, and disassemble in nearby operational platforms, parts can be produced directly wherever the need arises without having to wait for them to arrive from a supplier or central storage. Overall, a wealth

E. Puik et al. (Eds.): ICAD 2023, LNNS 849, pp. 15–25, 2024.
https://doi.org/10.1007/978-3-031-49920-3_2

of potential applications considering sustainability factors can be provided through the SMFs. A systematic literature review on modular and mobile facility location problems is done by Eduardo and Udo in [1]. According to [1], to provide a more efficient response to today's markets, more flexible networks have to be proposed by addressing the inclusion of modular units to consider fully mobile units. As the situation of flexibility in factories' planning horizon shows (see Fig. 1), flexibility directly depends on the degree of mobility [2]. After the idea of "factory in a box" as a solution to move toward SMFs (i.e., manufacturing small-scale components in a container, see Fig. 2), now it is time for emerging concepts for the additive manufacturing of prefabricated parts made of concrete or other materials for real-world industrial applications [3, 4].

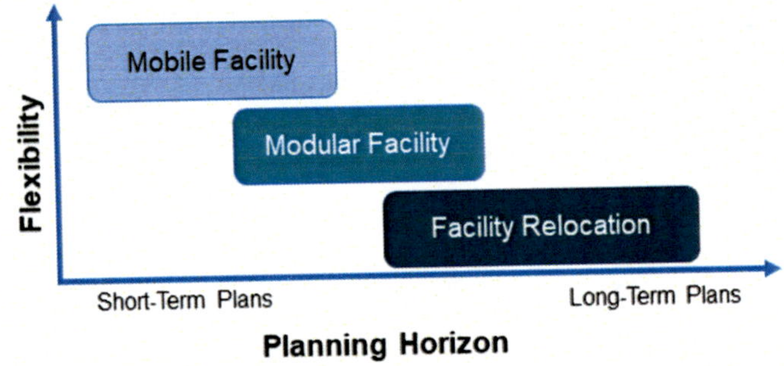

Fig. 1. Flexibility in factories' planning horizon [1]

Fig. 2. Factory in a box as a solution for SMFs [4]

In recent years, pandemic problems such as the COVID-19 crisis remarkably revealed supply chain vulnerabilities. The manufacturing industry strongly persists in promoting the expectations of previous years and a strong trend toward intelligent reindustrialization and local production can be seen these days. More and more companies are striving to alleviate supply chain difficulties due to geographically distant suppliers through SMFs [4]. At the same time, significant attention to the systems engineering field relying on a system thinking mindset will speed up systems design in a product lifecycle and it is expected to increase productivity through efficiency gains and thus gradually reduce the gap between design and manufacturing.

This research aims to illustrate how Model-Based Systems Engineering (MBSE) tools like Catia Magic can be used to decompose the functional requirements of complex systems. Therefore, producing infrastructure elements of the Hyperloop Transportation System (HTS) as an SMF is proposed.

In the following, first, a brief overview of the SMF of the HTS project at the Free University of Bozen-Bolzano is presented. Then, the problem definition and formulation as an Axiomatic Design (AD) are introduced in the next section. After that, the application of the Cameo Systems Modeler as part of Catia Magic in the automation of the AD process and related results are highlighted and proposed. The final section provides the conclusions of this research.

2 Mobile Smart Factory for Hyperloop Construction

The Hyperloop concept was born in 2013 when tech entrepreneur Elon Musk published a white paper on the subject [5] that focused on environmentally friendly goods and passenger transport. The Hyperloop's propulsion system is generated by a linear electric motor powered by renewable energy sources. Magnetic levitation is eco-friendly, consumes less energy, and causes no emissions. Eurotube Foundation [6] a non-profit research institution from Switzerland has developed a patent that envisages building the tube infrastructure using concrete instead of metal alloys. Within the joint research project Smart Mobile Factory for Infrastructure Projects (SMF4INFRA) between the Eidgenössische Technische Hochschule Zürich (ETH Zürich) and the Free University of Bozen-Bolzano, a prototype for a smart mobile factory to deliver material for the construction of hyperloop infrastructure is developed. Using a mobile factory in a linear construction site, with wide-ranging routes, allows for erecting the infrastructure sustainably. Moving the production factory of the individual pipe components while remaining close to the construction site's progression helps guarantee economic and ecological sustainability. Within the SMF4INFRA project, the physical mobile factory will be designed (Fig. 3) and its Digital Twin will be developed to ensure environmental sustainability during the construction of the hyperloop infrastructure project.

3 Axiomatic Design Decomposition

Axiomatic Design (AD) was developed by Nam P. Suh in the mid-1970s in the pursuit of developing a scientific, generalized, codified, and systematic procedure for design. AD uses the following four domains:

1) The customer domain where the customer wishes are described as so-called customer needs (CNs);
2) The functional domain where CNs are translated into functional requirements (FRs) as well as design constraints (Cs);
3) The physical domain where physical solutions (PSs) (or design parameters (DPs)) are derived that meet the previously defined functional requirements and

Fig. 3. First concept of the Smart Mobile Factory for Hyperloop construction

4) The process domain, where the DPs are transformed into real process variables (PVs).

The scientific theory gets its name from two axioms in AD that must be respected [7].

- The first is the Independence Axiom: Maintain the independence of the functional elements, i.e., avoid coupling in the system (e.g., avoiding dependencies between the DPs and other FRs).
- The second is the Information Axiom: Minimize the information content: select the solution with the least information content, i.e., that has the highest probability of success.

To apply these axioms, parallel functional and physical hierarchies are constructed, the latter containing the physical design solutions. The benefit of AD is that the designer learns how to construct large design hierarchies quickly that are more structured, thus freeing more time for mastering applications [8].

In the initial workshop on AD at Smart Mini Factory Lab. at Unibz, requirements and so-called CAs of the SMF4INFRA project were collected. Based on these inputs, FRs and Cs are defined and design parameters for a redesign were derived in an AD top-down decomposition and mapping process. The AD steps that have been carried out are as follows:

Step 1: Problem Formulation.
Step 2: Elaborate use cases into steps.
Step 3: Identify customer needs.
Step 4: Translate Needs and Use Case Steps to FRs and FRms.
Step 5: Generate Physical Solutions alternatives.
Step 6: Design decomposition – choose PS to achieve FR.

Figure 4 presents the result of these six steps for decomposing the design of a smart mobile factory for hyperloop infrastructure into 4 levels (Level 0 to Level 3). The design team has checked the independence axiom using the design matrix for each level to achieve an uncoupled or at least decoupled design.

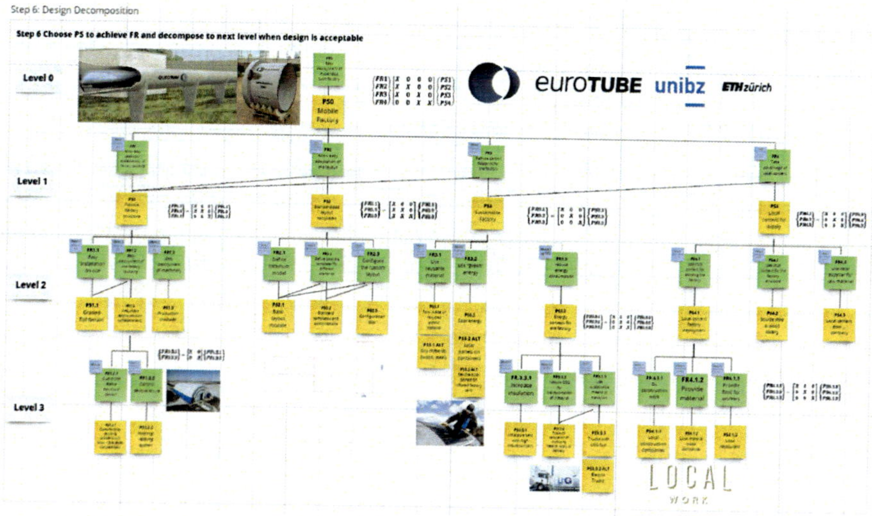

Fig. 4. Overview of the Axiomatic Design based decomposition of FR and PS

4 Model-Based Systems Engineering (MBSE) Using Catia Magic

Model-based systems engineering tools like Cameo System Modeler (developed by No Magic Inc. Which was purchased by Dassault Systems company in 2018 and is now part of Catia Magic) are suitable solutions for software architectures and operational processes. Requirement management is one of the features of this tool which provides capabilities as follows for users (see Fig. 5) [9]:

- Creating requirements
- Importing text-based requirements
- Requirements decomposition
- Requirements numbering
- Requirements gap and coverage analysis
- Tracing requirement changes in Teamwork Cloud
- Requirements verification
- Visualize and analyze.

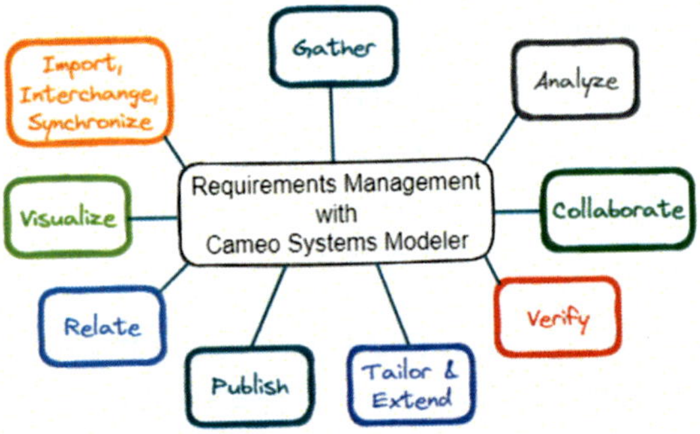

Fig. 5. The main features of Cameo Requirements Management [9]

Requirements can easily be visualized through the Requirement Diagram and Requirements Table by creating and importing them into the modeling tool. But before diving into the requirements, the structure of the problem can be modeled with blocks which here SysML Block Definition Diagram (BDD) plays an important role in this software. Using this part, you can see the problem's overall work breakdown structure and decide on decomposition and interaction between different blocks. In other words, system hierarchy from system to sub-systems and the specification of software, hardware, or human elements can be represented by blocks [9]. Figure 6 illustrates the SMF structure of the Hyperloop system using BDD, which comprises two Work Packages (WPs): the physical factory and its Digital Twin (DT).

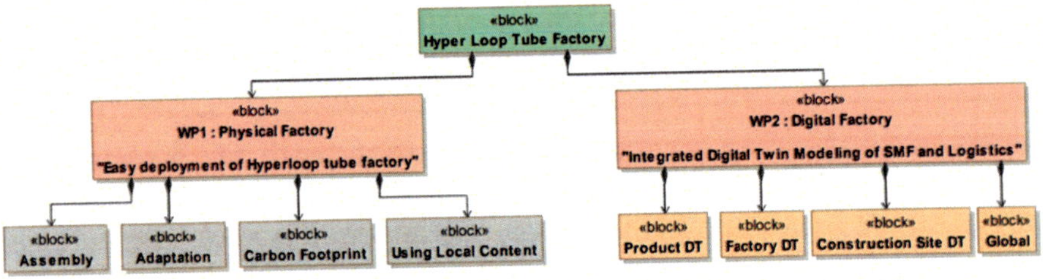

Fig. 6. The structure of the SMF of the Hyperloop project in the BDD

After creating the work breakdown structure of the problem from the system to the subsystem level, it is time to import predetermined FRs and PSs from Excel sheets to the Cameo software. This process can be done from the beginning in the software. However, importing and exporting requirements with different text-related software using Cameo is an advantage. All requirements can be easily updated in tables and diagrams just by copying and pasting them into the requirements table by a predetermined template. Figure 7 illustrates the FRs table based on related Excel sheet requirements. The requirement Diagrams are a valuable tool to provide a bridge between traditional requirement management tools and other SysML models. They are for demonstrating traceability from the requirements to the elements that are dependent on them. The FRs and PSs diagrams are represented in Fig. 8.

Such modeling can be done for PSs and finally, the relation between FRs and PSs can be shown and checked by providing a diagram including both (Fig. 9). One of the advantages of a requirements diagram like Fig. 9 is that the user can create any FRs and PSs and just link them together and by updating the software, the changes can be saved in other tables that could be exported for other usages.

We can use the Requirement Containment Map (RCM) and Requirement Derivation Map (RDM) to review, analyze, and decompose the Requirements. In these decompositions as trees, the RDM displays the decomposition of requirements related to the derived relationship. Figures 10 and 11 show the RDM and RCM of the SMF of the Hyperloop infrastructure project respectively. The user can determine the level/depth of decomposition to display results.

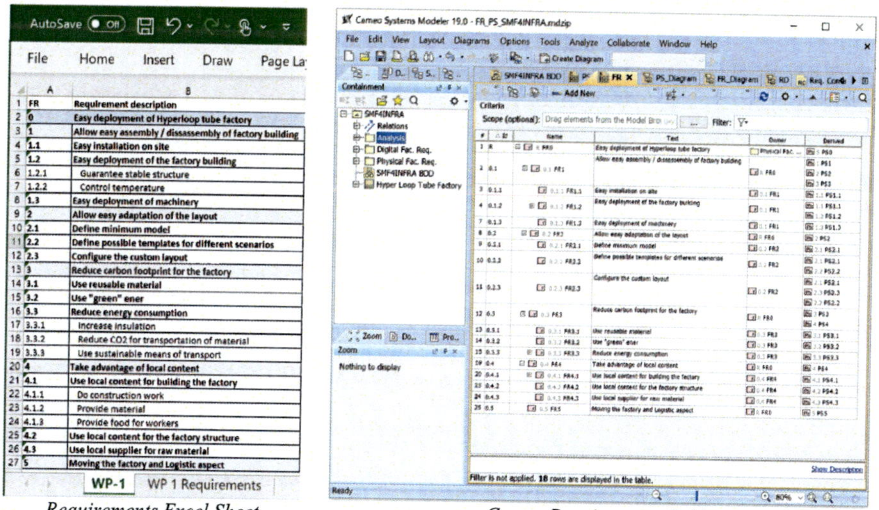

Requirements Excel Sheet *Cameo Requirements Table*

Fig. 7. The FRs table in Cameo

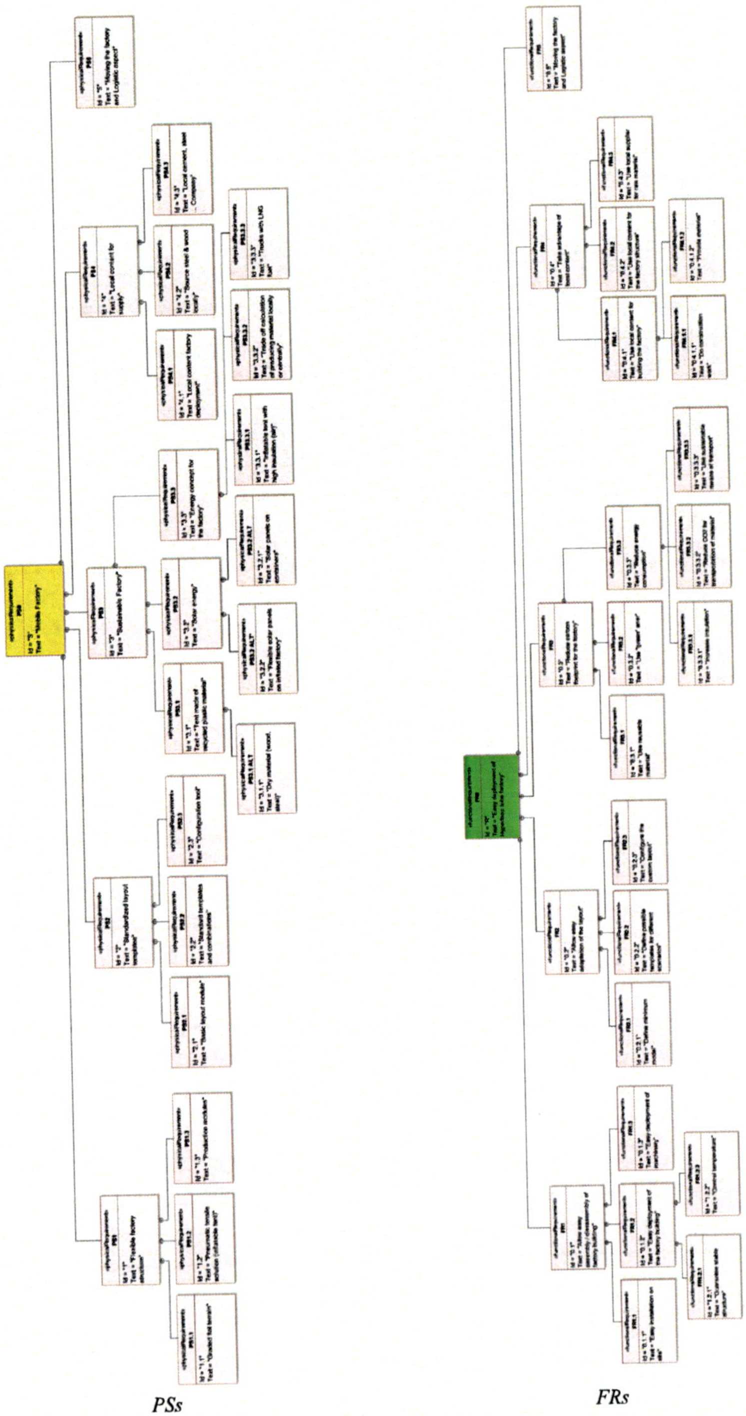

Fig. 8. The FRs and PSs diagrams in Cameo

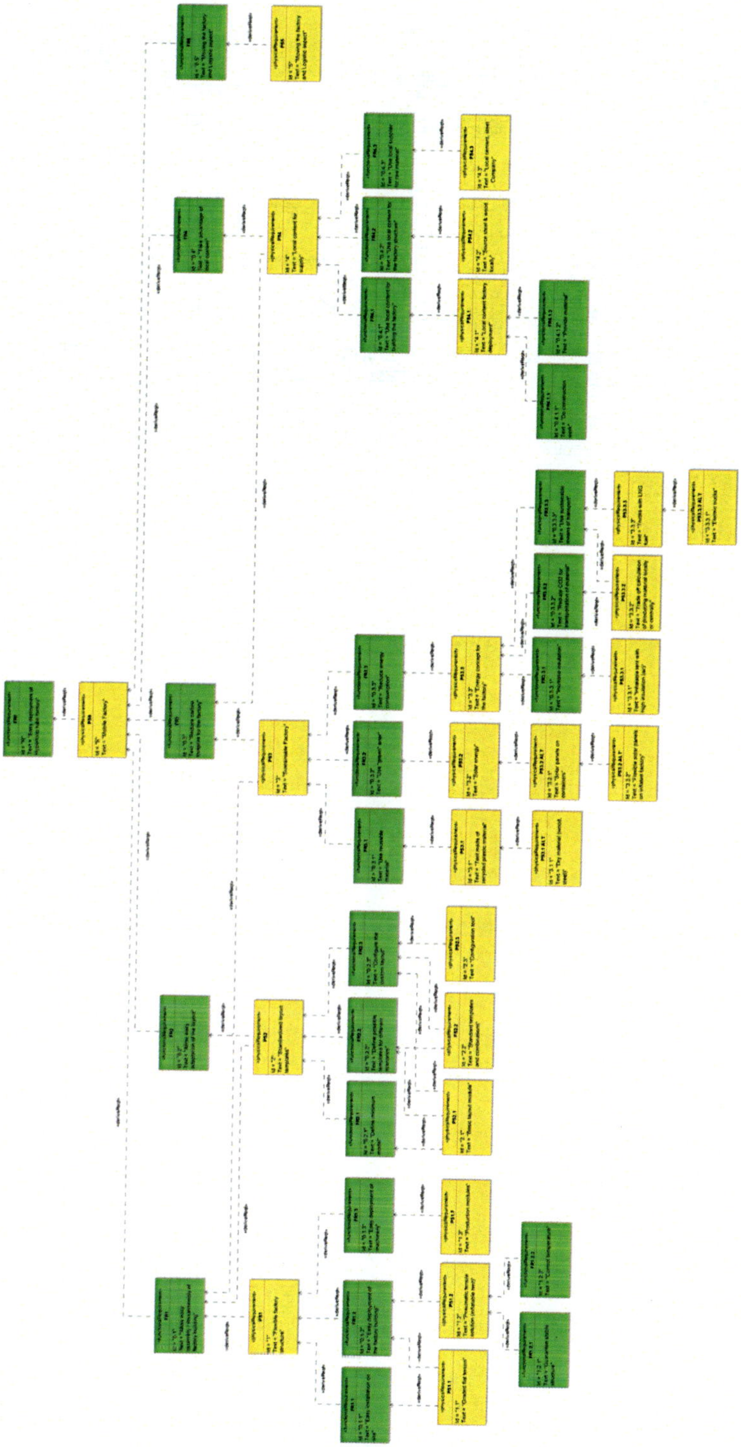

Fig. 9. The relation between FRs and PSs in the Cameo requirements diagram

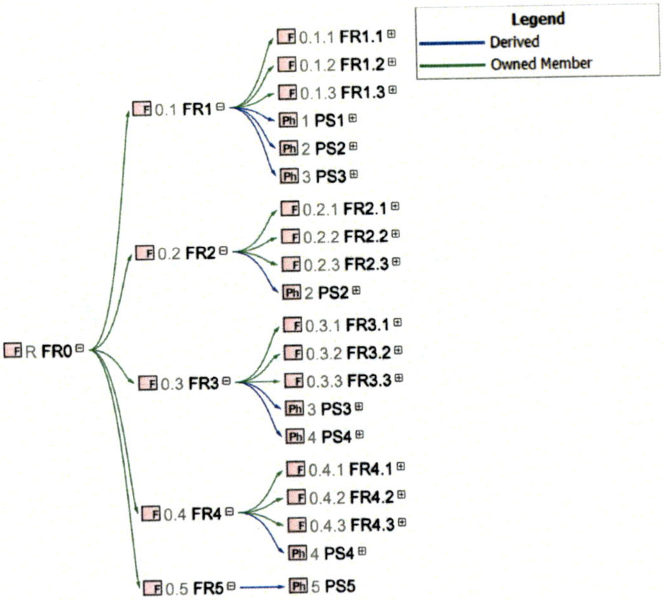

Requirement Derivation Map (depth 2)

Fig. 10. The requirement derivation map of SMF4INFRA in Cameo

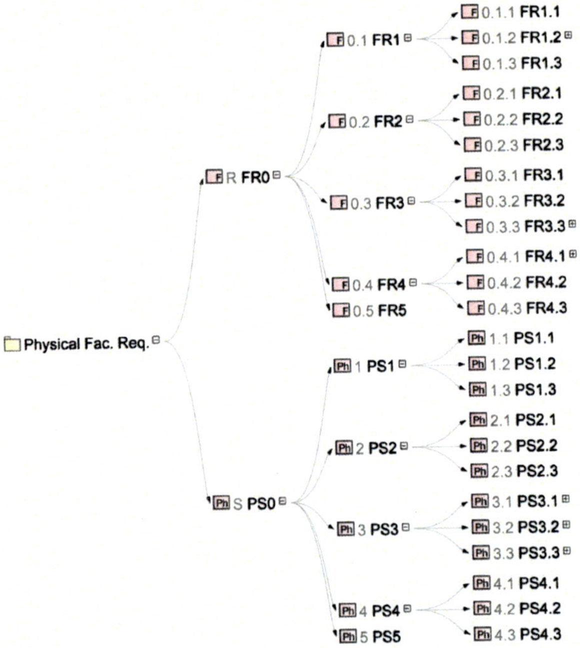

Requirement Containment Map (depth 3)

Fig. 11. The requirement containment map of SMF4INFRA in Cameo

5 Conclusion and Outlook

In this paper, we have introduced model-based systems engineering as an effective tool for alleviating difficulties with axiomatic design problems in dealing with real-world problems which include remarkable and inevitable interdisciplinary interactions in different levels of the system of systems, systems, and sub-systems. To further illustrate the capabilities of such software, the application of Cameo Systems Modeler as part of CATIA MAGIC software for AD of an SMF is proposed. In this respect, one of the projects (i.e. SMF4INFRA) that is defined between ETH Zürich and Unibz to deliver material for the construction of hyperloop infrastructure is presented and the final result of the AD which has been done in a workshop at Unibz is demonstrated. The automation of the AD process using MBSE tools helps managers and systems engineers to easier gather requirements through teamwork procedures and update, trace, and analyze them online. Having worked closely with the requirements management methodology presented herein, it could be expanded for the digital twin work package of the SMF4INFRA and bring the digital model as close as possible to the physical model.

Acknowledgments. The research presented in this article was carried out within the research project "Smart Mobile Factory for Infrastructure Projects (SMF4INFRA)", which has received funding from the Autonomous Province of Bolzano-Bozen as a Joint-Project South Tyrol – Switzerland 2021. The authors would also like to thank the involved company Eurotube for its valuable contribution from a practitioner's perspective to the research project.

References

1. Alarcon-Gerbier, Eduardo, Udo Buscher. Modular and mobile facility location problems: A systematic review. Comput. Ind. Eng. (2022): 108734
2. Fergani, C., Idrissi, A. E. B. E., Marcotte, S., Hajjaji, A., Optimization of hyperconnected mobile modular production toward environmental and economic Sustainability. Environ. Science Pollut. Res., (2020), **27**(31), 39241–39252
3. Gerrit Placzek, Veränderungen in der Bauproduktionsstrategie durch die Integration der additiven Fertigung, Beiträge zum 31. BBB Assistent:innentreffen vom 12. bis 14. July 2022 in Innsbruck
4. 3DStartup: Mobile Smart Factory Production Module Promotes Local Manufacturing, 3D printing news Interviews, April 13, 2021. https://www.3dnatives.com/en/3dstartup-the-mobile-smart-factory-production-module-130420214/
5. https://www.tesla.com/sites/default/files/blog_images/hyperloop-alpha.pdf
6. https://eurotube.org/
7. Suh, H.P.: Axiomatic Design: Advances And Applications Mit-Pappalardo Series In Mechanical Engineering. Oxford University Press, USA (2001)
8. Brown, C.A.: Teaching axiomatic design to engineers—Theory, applications, and software. J. Manuf. Syst. **24**(3), 186–195 (2005)
9. Cameo Systems Modeler 19.0 LTR, SP4, User Manual, No Magic, Inc., a Dassault Systems company, 2020

Optimizing Automated Manufacturing Processes Using Axiomatic Design Methods

Michael J. Browne(✉)

Department of Robotics Engineering, Worcester Polytechnic Institute, 100 Institute Road, Worcester, MA 01609-2280, USA
mjbrowne@wpi.edu

Abstract. Automating industrial manufacturing processes is a task that is often easier said than done. Due to emergent behaviors in both the end item being assembled, and in the robotic assemblers themselves, it is not uncommon for a change in one aspect of the design of the combined system (product + robot assembling the product) to have unintended impacts in seemingly unrelated areas of the overall system. These emergent behaviors are usually the result of poor or incomplete mapping of all the interactions between all the characteristics of the system. However, the axiomatic method provides the tools necessary to not only begin mapping these interactions, but to also confirm that all the requirements of the system have been met and are organized in an optimal way.

This paper aims to objectively analyze a hypothetical automated manufacturing environment, and all the aspects of its design that will be necessary for it to succeed in its mission of generating profit for the company that operates it. Currently, factories are often designed after-the-fact, after a product has been developed, and all manufacturing processes are tailored to suit it. Any defects or inefficiencies in a process are dealt with reactively, after they have already had a financial impact on the company. Instead, this paper proposes designing product and manufacturing process concurrently by utilizing the axiomatic design method, and by doing so, it becomes possible for interactions to be fully mapped and understood before anything—product or manufacturing tools—is built. By doing things this way, this paper shows that it then becomes possible to better utilize available robotic manufacturing tools & processes.

Keywords: Manufacturing · Automation · Robotics · Axiomatic Design · Process Design

1 Introduction

Manufacturing is an inherently complicated endeavor. Previous efforts by academia to contribute to the body of knowledge regarding manufacturing are often ignored by those working in a factory, if not outright rejected. While there

E. Puik et al. (Eds.): ICAD 2023, LNNS 849, pp. 26–65, 2024.
https://doi.org/10.1007/978-3-031-49920-3_3

have been methods proposed to increase the collaboration between academia and industry, and there are benefits to be had from such collaborations [1], a successful and deep collaboration between established for-profit companies and non-profit universities remains the exception, rather than the rule. Instead, factories tend to look inwards when solving their problems, and if they feel the need to seek outside information and expertise, they reach for a trade journal before they reach for an academic one. When developing best practices in the factory, empirical observations are used almost exclusively, and any proof they may have is based entirely on statistics of past events. This means that any practices developed this way are only "best" until another corner case is discovered or a new, more efficient method is developed. Methods developed in this way are purely reactionary, and while these observation-based methods can be made to work with manual manufacturing processes, where a human is involved in every step of the process, they begin to break down as humans are removed from the manufacturing cycle. The problem is that robots and other automated manufacturing methods can only do what they are told to do, and this requires the task to be automated to be fully defined in advance (including all corner cases).

The goal of this thesis is to lay out the argument in favor of utilizing Axiomatic Design to facilitate the automation of manufacturing processes. To that end, this thesis has two prongs: 1. Manufacturing processes can be more efficiently designed with Axiomatic Design methods than they can be with existing methods that seek to improve established processes after the fact; and 2. Robots can be better designed via Axiomatic Design methods. Taken together, this thesis makes a case that when designing automated manufacturing processes, utilizing Axiomatic Design methods will yield better results than more traditional engineering design methods.

Axiomatic Design is a rigorous design method that can quantify all aspects of a problem, and identify how they interact with one another [2]. By using the Axiomatic Design method - ideally from product conception with the customer - all aspects of a product can be objectively quantified and related to one another prior to ever drawing, designing, or building anything. In turn, in the context of the factory, this allows for all production tools - including robotics - to be identified and designed alongside the product itself.

Automated manufacturing processes are extremely complicated systems, where the factory's hardware and software must be tuned to perfectly produce the specified product in a reliable and repeatable manner. This is much easier said than done. With manual production cycles, the human laborers at each step can unconsciously work around the small variability in the parts that arrive at their bench. With a manual process, if a hole is a fraction of a millimeter off from the specified location, but still aligns with the rest of the assembly overall, the laborer installs the screw without even noticing and moves on to the next step. With the same issue on an automated process, the robot may crash as it aims for a location where there is no hole, causing both lost time and product, as well as impacting management's perception about the advantages of automated production environments. In order to successfully automate a production process, all of the aspects of the process must be accurately and precisely quantified, including all tolerances and potential failure modes.

One potential way to rigorously quantify all aspects of a production process is to use Axiomatic Design to break down all production requirements into their smallest components [3], map them to their matching physical parameters, and identify all interactions between these requirements and parameters (both intended and unintended interactions). By developing this Axiomatic Design matrix of design aspects, the whole system can be objectively evaluated for faults and risks, and all in advance of any tools being built, purchased, or deployed. Axiomatic Design has the potential to eliminate (or at least reduce) the need for continuously improving a production cycle, and can be used to minimize continuous operating costs earlier in the product's lifetime. But Axiomatic Design is not without its drawbacks.

The primary challenges with Axiomatic Design are the required up-front buy-in from management on a new design and project management philosophy (over established and accepted ones, like Six Sigma), and the significant amount of time spent up-front on designing the system on paper. Axiomatic Design cannot be shoehorned in after the fact, not without a major redesign effort, and it does not do any good if the process is not followed through to final delivery. Unfortunately, this significant up-front investment of time and effort—with nothing to show but work on paper—represents a risk to modern business thinking: if a product design effort fails, then all this time and money is viewed as wasted, with no return on investment. Every business owner wants a product to sell at the end of the day. But Axiomatic Design actually is a method used to reduce risk.

However, by taking the time to identify all problems in advance, so that they may be solved in concert with one another (instead of 'in series' as is typical with a lot of design efforts), a design team can increase their odds at arriving at a successful solution. It becomes possible to not only understand the full scope of a design effort before any CAD or calculus is done, it becomes possible to identify which problems have a lot of room to maneuver their solutions, and which have very narrow paths to success (see Fig. 2 and its relevant explanation for more information). With all of this in mind, the objective of this thesis is to prove that an automated manufacturing process can be designed using Axiomatic Design methods, and that these methods can identify the challenges of automated manufacturing and how they interact with one another.

1.1 Customer Needs

1.1.1 Manufacturing The primary role of the factory is to build the products that make the company its money. Market forces determine what a product sells for, so the factory's role in maximizing profits is to minimize its own costs. This means minimizing downtime, minimizing material loss, minimizing rework, minimizing production cycle time, and maximizing the number of products that can be in-work simultaneously. More simply put: efficient management of a factory dictates that products should be built perfectly the first time, with as few interruptions and delays as possible.

Currently, factories achieve these minimizations by reacting to issues and failures as they are discovered. There are many different methods that can be used to react to production failures in a consistent way - Lean Six Sigma [4], Continuous Improvement (Kaizen) [5], Total Quality Management (TQM) [6], Plan-Do-Check-Act (PDCA) [7], and 5-Whys [8], among countless others - but all of them, by their very nature, are attempting to find their solutions after the fact. They are not capable of proactively improving or optimizing any production processes. In order to be proactive in the factory, the problem being faced must be completely quantified and defined so that an effective solution/improvement can be designed and deployed.

Alternatively, Axiomatic Design seeks to eliminate the need to improve at all, and instead 'deliver perfect' at the very start of production. To borrow terminology from manufacturing: production engineers seek to increase the "first pass yield" of their products, to build as many products successfully the first time as possible, and to do this, they are always looking to improve their processes; Axiomatic Design seeks to improve the improvement process itself. By aiming to improve the "first pass yield" of the improvement processes themselves, rather than the products, Axiomatic Design is able to get closer to the root of the problems facing production. It is able to do this because the Axiomatic Design method itself is very flexible; it can be applied to anything that can be designed, not just hardware and software, but methodology as well. While Axiomatic Design often requires a larger up-front investment of non-recurring engineering time, it can be used to either optimize the manufacturing cell structure itself to decrease intra-factory lead times, or it can be used to design the processes themselves, so that recurring time expenditures can be minimized [9,10].

$$\begin{bmatrix} a_1 & 0 & 0 \\ 0 & b_2 & 0 \\ 0 & 0 & c_3 \end{bmatrix} \quad (1) \qquad \begin{bmatrix} a_1 & 0 & 0 \\ b_1 & b_2 & 0 \\ c_1 & c_2 & c_3 \end{bmatrix} \quad (2) \qquad \begin{bmatrix} a_1 & a_2 & a_3 \\ 0 & b_2 & b_3 \\ 0 & 0 & c_3 \end{bmatrix} \quad (3)$$

An uncoupled matrix A decoupled lower-triangular matrix A decoupled upper-triangular matrix

The primary way that Axiomatic Design ensures that all interactions are accounted for is the use of linear algebraic matrices. Specifically, by organizing "Functional Requirement" and "Design Parameter" (FR-DP) pairs into either a diagonal matrix (ideal) or triangular matrix (acceptable), it becomes possible to prove mathematically that a design is viable - including to what degree it is viable. Because multiplying diagonal matrices is commutative (If A is diagonal, and B is diagonal, then $C = AB = BA$), and multiplying two like-triangular matrices results in a third like-triangular matrix (multiplying two upper-triangular matrices together results in a third upper-triangular matrix of identical dimensions, *or* two multiplying lower-triangular matrices together results in a third lower-triangular matrix of identical dimensions) [11]. This

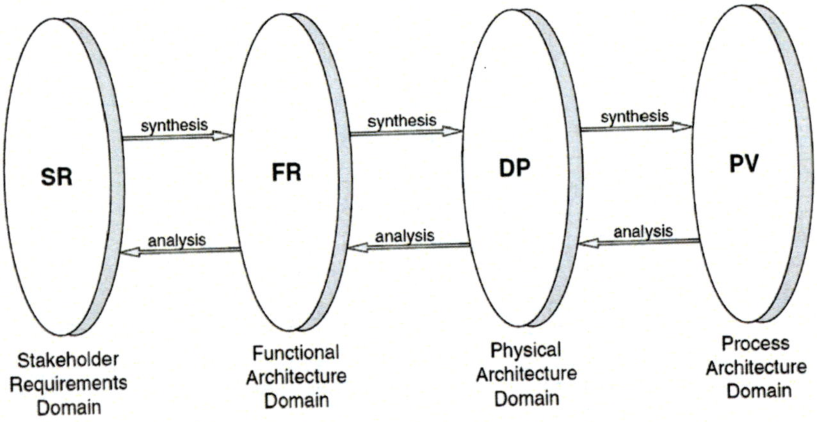

Fig. 1. Mapping the four domains of Axiomatic Design to one another [12]

means that by utilizing the Axiomatic Design method and organizing the overall design matrices for each domain in Axiomatic Design, as shown in Fig. 1, into either a diagonal matrix or triangular matrix, it becomes possible to calculate out all interactions from the definition of stakeholder requirements, all the way to process architecture, and mathematically prove that a design will work and is the optimum solution given all conditions. In Axiomatic Design, these matrices are referred to as "uncoupled" (diagonal, Eq. 1) and "decoupled" (lower and upper triangular matrix, Eqs. 2 & 3). Any other matrix is considered "coupled", and is not only undesirable in the Axiomatic Design method, but indicates that the whole design is caught in a feedback loop: any changes made to a design aspect are liable to spill over into other aspects, and eventually feedback in the originally changed aspect. A coupled matrix indicates that a design in its current state is unstable at best, and impossible at worst [2].

The challenge of Axiomatic Design is that it needs an early commitment from management, and a significant investment of time and energy from all team members in order to successfully execute it. All team members need to engage with the customers - both the internal customers and external customers - to make sure that every Design Parameter (DP) of the end product is identified, broken down into its smallest parts, quantified, and mapped to their relevant Process Variables (PVs). In order to properly do this, the DPs should also already be mapped to their respective Function Requirements (FRs), and the FRs should be mapped to their respective Customer Attributes (CAs)[1]. This will result in the four domains as shown in Fig. 1.

[1] Earlier works by Suh utilize the term "Customer Attributes" or "CAs" [2]. In later works, Suh began using the term "Stakeholder Requirements" or "SRs" [12], in place of Customer Attributes. This can be seen in Fig. 1. In both cases, the terms "CA" and "SR" can be thought of as the requirements of the system as defined by the end-user or 'investor'.

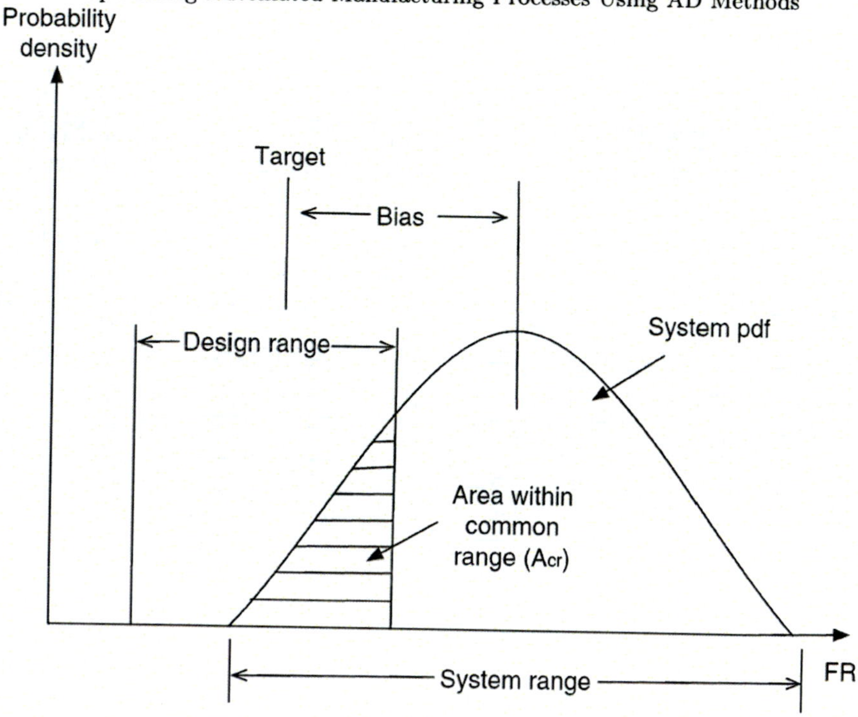

Fig. 2. The "Area within common range" represents the overlap between the design range and the system range, illustrating the probability of a design being able to satisfy the system's requirements [12]

Part of the reason why the initial investment in Axiomatic Design is so large is that, even after all the CAs/FRs/DPs/PVs have been identified, broken down, and mapped to one another, they need to be quantified in such a way that the overlap between the design range (what is needed in order for the system to function) and the system range (what the system is capable of physically achieving) needs to be identified for each interaction in the Axiomatic matrices. A visual of this can be seen in Fig. 2.

But this weakness is also its greatest strength. By mapping out every interaction from customer usage to factory production, and quantifying every interaction possible, it becomes possible to actually calculate things like system performance, production yields, and customer satisfaction in advance of investing in any tools or materials. This means that whether an endeavor will be financially successful can be rigorously evaluated after the design effort has been completed, but before any building takes place. And, if it looks like a product won't be profitable enough to warrant further investment, having all of the product mapped out can facilitate a re-evaluation of customer requirements to determine if there are any CAs that can be eliminated or relaxed in order to quickly and cheaply reduce product costs, with minimal sacrifice to capabilities.

1.1.2 Robotics When it comes to manufacturing, robotics can be sometimes viewed by management as more trouble than they are worth. Robots are inflexible tools. So long as the environment they exist within is consistent and within the designed expectations, they will do the same thing over and over, within a minimal amount of variability. When material or environments drift outside of design parameters, however, a robot is much less flexible than a human laborer. For example, if a robot's task is to install screws into prescribed locations in a certain order, but one particular assembly's screw hole locations are slightly out of alignment for one reason or another, then the robot will likely be unable to compensate, and at best will detect the error and 'call' a human for intervention, and at worst will crash and result in damaged product, lost time, and possible damaged tools as well. Alternatively, using a similar example of a screw, if an incorrect screw makes it into the hopper from which the robot is pulling, such as a screw with the incorrect thread pitch or damaged threads, it will similarly jam when the robot goes to install it. In both cases, a human laborer is very likely to identify the existence of the problem and document its details, all without causing damage to the product.

While robotics has the potential for significant improvements to all aspects of a manufacturing cycle, if it is not carefully and deliberately designed in all of its aspects, then it can turn into an unmitigated disaster for the company. In that regard, it has been shown that robot designs can be improved by Axiomatic Design methods [13], so if these same methods are applied to the design of manufacturing robotic systems, it stands to reason that their designs can be similarly improved.

However, improving the overall design of a robotic system is only one part of the problem. The other aspect of robotics is that the system's behavior also needs to be designed as well. Traditional methods rely on designers quantifying everything in the environment themselves. While this can result in very consistent and predictable behavior, it is also very rigid and does not leave much room for the system to adapt to unexpected interruptions and variability in its routine. Instead, there is potential that Axiomatic Design methods can be used for robotic motion planning in complex environments [14], by using Axiomatic Design combined with robotics algorithms to automatically analyze an environment for goals and obstacles, and generate the best path to achieve its goals while avoiding obstacles.

So, by utilizing Axiomatic Design methods, it should be possible to: 1. Design a cellularized manufacturing facility; 2. Design the robotic hardware and tools for an individual automated production cell; 3. Design robust behavior for the robotics in any given manufacturing cell.

2 Prior Art

The current state of the art in industry often revolves around so-called "trade matrices". This process involves coming up with multiple potential design candidates, assigning weights to design priorities (the greater the importance of a

Table 1. A demonstration of the trade matrix method; design candidate Charlie wins with the greatest total score of 115

		Design Candidates					
		Alpha		Beta		Charlie	
Characteristic	Weight	Score	Total	Score	Total	Score	Total
Strong	4	10	40	6	24	4	16
Fast	2	6	12	5	10	7	14
Cheap	5	2	10	6	30	3	15
User friendly	7	5	35	5	35	10	70
System Total			97		99		115

design characteristic, the greater the magnitude of the weight), scoring the design candidates on their ability to satisfy individual design priorities, and then multiply the design weights against the design scores to give a total product score. An illustrative example of what a trade matrix looks like can be seen in Table 1.

The trade matrix method is borrowed and adapted from Six Sigma. A lot of engineers trained in Six Sigma will also often stick to a 'multiple of 3' rule that helps to highlight and amplify differences in scores (not used in Table 1). Typically, weights and scores stick to a base-10 numerical system, but some will occasionally use weights that have a negative value (if there is an undesirable design characteristic that needs to be minimized or avoided). The main advantage of this method is that it allows the SMEs a lot of room to operate and do what they think is best, while still ensuring that all design options are evaluated in a consistent manner relative to one another. But there is a large drawback to this method: subjectivity. Both the characteristic weights and the design scores are assigned subjectively by the SMEs. The decisions may be informed by experience, but they are still subjective decisions, rather than objective ones. As long as a company is able to maintain an experienced workforce, it should be able to continue to succeed with this method of making design decisions. But if a company is newer, younger, or just less experienced in the area under study, it is possible that a 'wrong' weight or score may be assigned to either a characteristic or design candidate, which in turn could lead to the wrong design candidate being pursued.

3 Results/Experiments/Prototypes

3.1 Top Level FR-DP Pairs

The top-level FR/DP pair was identified as:

FR0: Maximize the ratio between revenue & expenses in the factory	**DP0:** A system that is flexible to market conditions

Ultimately, the goal of the factory is to maximize profits, while simultaneously minimizing the costs needed to achieve those profits. The costs in a factory also must be evaluated in reference to the profits as well, as they will increase as the volume of products moving through the factory also increases. So, when minimizing costs, care must be taken so that revenues are not simultaneously reduced. Or, if revenues are reduced, they are reduced by an overall smaller amount than what costs were reduced by. This is why FR0 is maximizing the ratio between revenue and expenses.

One key assumption in this thesis is that corporate strategy is not set by the factory. The factory is focused on beating its numbers from the previous quarter and year, and setting itself up to beat its current numbers next quarter and year. Longer-term planning is outside of the scope of this thesis, as this starts getting into business administration—and while Axiomatic Design can be used for coming up with a corporate strategy, that is not the goal of this thesis.

In order to satisfy FR0, it is not enough to simply reduce waste while expanding production. If a factory begins to over-produce, then demand for their products will begin to fall, leading to falling revenues. At the same time, if a factory fails to produce enough product, they may see the demand for their products skyrocket, leading to a spike in prices - but not every customer will be willing or able to purchase the products at the higher prices, and the factory starts to "leave customers on the table" that their competitors can snap up instead. So, to satisfy FR0, DP0 needs to be a system that monitors and reacts to market demands, both present and future.

Going deeper than the zeroth level, the following six pairs were identified using the Axiomatic Design method (Table 2):

Putting all of these into an Axiomatic matrix, and checking for interactions, a decoupled matrix was found, as shown in Fig. 3. The only off-diagonal pair in the top-level matrix is FR2-DP1; the interaction between minimizing production cycle complexity, and a system that evaluates market demand both present and future. If it weren't for this interaction, the top-level matrix would be uncoupled. However, even if the top-level were uncoupled, it would be possible that the other pair might have off-diagonal interactions after being decomposed. Just because a higher level is uncoupled, it does not mean that lower levels cannot be decoupled or even coupled.

As FR-DP pairs 3–6 cover more company logistics and human labor, and since they do not interact with FR-DP pairs 1 and 2, they only received a basic amount of study in this paper, and are left to the readers to evaluate further. Testing for interactions should be a simple exercise: simply compare the identified FR-DP pairs at the next lower level, and check for interactions off either side of the diagonal.

Going forward, the focus of this paper will be on FR-DP pairs 1 and 2, where much of the details of automated manufacturing were found to lay.

Table 2. First Level FR-DP Pairs

FR1: Match production rate with product complexity to meet current and future market demands	**DP1:** A system to evaluate market demand, both present and future
FR2: Minimize production cycle complexity	**DP2:** A system to evaluate products & processes for excessive complexities
FR3: Insure against potential supply shortages	**DP3:** An investment strategy that takes positions in the stock market that are inversed from material needs
FR4: Maintain worker safety at all times	**DP4:** A system that monitors injury occurrences, and correct root causes from the feedback
FR5: Attract the best talent available in the market	**DP5:** A program that actively engages with professionals - both young and experienced - and students, to maximize the bandwidth of the talent pipeline
FR6: Retain the best talent available in the market	**DP6:** Constant monitoring of market compensation packages, with proactive raises to match current market rates for all employees that meet or exceed performance goals

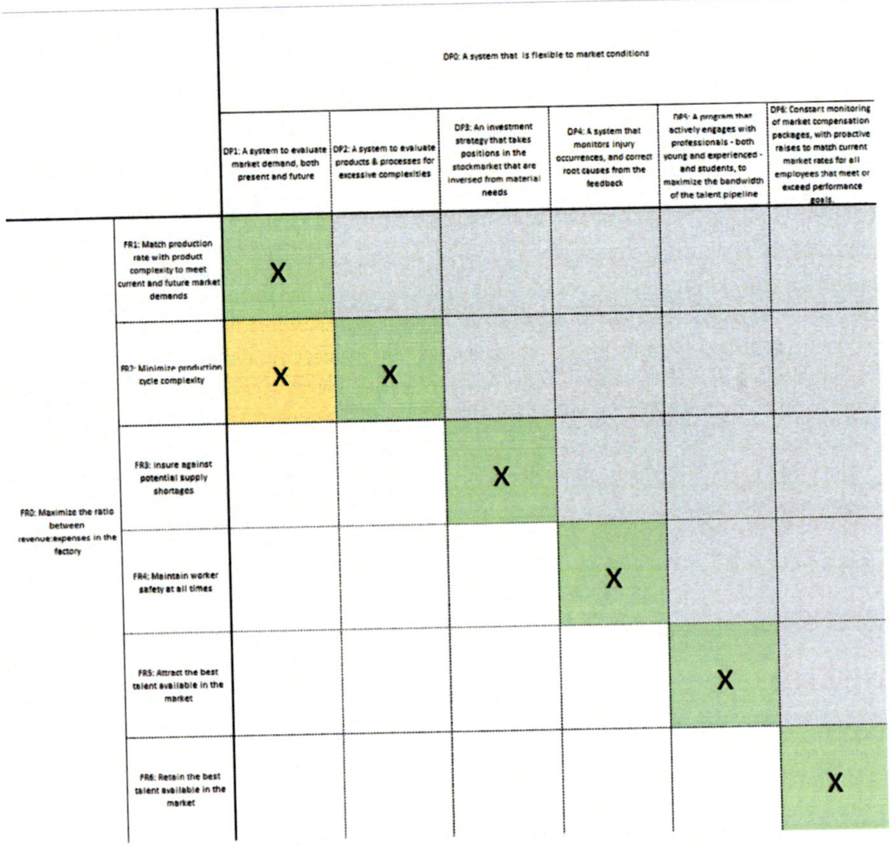

Fig. 3. The zeroth & first levels of the axiomatic matrix for a factory utilizing automated processes

3.2 FR1-DP1: Matching Production Rates to Market Demand

The first pair identified, over-production or under-production relative to product demand can easily impact the bottom line. If the manufacturing system fails to produce enough material to satisfy market demand, then sales are left uncaptured and revenues are smaller than they would be otherwise. If the factory system overproduces the amount of material, relative to market demands, then prices of its products may fall to a level where it is either no longer profitable to sell them, or the company could even be forced to destroy its own merchandise. So, the key to achieving this functional requirement is a system that can evaluate market demand for a product, both in the present and in the future.

FR1: Match production rate with produce complexity to meet current and future market demands

DP1: A system to evaluate market demand, both present and future

Table 3. FR1-DP1 Pairs

FR1.1: Automate as many production processes as possible	**DP1.1:** Robotic assembly processes
FR1.2: Minimize product complexity, while still achieving all customer requirements	**DP1.2:** Axiomatic product design
FR1.3: Minimize assembly process complexity	**DP1.3:** Axiomatic process design
FR1.4: Monitor current market demand for product(s)	**DP1.4:** Short-term (90 day) market survey mechanism
FR1.5: Forecast future market demand for product(s)	**DP1.5:** Long term (91–275 day) market survey mechanism

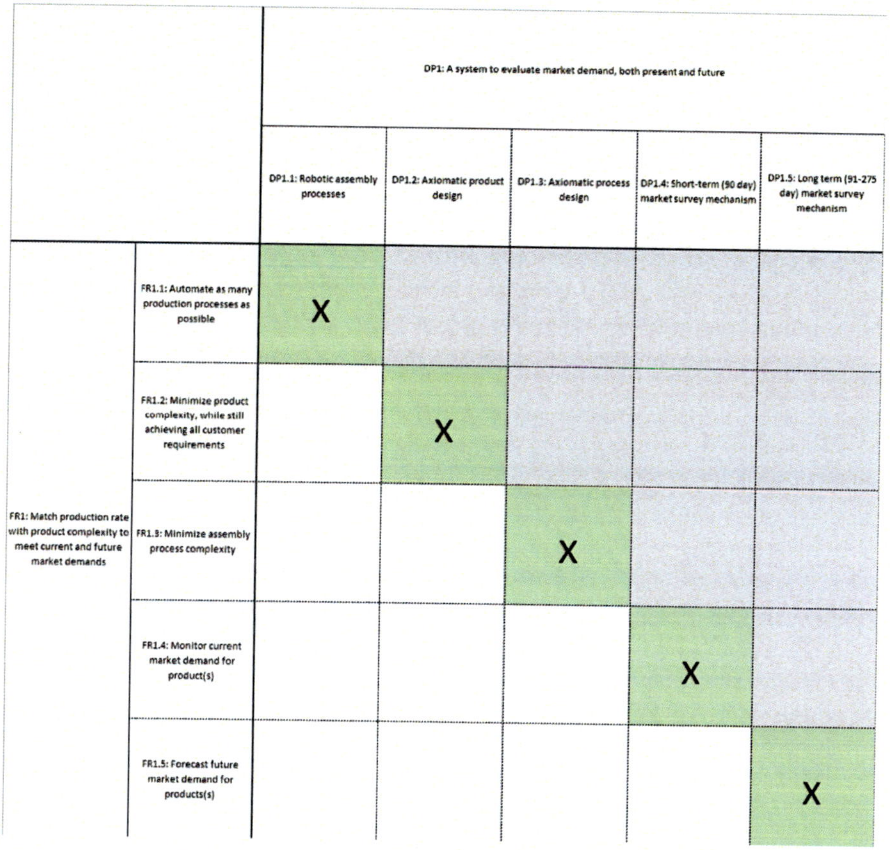

Fig. 4. FR1-DP1 Pairings

Decomposing this, the following FRs and DPs and their interactions can be seen as uncoupled in Fig. 4 (Table 3).

3.2.1 FR1.1-DP1.1 FR1.1 and DP1.1 is the first decomposition of the FR1:DP1 pair. They focus on automation, as the more the manufacturing process is automated, the greater the control over the overall system that can be exerted.

FR1.1: Automate as many production processes as possible

DP1.1: Robotic assembly processes

Decomposing further, the system begins to reach the limits of how far it can be broken down for this particular branch. The following two pairs of FRs and DPs are uncoupled in Fig. 5. This means that FR1.1.1 only maps to DP1.1.1 and vice versa; and FR1.1.2 only maps to DP1.1.2, and vice versa. With this, it is possible to segment manufacturing processes separately from identifying which processes are repetitive (and thus can be automated). This further implies that manufacturing processes can be segmented with the intent of automating them; automated processes can be grouped around the manufacturing steps that are repetitive (Table 4).

3.2.2 FR1.2-DP1.2 While FR1.1-DP1.1 was more focused on manufacturing processes, FR1.2-DP1.2 instead focuses on product complexity. By reducing and minimizing product complexity, not only can the reliability and quality of end products be ensured, but manufacturing processes can be kept as simple as possible.

FR1.2: Minimize product complexity, while still achieve all customer requirements

DP1.2: Axiomatic product design

To help achieve this, Fr1.2-DP1.2 can be decomposed as such (Table 5).

However, due to the natures of FR1.2.2 and DP1.2.1, this matrix is only decoupled, as seen in Fig. 6. In this case, FR-DP1.2.1 and FR-DP1.2.2 pair together as expected, but FR1.2.2 also interacts with DP1.2.1. This is because the effort to minimize the number of physical components naturally interacts with a component's versatility. Ideally, a single part satisfies every functional

Table 4. FR1.1-DP1.1 Pairs

FR1.1.1: Segment manufacturing into process steps	**DP1.1.1:** Breaks in assembly where stops are possible & natural
FR1.1.2: Identify processes that can be automated	**DP1.1.2:** Repetitive motions with predictable dimensions

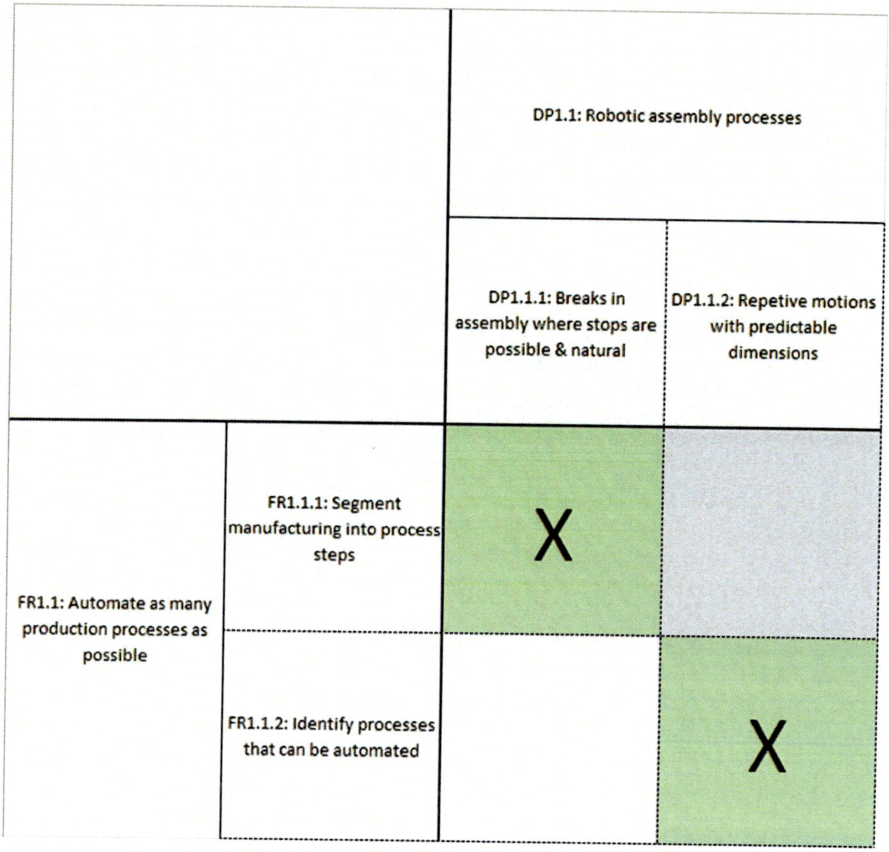

Fig. 5. FR1.1-DP1.1 Pairings

requirement - thus the interaction. In practice, this is not easy to achieve, and is sometimes outright impossible. Still, this interaction indicates that components should be as versatile as possible, without introducing extra functions that are not called for in the design.

3.2.3 FR1.3-DP1.3 Similarly to FR1.2-DP1.2, FR1.3-DP1.3 focuses on minimizing complexity, however, it focuses on manufacturing process complexity.

FR1.3: Minimize assembly process complexity

DP1.3: Axiomatic process design

From the very outset of a design effort, the manufacturing processes need to be considered. It does not matter if something can be achieved mathematically on paper if it cannot be achieved with tools in 3D space. With that in mind, the less complex a manufacturing process is, not only will the factory see better yields and shorter cycle times, but it will see a shorter on-ramp to the introduction of

Table 5. FR1.2-DP1.2 Pairs

FR1.2.1: Maximize the number of functions each component satisfies	**DP1.2.1:** Versatile components
FR1.2.2: Minimize the number of physical components	**DP1.2.2:** Essential Components

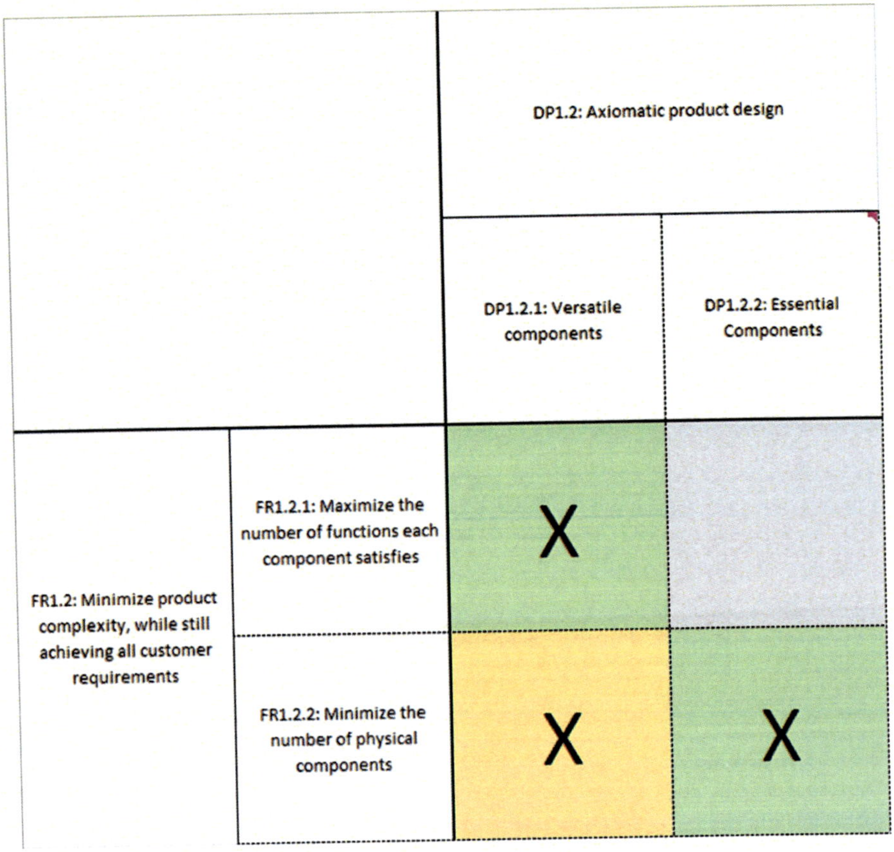

Fig. 6. FR1.2-DP1.2 Pairings

the new product and any future changes that may be made to it. More directly stated, do not cut two holes when the task can be achieved with one (Table 6).

Decomposing FR1.3-DP1.3, the following FR-DP pairs are shown as the uncoupled matrix shown in Fig. 7. In this matrix, we see that FR-DP1.3.1 only interacts with itself, and FR-DP1.3.2 also only interacts with itself. This proves that a combination of additive manufacturing whenever possible and appropriate has no impact on the number of fasteners in use. However, the minimization of fasteners and the utilization of additive processes (when viable) are both still desirable aspects per their parents FR1.3: minimize assembly process complexity.

Table 6. FR1.3-DP1.3 Pairs

FR1.3.1: Utilize additive manufacturing when possible & appropriate	**DP1.3.1:** Versatile processes
FR1.3.2: Utilize the minimum number of mechanical fastening steps	**DP1.3.2:** Essential process steps

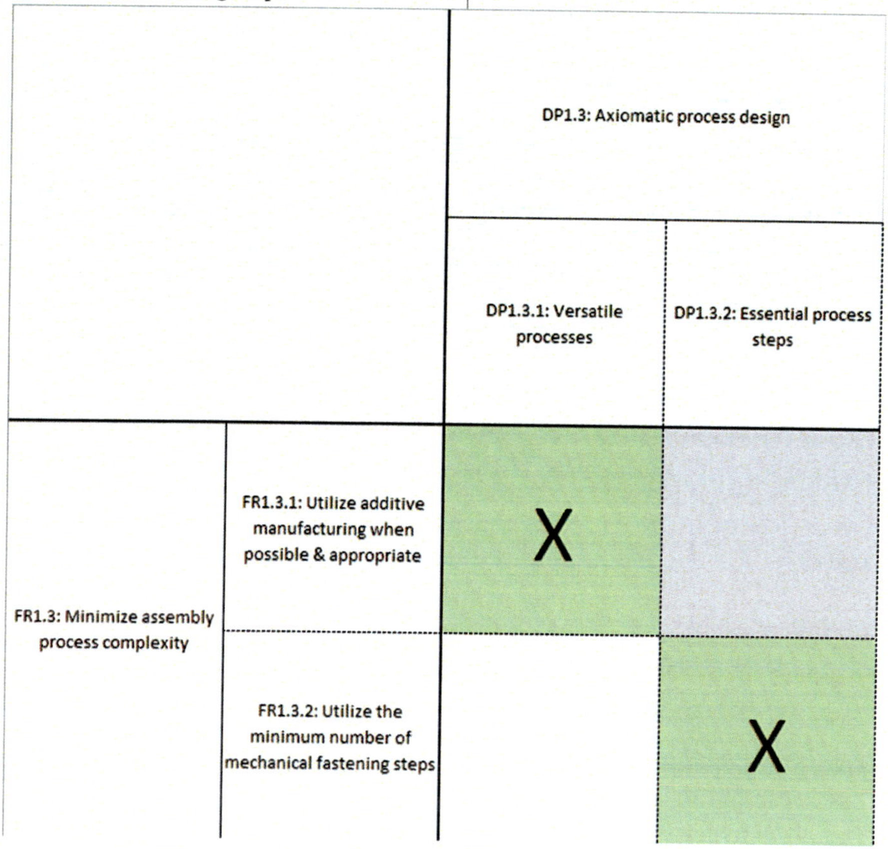

Fig. 7. FR1.3-DP1.3 Pairings

This may seem counter-intuitive at first, however, it becomes clearer when you consider that 3D printing not only can reduce the number of parts (via the designer combining them together), but it can also *increase* the number of parts, too, if the desired part cannot be fit into the available printer volume as a whole piece. How a product is put together is a task that is up to the designer. While 3D printing can enable novel ways of assembly (or completely eliminate the need for assembly at all, via print-in-place designs), it is not necessarily a guarantee of fewer assembly steps or fasteners, either. It is just another tool in the engineer's belt.

3.2.4 FR1.4-DP1.4 FR1.4 & DP1.4 are focused on immediate demand for the products of a company. They should be evaluated in the context of material movement within the company itself. Neither FR1.4 nor DP1.4 has any interactions with any other functional requirement or design parameter at the $1.x$ level. Additionally, looking at the highest matrix, we can see that while FR2 and DP1 do, in fact, interact with one another, as will be covered further in this paper, DP1.4 does not interact with any of the decomposed FRs of FR2. Thus, it can be concluded that neither FR1.4 nor DP1.4 will have any further interaction with any FRs or DPs outside of its own. FR1.4-DP1.4 is functionally independent of the rest of the Axiomatic matrix, indicating that the material inside of the factory—that this thesis is meant to analyze—can be moved freely to meet immediate market demand, without impacting the automated processes used to satisfy that demand. While scaling up beyond maximum capacity will still naturally require investment in additional tooling and personnel, this realization indicates that such a factory could be scaled *down* to match demand.

FR1.4: Monitor current market demand for product(s)

DP1.4: Short-term (90 day) market survey mechanism

3.2.5 FR1.5-DP1.5 Similar to FR1.4-DP1.4, FR1.5-DP1.5 is also focused on market demand. Unlike FR1.4-DP1.4, FR1.5-DP1.5 is focused on long-term demand and is intended to be used to look at a factory's *external* material position; supplier availability, material lead times, etc. Material needs to arrive at the factory with enough time left to still be turned into products that can meet time-dependent and cyclical demand.

Also, like FR1.4-DP1.4, FR1.5-DP1.5 does not interact with any other FR or DP at its own level, and is functionally independent because of it.

FR1.5: Forecast future market demand for product(s)

DP1.5: Long term (91-275 day) market survey mechanism

Because both FR1.4-DP1.4 and FR1.5-DP1.5 are both functionally independent - including from each other - and have little to do with automation, further decomposition, and more detailed analysis are being left as a future area of study.

3.3 FR2-DP2: Evaluating Production Cycle Complexity

While FR1-DP1 was primarily focused on material and tool management, FR2-DP2 is directly focused on production management. Specifically, it requires minimization of complexity in a production cycle. A simple production cycle minimizes movement, reduces manufacturing steps, and keeps waiting times during and between steps as short as possible. Part of the way this can be achieved is by saving repetitive tasks for automated tools (robots), as human error is one of the key drivers of rework and process variances. To this point, if the goal is to minimize the number of human laborers performing repetitive processes, and every product is assembled from a minimum (finite) amount of processes, then it would be logical to simultaneously maximize the number of repetitive processes needed to manufacture an item and ensure that enough automated systems existed to handle these repetitive processes. More directly put: automate as many process steps as cost-effective, and save the human labor for where it is really needed.

To this point, looking again at Fig. 3, we can see an interaction between FR2 and DP1, as it is this particular pairing where - after decomposing both - we see that production cycles begin to interact with market demand.

FR2: Minimize production cycle complexity	**DP2:** A system to evaluate products & processes for excessive complexities

Decomposing FR2-DP2, we get the following pairs, which produce the decoupled matrix shown in Fig. 8. The only off-diagonal pair that makes this decoupled is FR2.3-DP2.2, which indicated that the minimization of information content in overall product assembly processes also has a necessary interaction with the Axiomatic Design of the product itself. What this tells us, in plain terms, is that manufacturing processes must be considered and designed in parallel with the product design itself. A product cannot be delivered to a factory, for manufacturing processes to be figured out after the fact, and still be considered a product designed with Axiomatic Design process (Table 7).

Table 7. FR2-DP2 Pairs

FR2.1: Minimize the number of repetitive assembly processes performed by laborers	**DP2.1:** Automated simple & repetitive assembly steps
FR2.2: Minimize information content of product designs	**DP2.2:** Axiomatic Design of design parameters (DPs)
FR2.3: Minimize information content of overall product assembly process	**DP2.3:** Axiomatic Design of process variables (PVs)

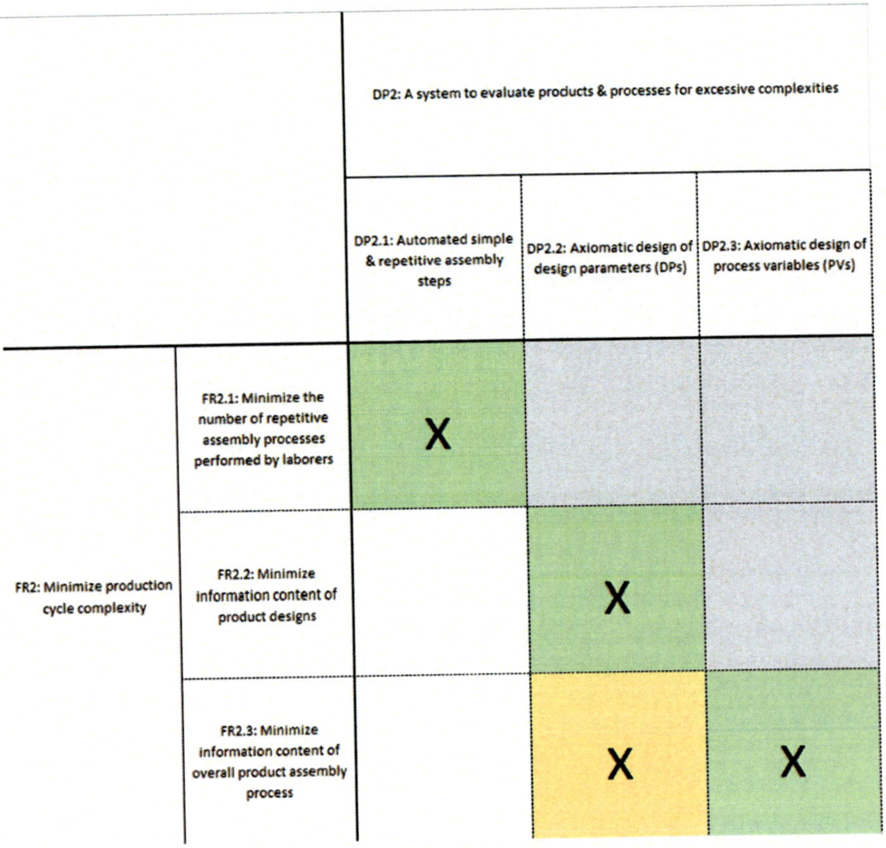

Fig. 8. FR2-DP2 Pairings

3.3.1 FR2.1-DP2.1 Decomposing FR2.1-DP2.1, we get the following pairs, expressed as a decoupled matrix in Fig. 9. The only off-diagonal pair that makes this decoupled is FR2.1.2 and DP2.1.1, which indicates that minimization of individual process step complexity has a direct interaction with any continuous improvement process to make a product and manufacturing process automation-centric. When looking to replace manual labor with an automated process, the complexity of the process must be both considered and minimized when possible, if it is to succeed in an automated environment (Table 8).

These pairs are primarily focused on keeping an assembly process as automated and automation-friendly as possible.

3.3.2 FR2.2-DP2.2 Decomposing FR2.2-DP2.2, and we get the following pairs, expressed as a decoupled matrix in Fig. 10 (Table 9).

These pairs are focused on minimizing the information content of the product design by ensuring that all the customer requirements are accounted for, with no

Table 8. FR2.1-DP2.1 Pairs

FR2.1.1: Replace a manual laborer with an automated tool wherever cost-effective	**DP2.1.1:** A continuing improvement process to improve product & process to be automation-centric
FR2.1.2: Minimize individual process step complexity	**DP2.1.2:** Minimum number of actions to complete step
FR2.1.3: Utilize all available automated assembly tools	**DP2.1.3:** Minimum automated tool downtime

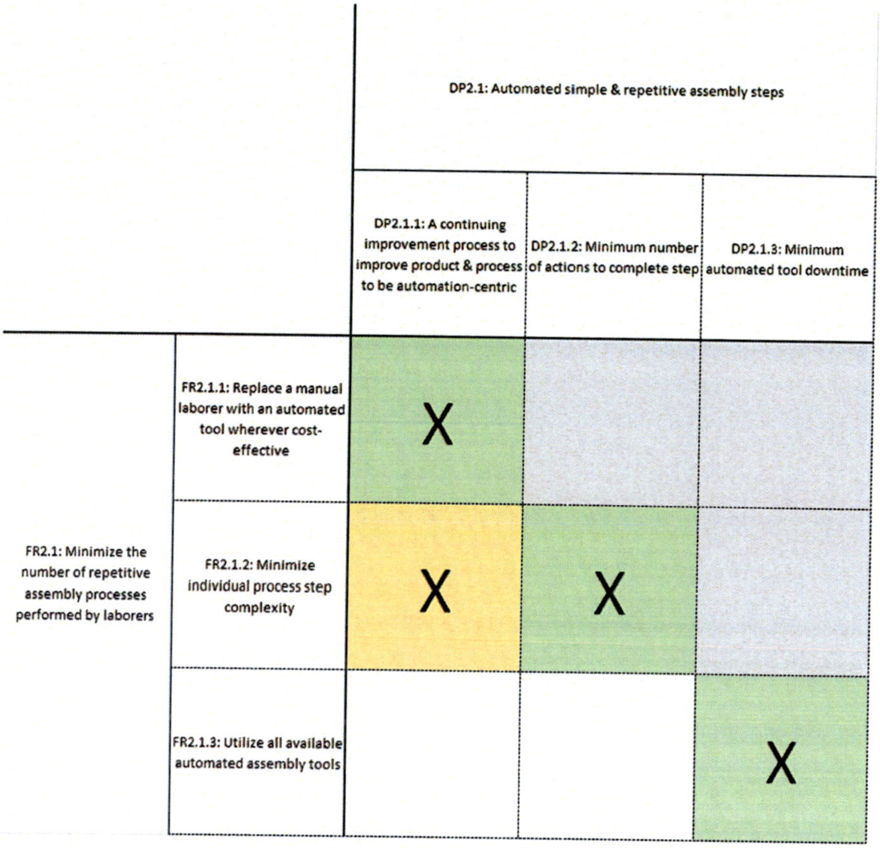

Fig. 9. FR2.1-DP2.1 Pairings

cases of extra features being included 'just because'. The only off-diagonal interaction making this particular sub-matrix decoupled is FR2.2.2-DP2.2.1, which is the interaction between excluding FRs that do not map to a customer attribute, and the list of customer attributes itself. This basically states that the engineer cannot be tempted to help by introducing FRs that the customer did not ask for. To do so could potentially destabilize the design in unpredictable ways. Design scope creep should be avoided under all circumstances, unless directly requested by the stakeholder.

3.3.3 FR2.3-DP2.2 In Fig. 8, it is shown that there is an off-diagonal interaction between FR2.3 and DP2.2, and this is what makes FR2-DP2 decoupled instead of uncoupled. For convenience, FR2.3's and DP2.2's respective decompositions are listed here again. Figure 11, shows another decoupled matrix, with all interactions being off of the primary diagonal of the overall Axiomatic matrix.

Table 9. FR2.2-DP2.2 Pairs

FR2.2.1: Include only one FR per customer attribute	**DP2.2.1:** A list of all customer attributes
FR2.2.2: Exclude any FRs that do not map directly to a customer attribute	**DP2.2.2:** Essential features

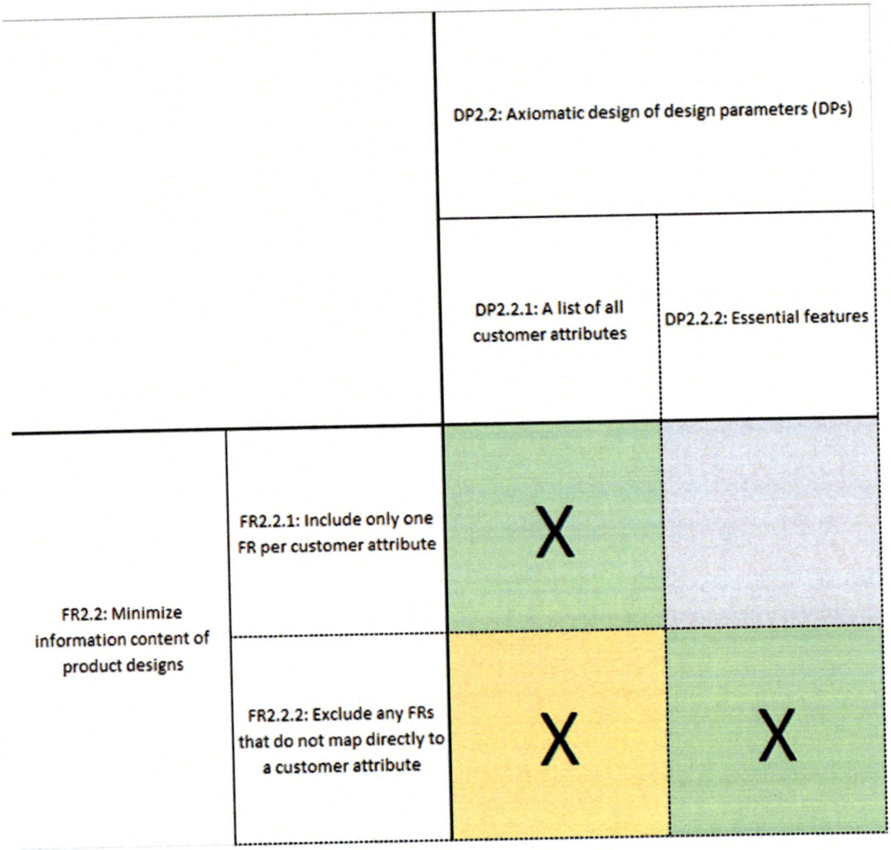

Fig. 10. FR2.2-DP2.2 Pairings

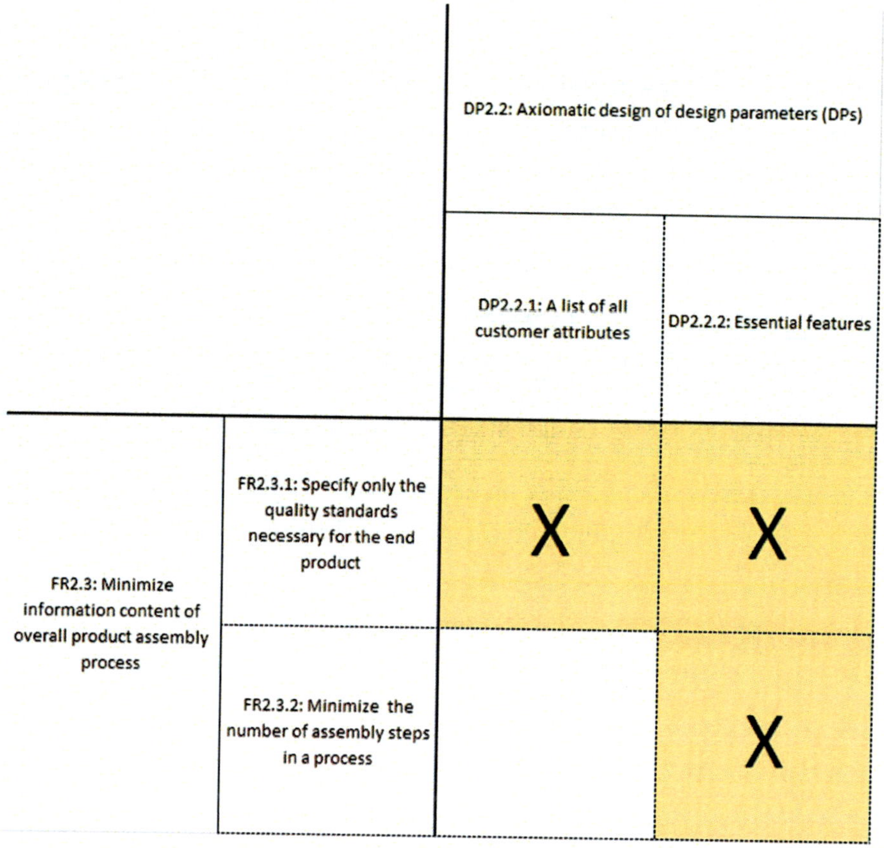

Fig. 11. FR2.3-DP2.2 Pairings

Table 10. FR2.3-DP2.2 Pairs

FR2.3.1: Specify only the quality standards necessary for the end product	**DP2.2.1:** An exhaustive list of all customer attributes
FR2.3.2: Minimize the number of assembly steps in a process	**DP2.2.2:** Features only as-specified, with no 'just because' extras

For FR2.3.1, it interacts with both DP2.2.1 and DP2.2.2, because all necessary quality standards should interact with all customer attributes and all essential features of a product. For FR2.3.2, minimizing the number of assembly steps in a manufacturing process will only interact with the essential features of a product - as the elimination of extra features will naturally eliminate extra assembly steps (Table 10).

3.3.4 FR2.3-DP2.3 Like FR2.2-DP2.2, FR2.3-DP2.3 is also focused on minimizing information content, but in this case, it is focused on minimizing the

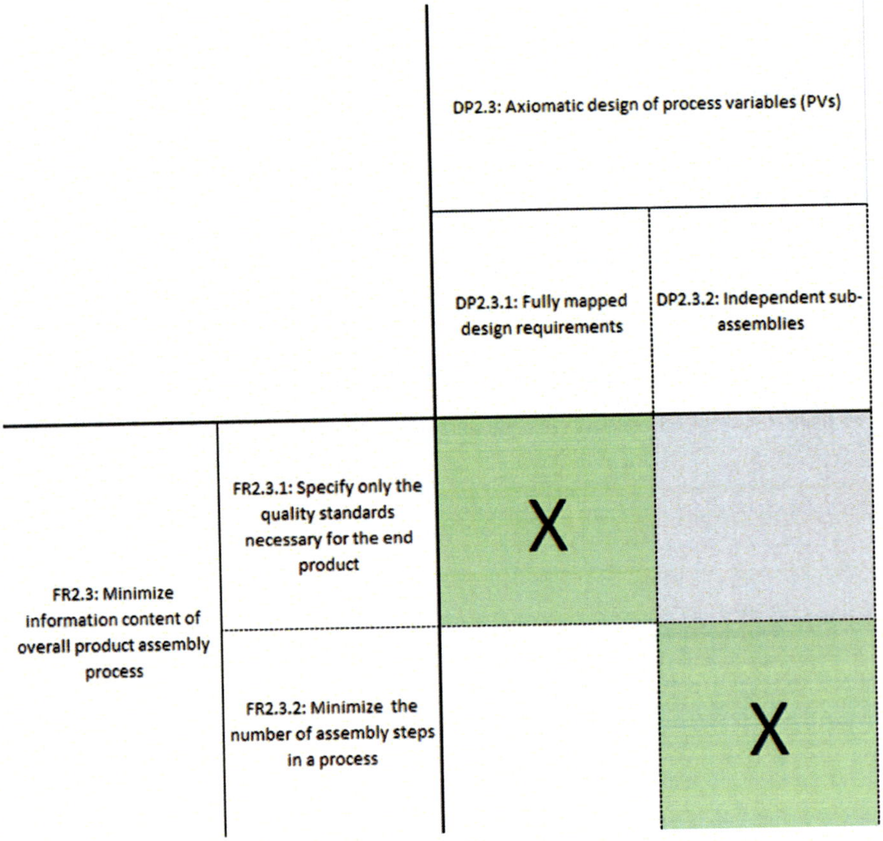

Fig. 12. FR2.3-DP2.3 Pairings

information content of manufacturing processes. Decomposing FR2.3-DP2.3, we get the following pairs, expressed as an uncoupled matrix in Fig. 12 (Table 11).

Table 11. FR2.3-DP2.3 Pairs

FR2.3.1: Specify only the quality standards necessary for the end product	**DP2.3.1:** Fully mapped design requirements
FR2.3.2: Minimize the number of assembly steps in a process	**DP2.3.2:** Products broken into manageable sub-assemblies

3.4 FR2-DP1: Production Cycle Complexity in Terms of Market Demand

With FR2-DP1, we start seeing interactions that are exclusively off-diagonal, in reference to the overall Axiomatic matrix for this design. The FR2-DP1 pairing is the primary driver keeping this design from being uncoupled, but it is not the only driver of it.

FR2-DP1 represents the interaction between production cycle complexity and material movement within the production environment Their decompositions are listed in Fig. 12, and the resulting matrix with all of their interactions is shown in Fig. 13. All of these interactions are about the way the minimization of information content in all aspects has interactions with the design and assembly processes, but no interactions with the supply chain itself.

FR2: Minimize production cycle complexity	**DP1:** A system to evaluate market demand, both present and future

As stated early, FR2 does not interact with DP1.4 or DP1.5 in any way. However, all decompositions of FR2 do interact with DPs 1.1, 1.2, and 1.3. FR2.1 interacts with DPs 1.1, 1.2, and 1.3, as minimizing repetitive labor performed by humans has interactions with robots performing repetitive tasks, as well as Axiomatic Design of both products and processes. FR2.2 only interacts with DP1.2, as both deal with product design, and FR2.3 only interacts with DP1.3, as both deal with process design (Table 12).

Table 12. FR2-DP1 Pairs

FR2.1: Minimize the number of repetitive assembly processes performed by laborers	**DP1.1:** Robotics performing repetitive assembly processes
FR2.2: Minimize information content of product designs	**DP1.2:** Axiomatic product design
FR2.3: Minimize information content of overall product assembly process	**DP1.3:** Axiomatic process design
	DP1.4: Short-term (90 days) market survey mechanism
	DP1.5: Long term (91–275 day) market survey mechanism

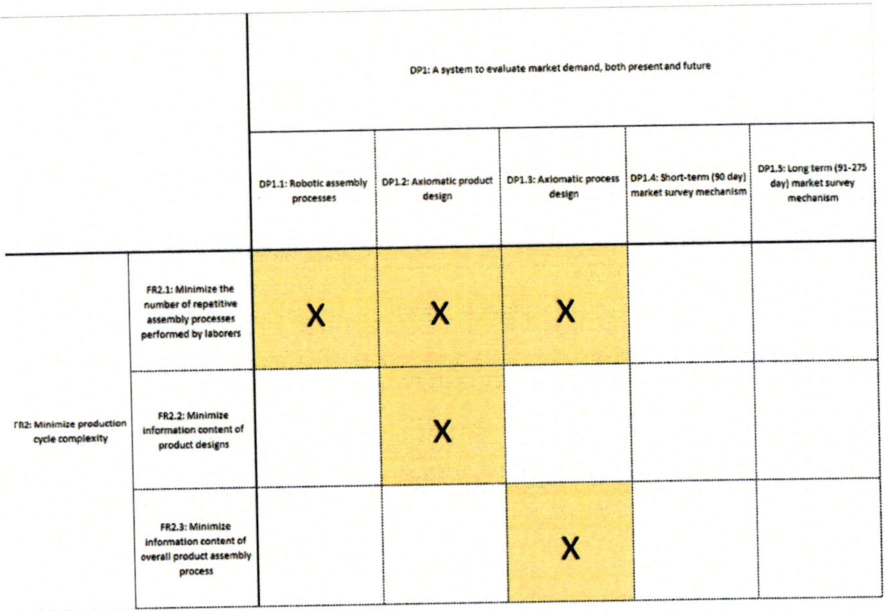

Fig. 13. FR2-DP1 Pairings

3.4.1 FR2.1-DP1.1 Diving deeper and looking at the decomposition of FR2.1-DP1.1, we get the following FRs and DPs, which combine to create the decoupled matrix seen in Fig. 14. In the process of minimizing the number of repetitive assembly processes performed by manual laborers (FR2.1), we see the only interactions with Robotic Assembly processes (DP1.1) are when replacing the manual laborer (FR2.1.1) interacts with breaks in the assembly process (DP1.1.1) and repetitive motions (DP1.1.2). For minimizing the process step complexity (FR2.1.2), we only see an interaction with the repetitive motions themselves (DP1.1.2) (Table 13).

Table 13. FR2.1-DP1.1 Pairs

FR2.1.1: Replace a manual laborer with an automated tool wherever possible	**DP1.1.1:** Breaks in assembly where stops are possible & natural
FR2.1.2: Minimize individual process step complexity	**DP1.1.2:** Repetitive motions with predictable dimensions
FR2.1.3: Utilize all available automated assembly tools	

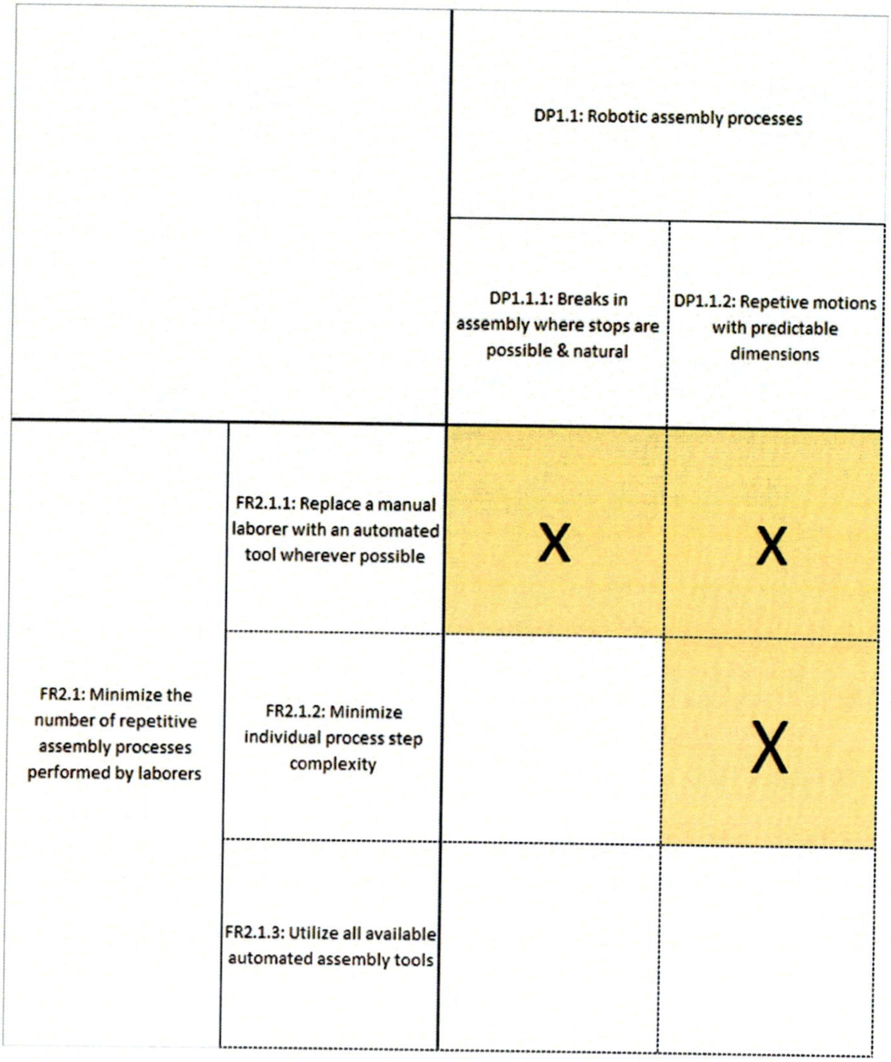

Fig. 14. FR2.1-DP1.1 Pairings

Table 14. FR2.1-DP1.2 Pairs

FR2.1.1: Replace a manual laborer with an automated tool wherever possible	**DP1.2.1:** Versatile components
FR2.1.2: Minimize individual process step complexity	**DP1.2.2:** No 'extra' parts
FR2.1.3: Utilize all available automated assembly tools	

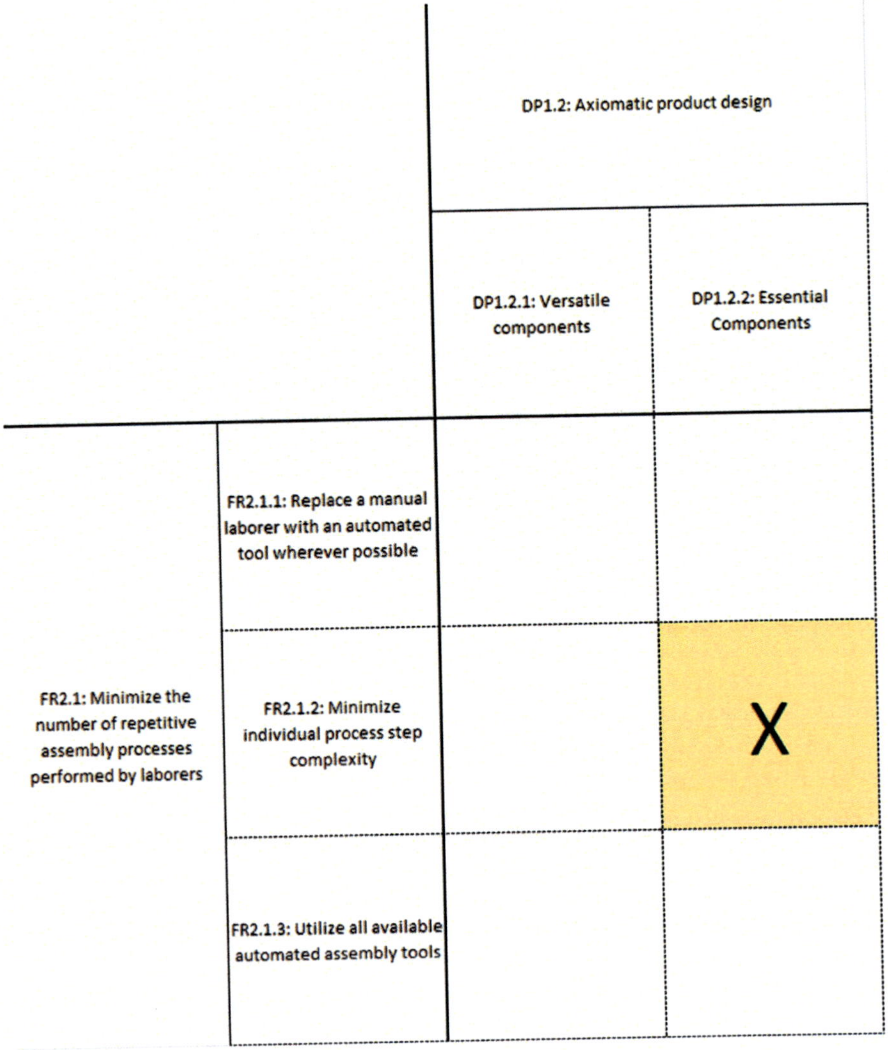

Fig. 15. FR2.1-DP1.2 Pairings

3.4.2 FR2.1-DP1.2 Looking at the decomposition of FR2.1-DP1.2, we get the following FRs and DPs, which combine to create the decoupled matrix seen in Fig. 15. In this case, there is only one interaction at this level: between FR2.1.2 and DP1.2.2. In the effort to minimize process step complexity, it will become necessary to consider which components are truly necessary and how they are necessary (Table 14).

3.4.3 FR2.1-DP1.3 Looking at the decomposition of FR2.1-DP1.3, we get the following FRs and DPs, which combine to create the decoupled matrix seen in Fig. 16. For this decomposition, we have two interactions: FR2.1.1 & DP1.3.1; and FR2.1.2 & DP1.3.2. For the first pair (FR2.1.1-DP1.3.1), when replacing a manual process with an automated one, the automated one should be as versatile as possible. This means that the automated process should be able to identify, and compensate for any reasonable part variabilities, and it should also be able to deal with a part that is out of spec on its own (ejecting a non-conforming part from the assembly line into a waste/scrap bin, obtaining a replacement, and continuing on without human interaction). For the second pair (FR2.1.2-DP1.3.2), this goes to keeping the overall assembly process as simple as possible. All individual steps should be as simple as possible, and it should use as few steps as necessary to complete the goal. More directly stated: the "keep it simple, stupid" (KISS) principle, and minimize product movement (Table 15).

3.4.4 FR2.2-DP1.2 Looking at the decomposition of FR2.2-DP1.2, we get the following FRs and DPs, which combine to create the decoupled matrix seen in Fig. 17. For this off-diagonal matrix, both FR2.2.1 and FR2.2.2 interact with just DP1.2.2. Both FR2.2.1 and FR2.2.2 deal with limiting scope creep, so both must interact with keeping a design limited to just its essential components.

Table 15. FR2.1-DP1.3 Pairs

FR2.1.1: Replace a manual laborer with an automated tool wherever possible	**DP1.3.1:** Versatile processes
FR2.1.2: Minimize individual process step complexity	**DP1.3.2:** No 'extra' process steps
FR2.1.3: Utilize all available automated assembly tools	

		DP1.3: Axiomatic process design	
		DP1.3.1: Versatile processes	DP1.3.2: Essential process steps
FR2.1: Minimize the number of repetitive assembly processes performed by laborers	FR2.1.1: Replace a manual laborer with an automated tool wherever possible	X	
	FR2.1.2: Minimize individual process step complexity		X
	FR2.1.3: Utilize all available automated assembly tools		

Fig. 16. FR2.1-DP1.3 Pairings

If a designer succeeds in only having one FR per customer attribute (which they should, if they are properly following the setup for Axiomatic Design), and excludes all FR that do not map directly to a customer attribute (at all levels), then all that should remain are the components essential to a design (Table 16).

Table 16. FR2.2-DP1.1 Pairs

FR2.2.1: Include only one FR per customer attribute	**DP1.2.1:** Versatile components
FR2.2.2: Exclude any FRs that do not map directly to a customer attribute	**DP1.2.2:** No 'extra' parts

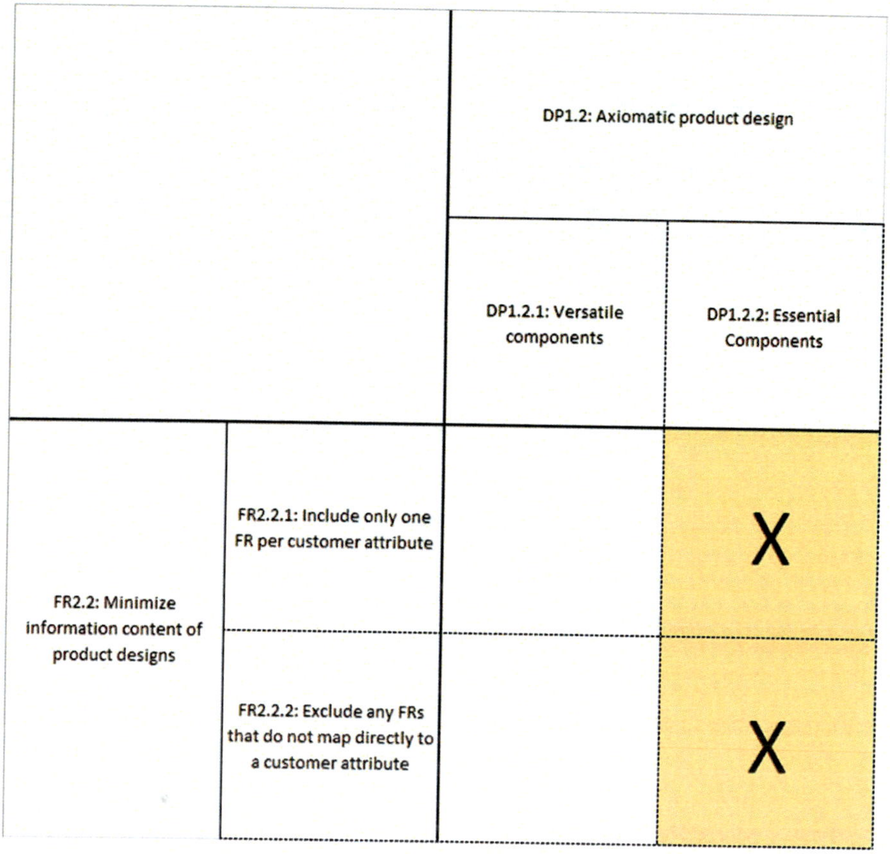

Fig. 17. FR2.2-DP1.2 Pairings

3.4.5 FR2.3-DP1.3 Looking at the decomposition of FR2.3-DP1.3, we get the following FRs and DPs, which combine to create the decoupled matrix seen in Fig. 18. Conversely, compared to FR2.2-DP1.2, FR2.3-DP1.3 is a situation where only FR2.3.2 interacts with the decomposition of DP1.3. In this case, it interacts with both DP1.3.1 and DP1.3.2. By minimizing the number of assembly steps in a manufacturing process, interactions with both the creation of versatile processes and utilizing essential process steps are seen. However, no interactions are seen between the quality standards, and how versatile or essential a process step is. This suggests that quality does not need to be sacrificed in order to

Table 17. FR2.3-DP1.3 Pairs

FR2.3.1: Specify only the quality standards necessary for the end product	**DP1.3.1:** Versatile processes
FR2.3.2: Minimize the number of assembly steps in a process	**DP1.3.2:** No 'extra' process steps

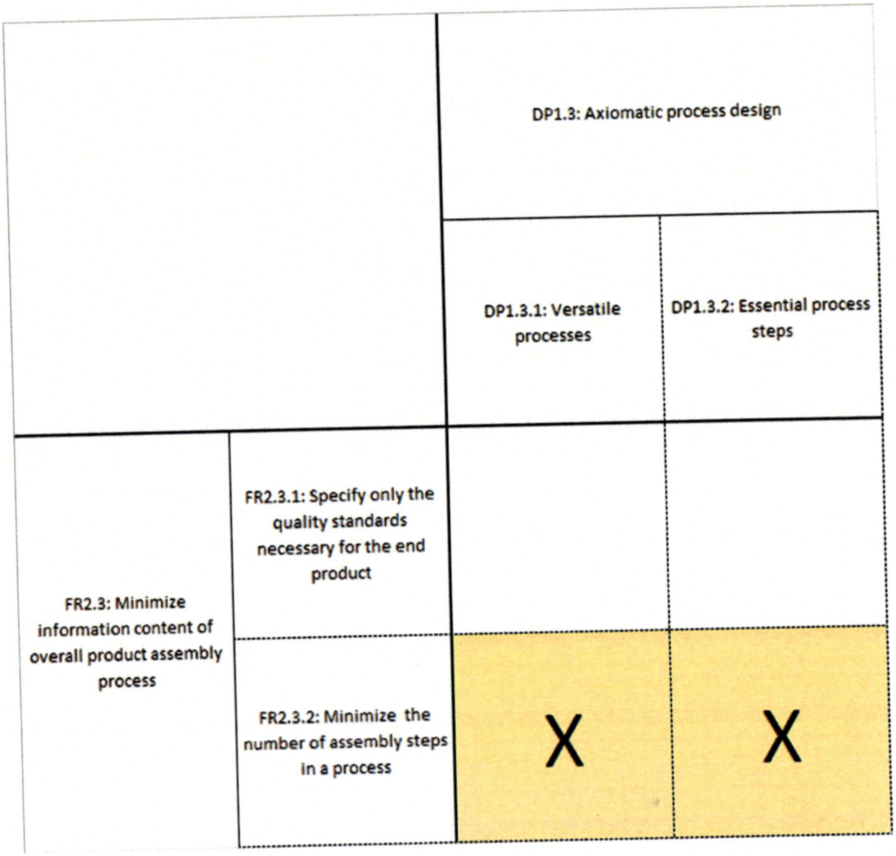

Fig. 18. FR2.3-DP1.3 Pairings

successfully design a manufacturing process with Axiomatic Design methods (Table 17).

3.5 FR3-DP3: Material Procurement Strategies

FR3-DP3 is uncoupled, at least to the levels that it was decomposed to. However, FR3-DP3 also deals with parts of the automated production cycle that cannot be completely ignored, but do not have much to do with automation itself; these fall outside of the scope of work, and were only included to complete the decomposition of FR0-DP0. It is possible that FR3-DP3 could also change from uncoupled to decoupled as it is decomposed. However, as long as each layer is decomposed correctly, it is unlikely that they will become coupled in this case.

Specific to FR3-DP3, the primary role of this pair is to financially insulate the company against supply chain shocks. A company can only control where they purchase its materials; it cannot control the market value of those materials. If raw material prices skyrocket, a company may not be able to afford to actually purchase the materials at a price that would allow production to remain profitable. However, if raw material values were to crater, a company may find itself in financial trouble if any stores of those materials were used to secure loans. So, as a way to help insure against such shocks, a strategy of commodity options contracts can be used as a way to offset risk. If material prices skyrocket, some call options contracts can allow for the purchase of materials at a lower price point. If material costs significantly decrease, put options contracts can be used to sell material at the older, higher price (potentially helping to cover the balance on a loan that was previously secured via the same material).

FR3: Insure against potential supply shortages	**DP3:** An investment strategy that takes positions in the stock market that are inversed from material needs

The decomposition of FR3-DP3 can be seen in Fig. 19 (Table 18).

3.6 FR4-DP4: Personnel Safety

FR4-DP4 deals with laborer safety. With very few exceptions, every factory needs human laborers. While there are some factories that can go "lights out" (no humans; fully autonomous machines building products in the dark), these are few and far between, and they require the product to be designed from the ground-up for 100% automated assembly. For every other factory, the introduction of robots represents a mixture of other risks to worker safety that needs to be accounted for and minimized, as well as a reduction of overall risk. While an individual robot represents a risk to the laborers around it - the same as a CNC machine

Table 18. FR3-DP3 Pairs

FR3.1: Hedge against raw materials in storage losing their value	**DP3.1:** Utilize Put Options contracts to take a 'short' position against all raw materials that must be kept on-hand
FR3.2: Hedge against price increases in raw materials needed to satisfy orders	**DP3.2:** Utilize Call Options contracts to take a 'long' position against all raw materials that must be purchased in the future to satisfy existing and forecasted orders
FR3.3: Hedge against outsourced component shortages	**DP3.3:** Utilize multiple sources of qualified component suppliers

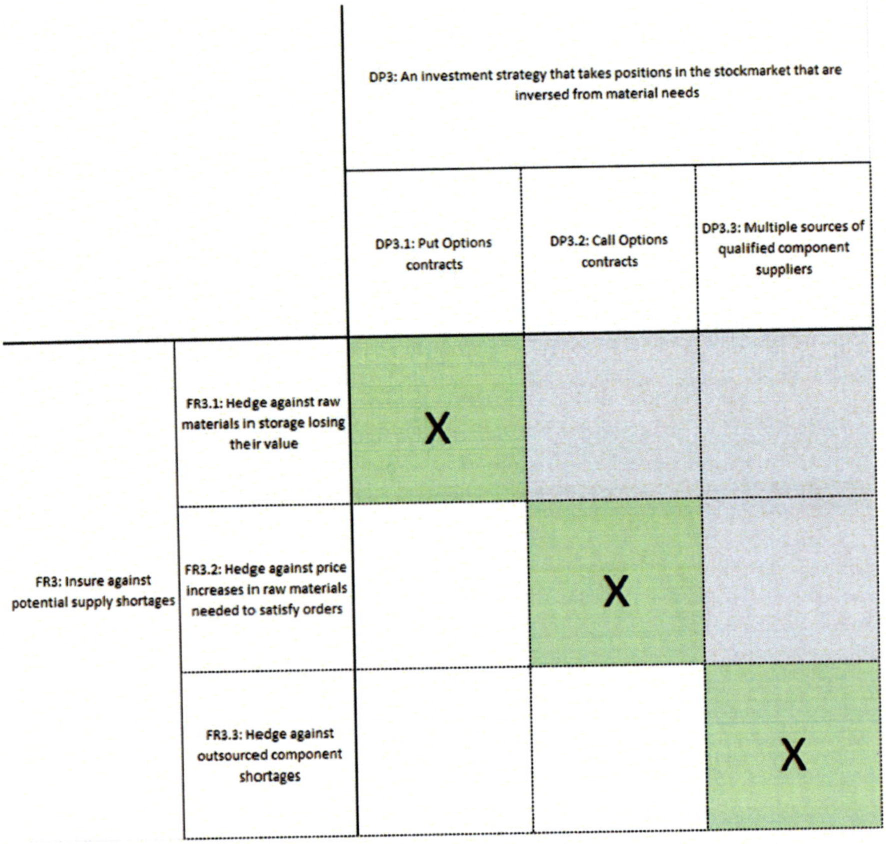

Fig. 19. FR3-DP3 Pairings

would, it also represents an elimination of risk by removing a human from the labor equation as well. The only way to 100% eliminate risk to a laborer is to remove that laborer from the work environment altogether. Robotics is one of the few technologies that can accomplish this. Meanwhile, when introducing a robot, care must be taken to install the appropriate barriers and interlocking systems to ensure that a laborer cannot be accidentally injured.

FR4: Maintain worker safety at all times

DP4: A system that monitors injury occurrences, and correct root causes from the feedback

Like FR3-DP3, worker safety (specific to how to keep them safe) is largely outside of the scope of this thesis. Care must be taken to design safe robotic manufacturing cells, but they do not play a role in employee attraction or retention when they are made safe to work around. It is likely that failing to design a safe robotic system will result in a negative impact on employee retention, however, this was not revealed in the decomposition in Fig. 3. This suggests that there is further decomposition to be made for both FR-DP4 and FR-DP6, or that the interaction may be revealed in an analysis of the CA-FR or DP-PV matrices.

Unlike FR3-DP3, FR4-DP4 is not an uncoupled matrix. There are interactions between FR4.4-DP4.3, and FR4.6-DP4.2. The decomposition of FR4-DP4 can be found in Table 19, and its matrix can be seen in Fig. 20.

3.7 FR5-DP5: Talent Attraction

Since this hypothetical factory cannot operate without human labor still, recruiting talent still needs to be considered for the factory. Even if all the manual tasks could be completely automated, there would still be a need for other support roles elsewhere in the company.

FR5: Attract the best talent available in the market	**DP5:** A program that actively engages with professionals - both young and experienced - and students, to maximize the bandwidth of the talent pipeline

The decomposition of FR5-DP5 can be found in Table 20, and its matrix can be seen in Fig. 21 (Table 20).

3.8 FR6-DP6: Talent Retention

With the attraction of talent comes the retention of talent. While the two may seem related at first glance, the reasons that people join a new company tend to be quite different from the reasons someone might leave their current company.

Table 19. FR4-DP4 Pairs

FR4.1: Capture all instances of recordable injuries	**DP4.1:** A consequence-free, injury reporting tool (reactive safety)
FR4.2: Determine root cause of recordable injuries	**DP4.2:** An independent accident & safety investigation team
FR4.3: Track injury rates relative to production areas	**DP4.3:** A tool for consistently logging data about accidents
FR4.4: Track injury rates relative to production tasks	**DP4.4:** A tool for feeding back safety data to process designers
FR4.5: Make feedback about injury data available to all employees	**DP4.5:** A system for disseminating statistics about safety & accident trends, and safe work practices
FR4.6: Create a system for anonymously and privately reporting safety concerns	**DP4.6:** A consequence-free safety-concern reporting tool (proactive safety)

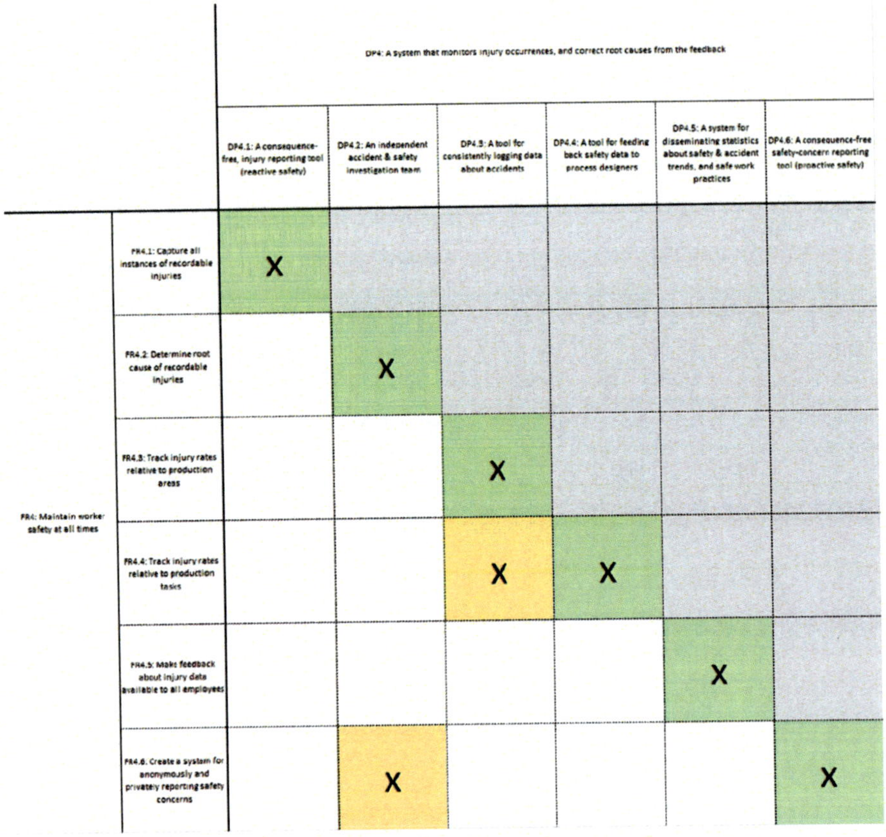

Fig. 20. FR4-DP4 Pairings

Management can't control why someone would want to leave their old role, so all that can be done is offer more money than other companies competing for the same talent, so that new talent may be more easily attracted. But management can make efforts to retain the talent they already have. Money is a large part of this as well, but in the case of retention, it also involves increasing the amount of money an employee receives each year - through direct pay, bonuses, and benefits - so that they do not feel any financial need to begin looking at what roles at other companies are listing for their salaries.

It should be noted that without further decomposition "raises" is a stand-in for the complicated topic of the relationship between labor and capital, a discussion that becomes even more complicated (and important) in automated manufacturing environments.

Table 20. FR5-DP5 Pairs

FR5.1: Offer average to above-average starting pay	**DP5.1:** A system to monitor average pay, relative to responsibilities, at direct competitors
FR5.2: Recruit top-performing employees from direct competitors	**DP5.2:** A program for collecting, publishing, and presenting the most technically interesting work currently being performed at the company by top-employees
FR5.3: Recruit from ABET accredited engineering schools	**DP5.3:** Co-op partnerships with programs teaching skills relevant to the business

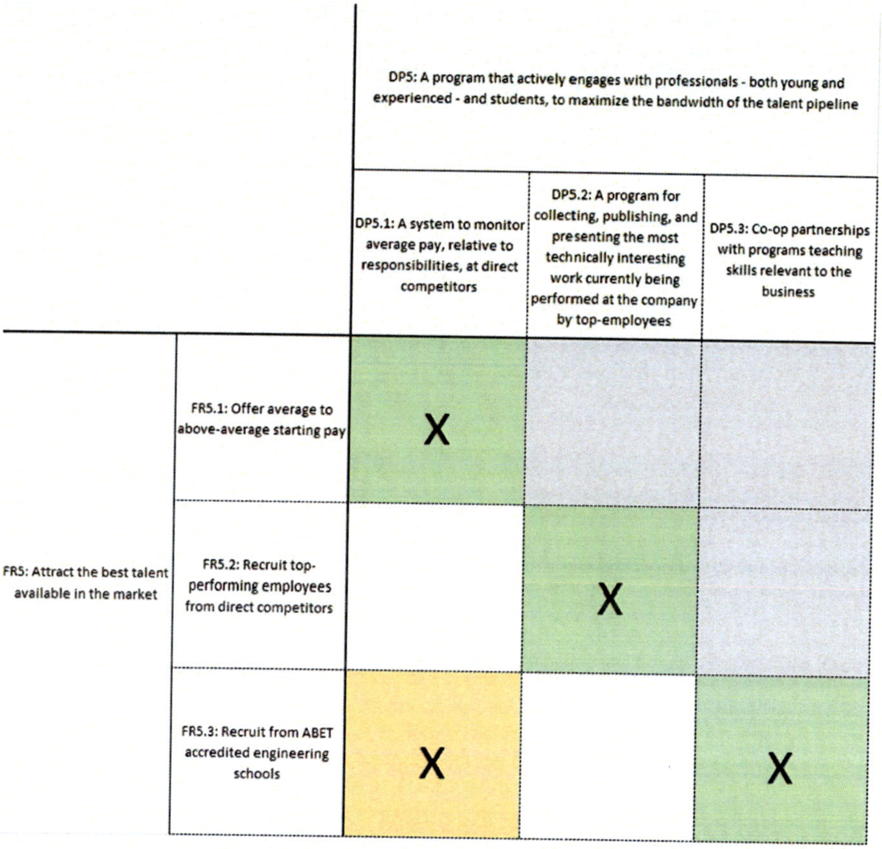

Fig. 21. FR5-DP5 Pairings

FR6: Retain the best talent available in the market

DP6: Constant monitoring of market compensation packages, with proactive raises to match current market rates for all employees that meet or exceed performance goals

The decomposition of FR6-DP6 can be found in Table 21, and its matrix can be seen in Fig. 22 (Table 21).

For FR-DP pairs 3, 4, 5, and 6, all of them are included through their first decompositions to ensure that FR0-DP0 is truly decoupled. However, none of them appear to directly interact with FR-DP pairs 1 or 2, where the primary focus of their thesis was: robotics and automation in a manufacturing environment. FR-DP pairs 3, 4, 5, and 6 all merit further study and likely can be decomposed into more layers.

Table 21. FR6-DP6 Pairs

FR6.1: Increase pay rate improvements to meet or beat competitor's	**DP6.1:** A system to monitor increases in compensation across the market
FR6.2: Sharing profits with employees	**DP6.2:** Bonuses paid out relative to profit goals
FR6.3: Make employees stakeholders in company ownership	**DP6.3:** Offer long stock option contracts to employees
FR6.4: Offer generous retirement plans	**DP6.4:** Offer employees employees generous plan contributions, and investment flexibility
FR6.5: Offer generous health plans	**DP6.5:** Keep employee out-of-pocket costs for medical expenses to a minimum
FR6.6: Hold managers accountable to their direct reports	**DP6.6:** A system for employees to review the performance of their direct managers, as a factor in the manager's performance regular performance reviews
FR6.7: Maintain a healthy work-life balance	**DP6.7:** Offer ample time off for life outside of work (child leave, PTO, sick time, flexible working schedules, etc.), and not only make is possible to utilize this time, but encourage them to

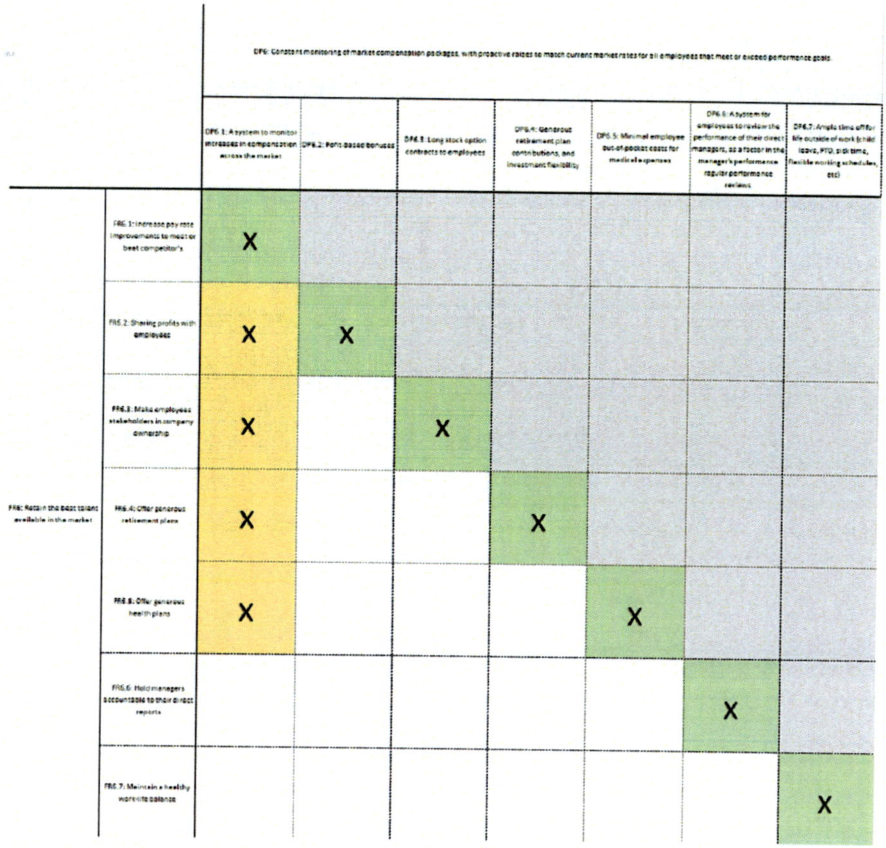

Fig. 22. FR6-DP6 Pairings

4 Discussion

By utilizing Axiomatic Design, not only can an entire automated factory be designed, but its supply chain can be made independent of its process cycle. It also becomes possible to determine which aspects of a product design are important to emphasize to help ensure the greatest financial success in the factory. Finally, using Axiomatic Design, becomes straightforward to identify and understand all the ways certain changes to both a product or a process could impact the overall yield and cycle time in the factory.

Interestingly, it seems that there are no interactions between the automated portions of the factory and the human portions, at least in terms of worker safety, attracting talent, and retaining talent. This was a surprising observation, and runs counter to the author's own experiences working in a large factory. There initially was an expectation to find an interaction between automated production cycles and the number of workers required on the fringes needed to support them - not unlike robots sitting inside of an imaginary volume and human laborers residing on the surface of that same volume, with both being necessary to successfully complete a production cycle.

A potential explanation for the lack of interactions between automation and worker safety, attraction, and retention is that by introducing robotics, you naturally eliminate the need for all three of these items for that particular position. If a task is automated, you do not need to attract nor retain talent for it. If a task is automated, there is no human present to be injured. Thus, it makes sense that there would not be any interactions between these three 'human' aspects of the Axiomatic Design matrix, and automation.

A possible limitation of this work was also identified upon peer review: it is possible that this design only works when a company already has a dominant position in its market. No consideration was made for the growth of the company in the Axiomatic Design matrix, only the growth of markets and a company's share of it. This is likely the result of author bias. It may be possible to eliminate this bias with additional work; through further decomposition, working with the other domains, or changing the overall design itself.

5 Conclusion

In conclusion, while there is still more work to be done, this thesis proves that it is possible to design at least a decoupled automated manufacturing process.

5.1 Future Work

This matrix still requires further study. Additional decompositions of FR4, FR5, FR6, and their matching DPs will likely reveal further information about automating a factory. There may be additional considerations in regards to all three of these FRs when it comes to laborers that are working in the periphery of an automated production cell, but all should be studied with the input of social scientists, as well as industry experts. FR3 also merits further decomposition to reveal more detail about the finances of running an automated factory, and those with experience in business administration should be engaged here. FR1 and FR2 can also be further decomposed, but doing so will likely require a specific manufacturing challenge to guide the decomposition process; an end goal (product) will need to be considered, so that its manufacturing process has a fixed set of CAs that FRs, DPs, and PVs can be designed for. The introduction of CAs and PVs could reveal interactions that are not visible in the FR-DP matrix.

References

1. Ahmed, F., Fattani, M.T., Ali, S.R., Enam, R.N.: Strengthening the bridge between academic and the industry through the academia-industry collaboration plan design model. Frontiers Psychol. **13** (2022). Accessed 19 Feb 2023

2. Suh, N.P.: Axiomatic Design: Advances and Applications. The MIT-Pappalardo Series in Mechanical Engineering. Oxford University Press, New York (2001)
3. Cochran, D.S., Reynal, V.A.: Axiomatic Design of Manufacturing Systems (1996). Accessed 30 Oct 2021
4. Six Sigma Definition - What is Lean Six Sigma? — ASQ. https://asq.org/quality-resources/six-sigma. Accessed 9 Feb 2023
5. Imai, M.: Kaizen (Ky'zen), the Key to Japan's Competitive Success. Random House Business Division, New York (1986). https://archive.org/details/kaizen00masa. Accessed 9 Feb 2023
6. Total Quality Management (TQM): What is TQM? — ASQ. https://asq.org/quality-resources/total-quality-management. Accessed 9 Feb 2023
7. PDCA Cycle - What is the Plan-Do-Check-Act Cycle? — ASQ. https://asq.org/quality-resources/pdca-cycle. Accessed 9 Feb 2023
8. Five Whys and Five Hows — ASQ. https://asq.org/quality-resources/five-whys. Accessed 9 Feb 2023
9. Durmusoglu, M.B., Satoglu, S.I.: Axiomatic design of hybrid manufacturing systems in erratic demand conditions. Int. J. Prod. Res. **49**(17), 5231–5261 (2011). https://doi.org/10.1080/00207543.2010.510487
10. Kulak, O., Durmusoglu, M.B., Tufekci, S.: A complete cellular manufacturing system design methodology based on axiomatic design principles. Comput. Ind. Eng. **48**(4), 765–787 (2005). https://doi.org/10.1016/j.cie.2004.12.006
11. Savov, I.: No Bullshit Guide to Linear Algebra, 2nd edn. v2.2 edn. Minireference Co., Montréal (2020). OCLC: 1310304974
12. Farid, A.M., Suh, N.P. (eds.): Axiomatic Design in Large Systems. Complex Products, Buildings and Manufacturing Systems. Springer, Cham (2016). https://doi.org/10.1007/978-3-319-32388-6
13. Qiao, J., Shang, J.: Application of axiomatic design method in in-pipe robot design. Rob. Comput. Integr. Manuf. **29**(4), 49–57 (2013). https://doi.org/10.1016/j.rcim.2012.10.007
14. Sui, Z., Xiang, L., Jenkins, O.C., Desingh, K.: Goal-directed robot manipulation through axiomatic scene estimation. Int. J. Rob. Res. **36**(1), 86–104 (2017). https://doi.org/10.1177/0278364916683444

Identifying a Device for Tracking the Evolution of Thermal Transfer in 3D Printed Parts Using Principles from Axiomatic Design

Adelina Hriţuc(✉), Oana Dodun, Petru Duşa, Andrei Mihalache, Gheorghe Nagîţ, and Laurenţiu Slătineanu

Gheorghe Asachi Technical University of Iaşi, 700050 Iaşi, Romania
adelina.hrituc@student.tuiasi.ro

Abstract. The evolution in the last decades of the manufacturing processes of parts by 3D printing has revealed the possibilities of changing the material properties of these parts using the values of some of the input factors in the 3D printing process. For the development of experimental research in such a direction, the requirement was formulated to design a device that would allow tracking the evolution of thermal transfer in parts manufactured by 3D printing. In this regard, it was proposed to use a test sample of a lamella made of polymeric material heated at one of the ends. The evolution of the thermal field affecting the sample could thus be followed using an infrared camera. The various components of the test sample support device were identified by applying some principles from axiomatic design. Analysis of the design matrix revealed that a decoupled design was reached. A principle solution was established for the device intended to track the evolution of the thermal field in the sample as a lamella.

Keywords: Heat Transfer · 3D Printed Part · Polymer Lamella · Experimental Device · Axiomatic Design

1 Introduction

Researching how heat is transmitted through parts made of different materials is of interest both from the point of view of obtaining a faster transfer of the heat released by a heat source and for the better characterization of insulating materials from a thermal point of view. The concept of *heat transfer* refers to the way in which the characteristics of heat are transmitted. In the case of solid bodies, heat transfer occurs by convection. If metals and metal alloys are generally appreciated as good heat-conducting materials, polymeric materials provide lower conditions for rapid heat transmission, being sometimes considered as insulating materials.

Parts made of polymeric materials can be obtained through various processes. The last two decades have highlighted an expansion of manufacturing parts from plastic materials through additive technologies. One such additive manufacturing technology is 3D printing. This technology allows the modification of the manufacturing conditions

E. Puik et al. (Eds.): ICAD 2023, LNNS 849, pp. 66–74, 2024.
https://doi.org/10.1007/978-3-031-49920-3_4

within wide limits, which facilitates obtaining materials with varied internal structures and characterized, as such, by different heat transmission capacities.

Different experimental research methods have been designed and applied to study the thermal properties of materials in 3D printed parts and therefore the ability of these materials to allow heat transfer.

Thus, de Rubeis et al. generated by 3D printing polylactic acid test samples with different geometries and free spaces inside [1]. Theoretical modeling was used, and, respectively, experimental research in which a heat flow meter and infrared thermography were used to evaluate the way heat is transmitted through these test samples. The thermal insulation capacities of the honeycomb structures were thus confirmed.

High-resolution infrared thermography was used by Muñoz – Codorníu for analyzing anisotropic heat flow in 3D porous architectural structures made of silicon carbide [2]. In this sense, they proposed using a device intended for laboratory applications.

Farzinazar et al. used infrared thermography to study thermal transfer in shape memory polymers embedded in 3D printed samples [3]. They appreciated that the main factors influencing thermal transfer are shape, solid volume fraction, and temperature.

It is noted that there is still research related to the approach of the heat transfer problem using axiomatic design [4–8].

The literature review led to the observation that an experimental investigation of the mode of heat transfer through 3D-printed polymer material parts could be performed using a test sample in the form of a lamella heated at one end. In the present paper, identifying a device solution for supporting the lamella-type test sample was considered using principles from axiomatic design.

2 Considered Experiment Scheme

The objective pursued was to ensure the conditions for visual highlighting and to evaluate the mode of heat transmission through test samples manufactured by 3D printing from a polymeric material. Different values of the input factors in the 3D printing process were used to manufacture the test samples.

It was preferred to use a lamella-shaped test sample manufactured from a polymeric material by 3D printing (Fig. 1). It was hypothesized that by heating the slide at one end and observing the lamella with an infrared camera, one could track how the test sample heats up over time. At the same time, the temperatures reached during heating in different areas of the lamella could be evaluated. In Fig. 1, three arrows indicate the propagation of heat from the support part to the lamella, and the other three arrows indicate heat dissipation by the lamella in the surrounding environment.

If, initially, the lamella will be at the temperature of the surrounding environment, once the heating process of one end of the lamella is initiated, there will be a transfer of heat to the rest of the sample. It is expected that a thermal flow will occur, and it will move from the heated end of the lamella to the rest of the material of the test sample, initially at ambient temperatures. Part of the heat will be transmitted from the lamella to the surrounding environment through the lamella's free surfaces. When the entire amount of heat that enters the lamella from the heated end of it passes through a part

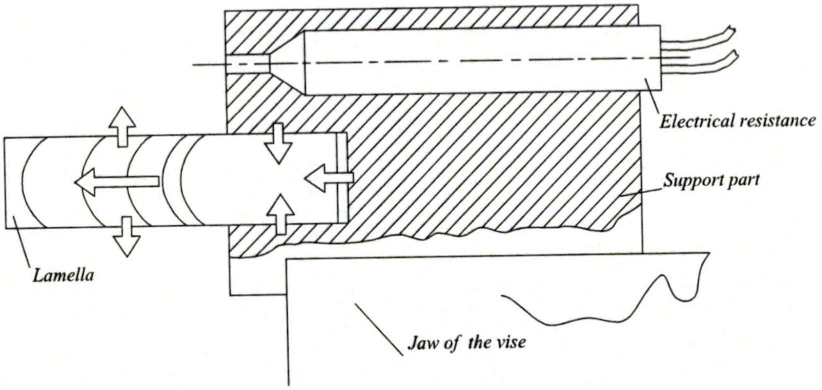

Fig. 1. Hypothesis on how to highlight the transmission of heat with the help of an infrared camera

of the test sample and is subsequently discharged to the outside environment, a state of thermal equilibrium will be reached.

The following factors or groups of factors could influence how the heat transfer occurs along the test sample:

- The nature and structure of the lamella material;
- The temperature at which the environment surrounding the free area of the test sample is located;
- Maximum temperature at the heated end of the test sample;
- The speed with which the temperature increases at the heated end of the lamella;
- The dimensions characterizing the cross-section of the test sample;
- The capacity of the test sample material to ensure heat transfer to the external environment;
- The ability of the external environment to absorb the heat transferred by radiation and convection from the test sample.

3 Using Some Principles from Axiomatic Design to Design the Device for Heat Transfer Research

Professor Nam Pyo Suh proposed the axiomatic design in the 70s of the previous century. Such a way of designing followed the improvement of manufacturing technologies, but nowadays the axiomatic design is applied to solve problems in very different fields. In principle, axiomatic design involves the use of two axioms: a) The Axiom of independence of functional requirements; b) The axiom of information, according to which the version of the project that requires a minimum of information will be used [6–9]. These principles aim to ensure that the functional requirements of a system are independent and the design is as simple as possible.

Designing a device for the study of heat transfer by using some principles from axiomatic design [9] makes it necessary to define a so-called *customer*. In this situation, it will be considered that the possible client is a Ph.D. student or a young researcher. He must develop research aiming at highlighting the influence of different factors on heat transfer through the materials incorporated in the parts manufactured by 3D printing.

A further sequence will follow the discovery of the customer's need. Following those mentioned above, it will be considered that there is only one customer requirement and that it could be formulated in the following way:

CN: Ensure the existence of a device that allows the study of the influence exerted by different factors on how heat transfer occurs in the material of a test sample manufactured by 3D printing.

Before moving on to the next step, the development of the zero-order functional requirement FR0, it is necessary to mention that the possibilities of using an infrared camera have been analyzed. It is thus known that such equipment (infrared camera) allows highlighting on a screen the temperatures reached in different areas of a body by receiving the infrared radiation emitted by these areas. Using the received radiation, an image is generated on a screen. In this image, areas with different temperatures are represented by different colors or shades of colors.

It is further necessary to develop the so-called *functional requirements FRs*, which should highlight the requests to which the device must respond to satisfy the customer's needs. The zero-order functional requirement may take the form of *FR*0: design a device capable of providing conditions for studying the influence exerted by different factors on the way thermal transfer is carried out in the material of a test sample manufactured by 3D printing. Following the principles of axiomatic design, for the zero-order functional requirement, the statement corresponding to the zero-order design parameter can be formulated: *DP*0: Device that ensures conditions for researching the influence exerted by different factors on how heat transfer occurs in the material of a 3D printed test sample. These design parameters will represent the technical features that can fulfill the FRs to gather them all together to obtain the proposed solution for the equipment.

The next stage of applying axiomatic design principles aims to decompose the zero-order requirement into first-order requirements. Subsequently, each zero-order functional requirement will be associated with a zero-order design parameter. Each *FR* will correspond to a specific *DP* as an applicable technical solution.

A review of the main requirements with the highlighting of the solution found as a design parameter that the tracked device will need to meet could be as follows:

*FR*1: Determine the shape of the test sample intended to allow the study of heat transfer evolution. It can be mentioned here that the results obtained through previous experimental tests that followed the thermal transfer, in more limited conditions, through some cylindrical polymer rods manufactured by 3D printing [10] were taken into account. The obtained results showed that it is more difficult to formulate general observations regarding the thermal transfer for such situations due to the cylindrical bar's relatively large thickness and the cross-section's specific circle shape. The design parameter that arose after formulating the functional requirement of *FR*1 was to use a plate-shaped test sample with a rectangular cross-section and sufficiently thin. Heating at one end of the test sample should allow a clearer image of how the heat is transmitted along the thin, constant-thickness lamella. Based on the second axiom of the axiomatic design [11, 12], the two types of test samples considered (the one in the form of a cylindrical bar with a circular section and, respectively, the one in the form of a lamella with a rectangular cross-section), the test sample variant in the form of a thin lamella, with a rectangular cross-section, was selected. There is a higher probability of success in the case of this variant to the first alternative since the lamella type test sample would allow obtaining an image closer to what the CN customer's requirement contains;

For the next *FRs*, the same approach was applied so that the dependence matrix FRs-DPs could be created at the end.

*FR*2: Provide a body of the device (support part) to which some of the other components will be assembled, such as those for locating and clamping the lamella, heating one end of the test sample, determining and adjusting the value of the heating temperature, etc. This device body could be of the monobloc type, as a body obtained by assembling distinct components could also be considered. For now, by also taking into consideration axiom two, a monobloc-type body was preferred;

*FR*3: Provide conditions for controlled heating of one end of the test sample. Among the various solutions to meet this requirement (flame heating, induction heating, heating using an electric resistance, etc.), the use of an electric resistance was preferred since it ensures simpler conditions for assembly and controlled heating of the part support and, through it, the end of the lamella-shaped test sample;

*FR*4: Provide conditions for supplying electrical current to the electrical resistance. It was assessed that it is the taking of energy from the outlet corresponding to the electrical network of the laboratory where the experimental research is carried out. The electric heating resistance will be supplied with electric current through a controller that allows the adjustment and maintenance of the temperature variation between certain limits to the temperature determined using a temperature sensor;

*FR*5: Provide possibilities for programming and controlling temperature values. The previously mentioned temperature controller and sensor could help meet this requirement;

*FR*6: Provide conditions for assembling the subsystem corresponding to the temperature sensor to the support part.

*FR*7: Provide conditions for placing the device on a locksmith table in the laboratory. The experience accumulated through previous research [10] highlighted the need to use lower temperatures (around 60–100 °C) to avoid reaching a temperature at which the material of the test sample could reach a state of plasticity. The solution identified for this purpose could involve the use of a vise. The previously mentioned reason and the relatively high thermal conductivity of the vise parts' metal material were considered. Since the support part will heat itself, it may be necessary to use an intermediate part made of a material with low thermal conductivity and placed between the support part and the vise. It was thus estimated that the vise components would be relatively low heating, making the use of the intermediate piece unnecessary.

As previously mentioned, the corresponding design parameters in the form of solutions found were also identified for each functional requirement. A synthetic presentation of the correlations between the first-order operating requirements and the first-order design parameters can be seen in Table 1.

For some of the highlighted functional requirements, it is possible to resort to the continuation of the decomposition activity by generating the second-order functional requirements.

For example, in the case of functional requirement *FR*2, the following could be considered:

Table 1. The matrix containing *FRi* functional requirements and *DPi* design parameters in the case of the equipment for the study of thermal transfer in the polymeric material of a test sample manufactured by 3D printing.

1		Design parameters	DP0: device to allow the study of the influence exerted by different factors on the way thermal transfer takes place in the material of a sample manufactured by 3D printing.						
2	Functional requirements		*DP1*: Proba de testare sub forma unei lamele	*DP2*: Monobloc support piece	*DP3*: Electrical resistance	*DP4*: The electrical network	*DP5*: Controller and temperature sensor	*DP6*: Subsystem for assembling the sensor	*DP7*: Vise
3 Column no. 1	2	3	5	6	7	8	9	10	11
4	Zero order functional requirement	1st order *FR* functional requirements	Highlighting the *DPi* design parameters corresponding to each *FRi* functional requirement						
5	*FR0*: design a device capable of providing conditions for studying the influence exerted by different factors on the way thermal transfer is carried out in the material of a sample manufactured by printing	*FR1:* Determine the shape of the specimen intended to allow the evolution of heat transfer	X						
6		*FR2* Provide for a device body (a support part) to which some of the other components will be assembled		X					
7		*FR3*Provide controlled heating of one end of the test sample		X	X				
8		*FR4*: Ensure the electrical current supply to the electrical resistance				X			
9		*FR5*: Provide programming and control of temperature values		X		X	X		
10		*FR6*: Provide conditions for assembling the subsystem corresponding to the temperature sensor to the support piece						X	
11		*FR7*: Secure the placement of the device on a locksmith table in the laboratory		X					X

*FR*2.1: Determine the basic shape of the support part;

*FR*2.2. Determine the material of the support part;

*FR*2.3. Ensure the locating and clamping of the test sample in the support part;

*FR*2.4. Ensure the shape and position of the clearance for the temperature sensor.

These functional requirements could be assigned the following *DPi* design parameters:

*DP*2.1. Support part in the form of a plate, with the removal of material from areas where it is not needed;

DP2.2. Aluminum. It could also be considered to use copper to make the support part, but this material is more expensive;

*DP*2.3. A clearance to facilitate the self-clamping of the test sample, preferably a clearance whose flat walls form an angle of about 20°. It is noted, at this moment, that exploiting the zig-zagging facilities [12], it is necessary to modify the design parameter *DP*1: the test sample must have the shape of a lamella, but the end that will be used to locate and clamp the test sample will have to have flat walls arranged at an angle of 20° (Fig. 2, a). Other ways of locating and clamping the test sample could be considered, for example, by using a parallel-walled clearance with some elasticity to immobilize the test sample (Fig. 2, b), or the clamping could be by using screws and nuts (Fig. 2, c). It was appreciated that the angular release would allow a more efficient transfer of heat through the inclined walls, and at the same time, it is simpler than the variant that involved the use of screws and nuts. Therefore, the requirement corresponding to the second axiom seems to consider the variant with angular release as having a higher probability of success.

*DP*2.4: Threaded hole located perpendicular to the axis of the cylindrical surface corresponding to the electrical resistance subsystem (the latter being a commercially available component).

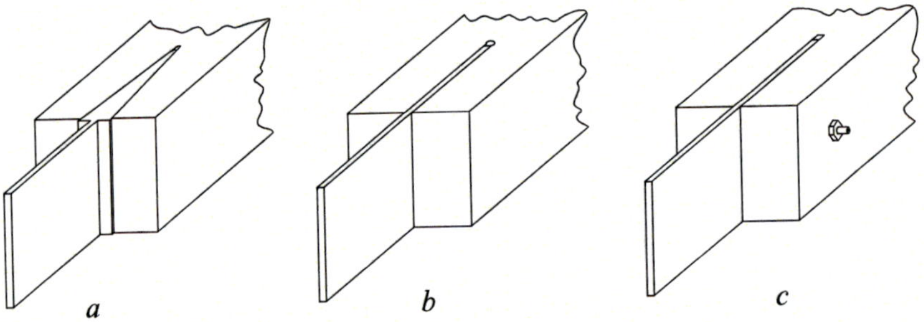

Fig. 2. Variants taken into account when establishing the design parameter *DP*2.3.

The matrix equation corresponding to the functional requirement *FR*2 can now be written:

$$\begin{Bmatrix} FR2.1 \\ FR2.2 \\ FR2.3 \\ FR2.4 \end{Bmatrix} = \begin{bmatrix} X & 0 & 0 & 0 \\ X & X & 0 & 0 \\ X & 0 & X & 0 \\ X & 0 & 0 & X \end{bmatrix} \cdot \begin{Bmatrix} DP2.1 \\ DP2.2 \\ DP2.3 \\ DP2.4 \end{Bmatrix} \quad (1)$$

The analysis of the information in Table 1 and in Eq. (1) highlights that in both situations, we are dealing with *a decoupled design* since fulfilling some functional requirements requires the involvement of two design parameters. The obtained result can be represented in the form of an upper triangular matrix. It can be seen that revealing the correlations between the functional requirements *FRs* and the design parameters *DPs* led to the placement of "X" symbols below the descending diagonals of the matrix representation.

Certain constraints were also considered in the design of the schematic diagram of the device. Such constraints referred to the need to identify a solution that is not too complex and has dimensions that allow the support piece to be fixed in an existing laboratory vise. Another constraint was that the components of the device could be purchased from the trade or require only processing that can be done on the machine tools of the laboratory.

4 Proposed Solution

The solution, whose schematic representation can be seen in Fig. 3, was proposed by considering the results of applying some axiomatic design principles. It is noted that the area of the lamella that protrudes outside the support part can be examined and filmed using an Infrared camera. The heating of the support part is made using electric resistance. Achieving and maintaining a low variation of a pre-set temperature by the support part is possible using a temperature sensor and controller.

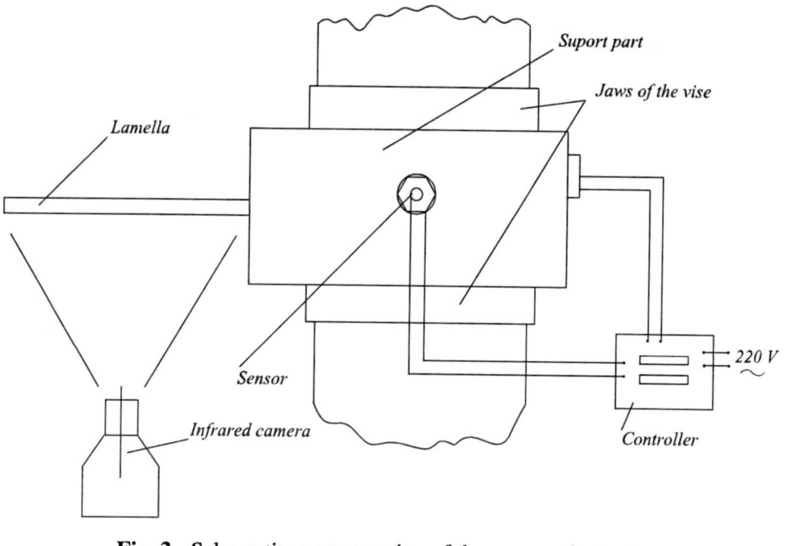

Fig. 3. Schematic representation of the proposed solution.

5 Conclusions

The need to design a device that allows the investigation of the influence exerted by some factors on the evolution of thermal transfer inside some test samples of polymeric material manufactured by 3D printing was taken into account. It has been appreciated that a lamellar specimen provides general information on how heat is transmitted through the test sample when one end of the test sample is heated. Some axiomatic design principles were used to identify a solution corresponding to the device for supporting and heating the test sample. In this sense, the main functional requirements of the first order were formulated, to which the corresponding design parameters were attached. When a functional requirement could be met in more than one way, the second axiom of the axiomatic design was used to select the alternative that could provide the highest probability of successful use of the device. The development of the design matrix led to the observation that the proposed solution corresponds to a decoupled design. By applying the axiomatic design, a device was proposed that would allow the highlighting of the heat propagation way inside a parallelepiped plastic test sample manufactured by 3D printing. The intention is to realize and experimentally test the proposed device. Later, research will be carried out regarding the behavior of different plastic materials and distinct internal structures of the test samples made by 3D printing from the point of view of heat propagation through thermal conduction.

References

1. de Rubeis, T., Ciccozzi, A., Giusti, L., Ambrosin, D.: The 3D printing potential for heat flow optimization: influence of block geometries on heat transfer processes. Sustainability **14**, 15830 (2022)
2. Muñoz Codorníu, D., Moyano, J.J., Belmonte, M., Osendi, M.I., Miranzo, P.: Thermal conduction in three-dimensional printed porous samples by high resolution infrared thermography. Open Ceram. **4**, 100028 (2020)
3. Farzinazar, S., Wang, Y., Owens, Abdol-Hamid, C., Yang, C., Lee, H., Lee, J. : Thermal transport in 3D printed shape memory polymer metamaterials. APL Mater 10, 081105 (2022).
4. Soderborg, N.R., Crawley, E.F., Dori, D.: System definition for axiomatic design aided by objectprocess methodology, In: Proceedings of ICAD2002 Second International Conference on Axiomatic Design, Cambridge (2002)
5. Heo, G., Lee, T., Do, S.-H.: Interactive system design using the complementarity of axiomatic design and fault tree analysis. Nucl. Eng. Technol. **39**(1), 51–62 (2007). https://doi.org/10.5516/NET.2007.39.1.051
6. Heo, G., Jeong Y.H.; Design of safety injection tanks using axiomatic design and TRIZ. In: International Youth Nuclear Congress 2008, Interlaken, Switzerland, Paper No. 200, (2008)
7. Wang, Y., Liu, A.: A student's axiomatic design application example of battery thermal controller for high altitude balloon. MATEC Web Conf. **223**, 01017 (2018)
8. Torabi, A., Abedian, A., Farsi, M.A., Kalantarinezhad, R.: Axiomatic design of a reflective multilayer high-temperature insulator. Proc. Inst. Mech. Eng., Part G: J. Aerosp. Eng. **233**(2), 457–471 (2019)
9. Suh, N.P.: Axiomatic Design: Advances and Applications. Oxford University Press, New York (2001)
10. Hriţuc, A., Mihalache, A.M., Slătineanu, L., Dodun, O., Nagîţ, G.: Heat transfer in 3D-printed polymer cylindrical parts. Paper accepted for publication in Macromol Symp. (2023).
11. Fradinho, J., Gonçalves-Coelho, A.: The Information Axiom and Robust Design. In: Suh, N.P., Cavique, M., Foley, J.T. (eds.) Design Engineering and Science, pp. 307–325. Springer, Cham (2021). https://doi.org/10.1007/978-3-030-49232-8_11
12. Cavique, M., Gabriel-Santos, A., Mourão, A.: Mapping in Design. In: Suh, N.P., Cavique, M., Foley, J.T. (eds.) Design Engineering and Science, pp. 255–284. Springer, Cham (2021). https://doi.org/10.1007/978-3-030-49232-8_9

Sustainable Manufacturer Engineering for Industry 6.0

Janne Heilala(✉) and Jussi Kantola

Department of Mechanical and Materials Engineering, University of Turku Technology, Turku, Finland

janne.p.heilala@utu.fi

Abstract. This is axiomatic design (AD) research of Finnish Manufacturers practices to optimize design principles in European-level. We dive into the deep end of the pool by leveraging cutting-edge AD methodology by partnering with businesses surveying to see best productivity. The companies are integrated from simplified industry 6.0 perspective, while some industries technology readiness is benchmarked between 4.0–5.0 compliant. In this paper we take new insights into consideration from the supply chain, understanding contracts, competitiveness, and profit-seeking companies' sustainability strategies. We discover high-order axioms and their complex, multidimensional modeling, refined through supplier selection models. Hold tight as we focus on organizational planning based on sustainable process management. Observe how outsourcing simplifies organizational structures and how the complex becomes simple when the production organization aligns to maximize revenue, control, and management. Witness the creation of a decoupled AD for higher-level manufacturing organizations design suitable for horizon financing activities.

Keywords: axiomatic design · organizational concepts · European manufacturing survey

1 Introduction

AD principles have gained traction, with researchers exploring new applications beyond traditional engineering design, including manufacturing engineering [1]. Emphasizing manufacturing engineering leads to successful operations through effective organizational design, which can transform supply chains and create ripples in society. Crucial to any manufacturing capacity, successful operations necessitate efficient manufacturing organizations [2]. To address uncertainties, semi-structured surveying methods have been combined with Axiomatic Design (AD) [3].

Historically, organizational practices focused on manual labor for added-value jobs divided into specialized tasks. However, recent decades have seen industrial engineers shift their focus towards systematic thinking in human-centered design (HCD) to improve productivity [4, 18]. This approach addresses the integration of various human systems in response to technological challenges in industrial markets. In this study, AD been

E. Puik et al. (Eds.): ICAD 2023, LNNS 849, pp. 75–87, 2024.
https://doi.org/10.1007/978-3-031-49920-3_5

employed in research and development methodologies within small and medium-sized enterprises (SMEs) to create flexible and agile manufacturing system integration. While other design principles share similarities with AD, many is less relevant in effective generation of design domains [5]. Consequently, AD will be used in this study to examine new organizational design.

A four-layer representation of the production system consists of enterprise organization, plant organization, production organization, and operation. The system depends on resources or technologies within the production layout, with innovation contributing to the enterprise's value from an operational standpoint. The entire organizational scope, from goal-setting to strategy development and process execution, defines the enterprise's value. Production encompasses various stakeholders, including customers and businesses (e.g., distributors). Improving organizational design in production can be divided into three managerial tasks: problem-solving in design and engineering, information management, and resource transformation [6].

This research investigates the applicability of two fundamental axioms in organizational design. Axiomatic domain mapping for the production system design levels is achieved using empirical findings from the business portfolio. Axiom 1 maintains the independence of functional requirements (FRs), while Axiom 2 minimizes the design's information content. These axioms form the basis of design domain relationships, addressing the complexity of interconnections between customer domains, constrained FRs, and design parameters [6, 11–13].

Axiomatic vector spaces provide the foundation for finite-dimensional prototype matrices [1]. Real number scalars denote a collection of vectors within these matrices, allowing linear algebra to effectively represent relationships between various entities [2]. In this study, we utilize empirical data obtained from Finnish manufacturing companies participating in the European Manufacturing Survey (EMS) [3]. By characterizing the primary production of these enterprises through ideal organizational practices associated with growth companies, we derive insights from the connections and magnitudes of the measured observations [4].

2 Empirical Method and Material

This study utilizes manufacturing research measurements from the Finnish manufacturing industry. The Delphi approach, along with axiomatic theories, has been suggested in previous research as a suitable method for data analysis (e.g., [7]).

2.1 EMS Data Source

Data were collected from Finnish manufacturing companies using the EMS research instrument. Company representatives responded to coded arguments (abbreviated as coding) [8]. Z-score normalization was applied to the data using IBM's statistical package for social science analysis. The sample comprised of supply chain contract (SCC) companies, including operating manufacturers (MFR, m03a1-m03a3), contracted suppliers (SPLR, m03a4-m03a5), and contract manufacturers (CM, m03a6) [8].

The study investigates the development of competitiveness and employment situations from the perspectives of turnover, employees' salaries, and capital utilization in SCC companies [8]. Selected variables include annual turnover (AT, m23a1), number of employees (NEs, m23b1), manufacturing capacity utilization (MCU, m23h), and return-on-sales (ROS, m23i1-m23i5) [8].

Organization concepts (OCs) consist of surpassing Industry 4.0 readiness to emphasized human creativity and innovativeness to Industry 5.0, while partially we use the sample from integration of industry 6.0 perspective:

a. Organizing production (OP), which involves planning (OP1, m06a1), customer- or product-oriented lines/cells organization (OP2, m06b1), the pull principle (OP3, m06c1), change-over time optimization or set-up time reduction (OP4, m06d1), and standardized work instructions organization (OP5, m06e1) for Industry 4.0–5.0.
b. Production management and control (PMC), which includes visual management (PMC1, m06f1), quality standard-based manufacturing (PMC2, m06g1), employee involvement in manufacturing and innovation (PMC3, m06h1), bonus systems for outstanding performances (PMC4, m06i1), environmentally conscious manufacturing (PMC5, m06k1), and energy management (PMC6, m06l1) for Industry 4.0–5.0.
c. Task- (TCD1, m17a1), cross-functional- (TCD2, m17b1), digital product/system implementation support- (TCD3, m17c1), and data security/compliance-related (TCD4, m17d1) TCD key measures [8] for industry 5.0.

SPSS was used to process the observed variables, yielding an outcome space for each organization ($n = 31$). Table 1 presents the descriptive matrix, while Table 2 displays the correlations.

2.2 Extracting Statistics into Axiom Testing

The integration in systems engineering is examined through cross-tabulation, a method of data analysis that scrutinizes the relationships between variables with the aid of convenience sampling. This approach yields numerical values that reflect relationships with syntactical and semantical significance, thus facilitating supervised learning in pairs, and offering insights into the data reliability among factors.

From a systems design perspective, challenges are identified, milestones towards goals are set, and concepts evaluated, eventually leading to the selection of an optimal function as a system function. The axiomatic perspective allows for the evaluation of the design quality. Systems designed to address challenges underscore the importance of hierarchy in the design of subsystems, with the creation and evaluation of processes based on manufacturability or maintenance. Successful strategies are managed then with AD.

Table 1. The descriptive matrix according to the observed variables for the production organizations [8].

	MIN	MAX	M	MED	MOD	STD	SKEW	KURT	Range	Valid
AT21	.40	220.0	39.324	31.229	6.0	58.03	1.973	3.955	219.60	20.0
AT19	.10	250.0	36.372	7.00	6.0	63.20	2.542	6.993	249.90	19.0
NE21	4.0	600.0	129.86	70.00	4.0	157.608	1.993	3.693	596.0	22.0
NE19	3.0	500.0	113.55	50.00	50.0	138.699	1.816	2.648	497.0	22.0
MCU21	0.0	100.0	68.00	80.00	80.0	30.986	-1.322	.824	100.0	18.0
MCU19	0.0	100.0	65.00	79.00	80.0	32.820	-.927	-.467	100.0	15.0
ROS16	1.0	5.0	5.0	5.0	5.0	1.619	-1.026	-.583	4.0	19.0
OCS	.09	1.00	.531	.545	.545	.262	.191	-.567	.91	31.0
OP	.0	1.0	.497	.400	.4	.3049	.244	-.685	1.0	31.0
PMC	.00	1.00	.56	.67	.167	.312	-.230	-1.277	1.0	31.0
TCD	.0	1.0	.633	.600	4.0	.2928	-.450	-.401	1.0	30.0
OP1	0.0	1.0	.55	1.00	1.0	.506	-.204	-2.098	1.0	31.0
OP2	0.0	1.0	.48	.00	0.0	.508	.068	-2.138	1.0	31.0
OP3	0.0	1.0	.52	1.00	1.0	.508	-.068	-2.138	1.0	31.0
OP4	0.0	1.0	.32	.00	0.0	.475	.798	-1.462	1.0	31.0
OP5	0.0	1.0	.61	1.00	1.0	.495	-.487	-1.889	1.0	31.0
PMC1	0.0	1.0	.45	.00	0.0	.506	.204	-2.098	1.0	31.0
PMC2	0.0	1.0	.87	1.00	1.0	.341	-2.327	3.648	1.0	31.0
PMC3	0.0	1.0	.58	1.00	1.0	.502	-.344	-2.017	1.0	31.0
PMC4	0.0	1.0	.71	1.00	1.0	.461	-.972	-1.134	1.0	31.0
PMC5	0.0	1.0	.55	1.00	1.0	.506	-.204	-2.098	1.0	31.0
PMC6	0.0	1.0	.19	.00	0.0	.402	1.631	.702	1.0	31.0
TCD1	0.0	1.0	.83	1.00	1.0	.379	-1.884	1.657	1.0	30.0
TCD2	0.0	1.0	.50	.50	0.0	.509	.000	-2.148	1.0	30.0
TCD3	0.0	1.0	.47	.00	0.0	.507	.141	-2.127	1.0	30.0
TCD4	0.0	1.0	.83	1.00	1.0	.379	-1.884	1.657	1.0	30.0
TCD5	0.0	1.0	.53	1.00	1.0	.507	-.141	-2.127	1.0	30.0

2.3 Advanced Engineering and Technology Solutions

AD, particularly the sequential zig-zag approach, elucidates compact system structures with an emphasis on the design life cycle [9]. Complex systems should maximize functional independence and minimize information content. Ideally, an uncoupled model is preferred over a coupled model [10]. Stability in the long term can be achieved by a simplified system version [9 adapted to 10].

In the context of organizational design, multi-level components originating from EMS research precede the empirical evidence of each variable, enabling axiom testing. The tangible and intangible elements contributing to an organization's operations are emphasized on the practical side, while the empirical side relies on collected samples. Given the range of policies and integration domains, it's an opportunity to examine sociotechnical systems, amalgamate their organizational structures and practices, and select processes from unique systems for conversion to individual cultures, based on integration optimization.

Table 2. The correlation matrix according to the observed variables for the production organizations [8].

	AT21	AT19	NE21	NE19	MCU21	MCU19	ROS16	OCS	OP	PMC	TCD	OP1	OP2	OP3	OP4	OP5	PMC1	PMC2	PMC3	PMC4	PMC5	PMC6	TCD1	TCD2	TCD3	TCD4	TCD5
AT21	1
AT19	.987**	1
NE21	.917**	.875**	1
NE19	.950**	.944**	.978**	1
MCU21	.261	.297	.140	.123	1
MCU19	.231	.298	.344	.341	.953**	1
ROS16	.433	.409	.458*	.467*	.434	.591*	1
OCS	.720**	.671**	.703**	.671**	.296	.205	.598**	1
OP	.492*	.463*	.485*	.454*	.293	.055	.410	.809**	1
PMC	.703**	.662**	.689**	.665**	.204	.245	.591**	.878**	.428*	1
TCD	.475*	.420	.522*	.498*	.325	.355	.315	.302	.129	.368*	1
OP1	.216	.186	.210	.163	.188	.081	.249	.544**	.595**	.351	.015	1
OP2	.142	.148	.192	.169	.204	-.004	.194	.397*	.677**	.059	.077	.360*	1
OP3	.158	.177	.189	.178	.033	-.240	.292	.333	.657**	-.024	.062	.029	.679**	1
OP4	.628**	.608**	.656**	.644**	.153	.062	.397	.673**	.698**	.467**	.016	.349	.160	.254	1
OP5	.484*	.440	.384	.379	.369	.278	.224	.540**	.433*	.477**	.232	.077	-.158	.026	.407*	1
PMC1	.357	.318	.344	.290	.098	.063	.522*	.670**	.485**	.635**	.179	.693**	.159	-.029	.344	.322	1
PMC2	.480*	.426	.447*	.405	.204	.107	.360	.598**	.274	.696**	.522**	.017	.169	.224	.167	.264	.246	1
PMC3	.509*	.483*	.515*	.520*	-.208	.034	.495*	.642**	.254	.779**	.126	.276	-.092	-.050	.441*	.221	.438*	.465**	1
PMC4	.268	.234	.245	.262	.094	.254	.172	.386*	.060	.545**	.114	.037	-.012	-.180	.060	.287	.156	.258	.390*	1
PMC5	.596**	.539*	.513*	.496*	.412	.320	.542*	.727**	.336	.844**	.294	.218	-.159	-.100	.488**	.610**	.433*	.542**	.562**	.424*	1
PMC6	.842**	.866**	.869**	.879**	.288	.295	.333	.518**	.277	.571**	.197	.116	.179	-.016	.361*	.222	.212	.251	.313	.189	.445*	1
TCD1	.252	.225	.200	.174	.453	.142	.095	.309	.264	.264	.487**	.120	.060	.120	.316	.217	.211	.331	.098	-.175	.299	.200	1
TCD2	.467*	.425	.473*	.457*	.223	.295	.275	.176	-.022	.295	.764**	-.134	.000	.000	-.141	.208	.067	.471**	.073	.000	.267	.268	.268	1
TCD3	.351	.296	.408	.373	.166	.343	-.007	.053	.000	.084	.681**	.071	.063	-.063	-.094	.018	-.009	.279	.029	.170	-.063	-.060	.060	.401*	1
TCD4	.259	.227	.272	.275	.223	.198	.172	.152	-.029	.264	.611**	-.060	-.120	-.060	-.063	.217	.030	.150	.098	.351	.299	.200	.280	.268	.418*	1
TCD5	.270	.257	.320	.320	.103	.253	.542*	.299	.219	.286	.619**	.063	.205	.196	.094	.120	.279	.396*	.117	.026	.196	.060	.120	.401*	.205	.120	1

A sample of respondents from the manufacturing sector provided products as a linear combination represented by (2a + 10b + 5c + 4d + 2e + 2f + 2g + 2h + 3i = 0), and further interconnectedness of various technologies $[f(2a, 10b, 5c, 4d, 2e, 2f, 2g, 2h, 3i) = 0]$, Manufacturers' generalized views encompass other services. While the main product is recognizable to another sector, it must still relate to the precise manufacturing and operational needs with design parameter (DP) representing the design layout. These various designs were based on the design for renewable energy solutions (a, DP1); metal fabrication and construction (b, DP2); electronics and communication systems (c, DP3); element products (d, DP4); electromechanical systems (e, DP5); controlled environment solutions (f, DP6); ship engineering (g, DP7); software development and integration (h, DP8); machinery and hydraulic systems (i, DP9) [8]. The domains require mapping on unit vectors to reflect the systems' integrationist perspective from advanced engineering and technologies. By organizing the concepts in this manner, it becomes easier to understand the relationships between them and how advancements in one domain might influence or be influenced by those in another aligned with the design matrix (DM) and design parameters (DP) has to be aligned [11], represented in (1) as $\boldsymbol{FR} = [\boldsymbol{DM}] \times \boldsymbol{DP}$, where $\boldsymbol{DM}_{ij} = \frac{\partial \boldsymbol{FR}_i}{\partial \boldsymbol{DP}_j}$.

$$\{FR\} = \left| \begin{array}{ccccccccc} 2a & & & & & & & & \\ & 10b & & & & & & & \\ & & 5c & & & & & & \\ & & & 4d & & & & & \\ & & & & 2e & & & & \\ & & & & & 2f & & & \\ & & & & & & 2g & & \\ & & & & & & & 2h & \\ & & & & & & & & 3i \end{array} \right| \times \left\{ \begin{array}{c} DP1 \\ DP2 \\ DP3 \\ DP4 \\ DP5 \\ DP6 \\ DP7 \\ DP8 \\ DP9 \end{array} \right\} \quad (1)$$

The need for mapping investigation from an intangible process domain is suggested by [11, 258]. The {FR} domain levels for concurrent engineering can be expressed as in [12].

The design for various sectors such as engineering offices, metal, electrical/electronic, construction, software, chemical, marine, machine, and supplier systems, can be established by mapping the {FR}, [DM] and {DP} character vectors.

The sample also represents customer domains of enterprises, which include engineering offices that consult and manage product manufacturing processes according to customer requirements. The metal industry deals with large-scale production using additive and subtractive manufacturing techniques to create end products. The electrical industry focuses on the production of new electronics and the complex installation from a construction viewpoint, extending the design from an engineering facilities perspective.

The intangible viewpoint aligns with the perspective of software producers, extending the design to chemical manufacturing, which produces industrial chemicals as resources. Recreational manufacturing, concerning the manufacture of ships, considers the differences between machinery design manufacturing and suppliers, which are crucial across all industries. This viewpoint, particularly relevant to general suppliers, enhances the distribution characteristics that respond to customer-specific solutions.

The performance of these companies concerning their organizational practices was chosen for testing to offer each company its axiomatic optimum, such as directing towards industry 6.0 systems integration, which may be yet fictious but as research initiative. A core model was selected from these companies based on the integrability of the products and the use of axiomatic theory was expanded to include the variants emerging from recent literature, indicating the research popularity of this area. A successful transition requires continuous knowledge exchange between and within different design domains, e.g., [12].

3 AD of a Manufacturing Organization

AD offers a solution for achieving design independence early in the program phase. This chapter discusses the formulation of assumptions based on hierarchical decomposition and determining if an organization requires specific practices or if it can be adaptable.

Prior conceptual design reflection chapter concluded the used enterprises and noted the functionality basis of the design of a system. Herein the design of the manufacturing systems is human centered. Generally, a manufacturing system comprises a series of processes surrounding the business owner, namely, design, materials, refining and assembly. Modeling the entire manufacturing structure supports organizational productivity and must be viewed because the components are connected. The AD introduced previously encouraged independence and information number representation. As per the following optimization, we are applying an axiomatic approach by advancing axioms to each component from the production side (adapted to [13]).

AD of a manufacturing organization can be summarized by applying an axiomatic approach to each component from the production side, based on the functionality basis of the system design. The manufacturing systems are human-centered and consist of processes like design, materials, refining, and assembly. The AD encourages independence and information number representation. The following equations represent the observed relationships as $\boldsymbol{F}(\boldsymbol{X},\boldsymbol{Y},\boldsymbol{Z}) = \boldsymbol{h}(\boldsymbol{x}_1,\ldots,\boldsymbol{x}_n), \boldsymbol{h}(\boldsymbol{y}_1,\ldots,\boldsymbol{y}_n), \boldsymbol{h}(\boldsymbol{z}_1,\ldots,\boldsymbol{z}_n)$ in (1). In the Finnish manufacturing domain, the establishment of innovative thematics for organization concepts (OCs): organizing production (OP) design procedures, and production management/control (PMC) and training & competence development (TCD) that are represented as partial tensors from profitable utilization both sides, two parts representing labor market turnover (LMT, j), and dollar utilization (DU, k) as minimized (A's criterion) as continuing to (2).

$$A = F(\boldsymbol{X},\boldsymbol{Y},\boldsymbol{Z}) = \begin{vmatrix} \boldsymbol{h}(\boldsymbol{x}_1,\ldots,\boldsymbol{x}_n) \\ \boldsymbol{h}(\boldsymbol{y}_1,\ldots,\boldsymbol{y}_n) \\ \boldsymbol{h}(\boldsymbol{z}_1,\ldots,\boldsymbol{z}_n) \end{vmatrix} \begin{pmatrix} j \\ k \end{pmatrix} = \mathrm{OC}\begin{pmatrix} \mathrm{OP} \\ \mathrm{PMC} \\ \mathrm{TCD} \end{pmatrix}\begin{pmatrix} \mathrm{LMT} \\ \mathrm{DU} \end{pmatrix} = \mathrm{hf}\left(\mathrm{x},\mathrm{y},\mathrm{z}\begin{pmatrix} \mathrm{j} \\ \mathrm{k} \end{pmatrix}\right) \tag{2}$$

where the data processor takes following parameters,

	x_1= integration of tasks
	x_2=Customer-/product-oriented lines/cells
	x_3=Pull principle production control
	x_4=Change-over time optimization
	x_5=Standardized work instructions
	y_1= Visual management & monitoring
	y_2= Quality assurance methods
	y_3= Employee innovation involvement
	y_4= Employee performance bonus system
	y_5= Certified environmental mgt. (ISO 14001/EMAS)
	y_6= Certified energy mgt. (ISO 50001)
	z_1= Task-specific focus
	z_2= Cross-functional focus
	z_3= Digital production technology support
	z_4= Data security & compliance
	z_5= Creativity & innovation focus
	z_6= Project management
	$\boldsymbol{j}_1$= annual turnover
	$\boldsymbol{j}_2$= number of employees
	$\boldsymbol{k}_1$= manufacturing capacity utilization
	$\boldsymbol{k}_2$= profit

The final coupled matrix for the organization, based on a priori optimization, is shown in Eq. (3).

$$\boldsymbol{OCi} = \boldsymbol{DPI} \tag{3}$$

Customer requirements $\{\boldsymbol{CR}\}$ correspond to the space chosen from the Finnish equivalents of European manufacturing research $\{\boldsymbol{FR_i}\}$. The aim is to respond to these from the axiomatic perspective of the $\{\boldsymbol{FR}\} = \{\boldsymbol{A}\}\{\boldsymbol{DP}\}$, when the simplified form of the matrix prototype becomes (4), which information criteria (IC) is minimized [14]:

$$\boldsymbol{FR} = \{\boldsymbol{A}\} \times \{\boldsymbol{DP}\} by \text{ minimized } IC \left(\mathbf{log_e}\frac{\mathbf{1}}{\mathbf{p_s}}\right) \tag{4}$$

While specifying system terms, practices and maintenance must be maintained to a corresponding firm related to the decoupling given in (5).

$$
\begin{Bmatrix} FR1 \\ FR2 \\ FR3 \\ FR4 \\ FR5 \\ FR6 \\ FR7 \\ FR8 \\ FR9 \\ FR10 \\ FR11 \\ FR12 \\ FR13 \\ FR14 \\ FR15 \\ FR16 \\ FR17 \end{Bmatrix} = \begin{vmatrix}
X\,0\,0\,0 & 0\,0\,0 & 0\,0\,0 & 0\,0\,0 & 0\,0\,0\,0 \\
0\,X\,0\,0 & 0\,0\,0 & 0\,0\,0 & 0\,0\,0 & 0\,0\,0\,0 \\
0\,0\,X\,0 & 0\,0\,0 & 0\,0\,0 & 0\,0\,0 & 0\,0\,0\,0 \\
0\,0\,0\,X & 0\,0\,0 & 0\,0\,0 & 0\,0\,0 & 0\,0\,0\,0 \\
0\,0\,0\,0 & X\,0\,0 & 0\,0\,0 & 0\,0\,0 & 0\,0\,0\,0 \\
0\,0\,0\,0 & 0\,X\,0 & 0\,0\,0 & 0\,0\,0 & 0\,0\,0\,0 \\
0\,0\,0\,0 & 0\,0\,X & 0\,0\,0 & 0\,0\,0 & 0\,0\,0\,0 \\
0\,0\,0\,0 & 0\,0\,0 & X\,0\,0 & 0\,0\,0 & 0\,0\,0\,0 \\
0\,0\,0\,0 & 0\,0\,0 & 0\,X\,0 & 0\,0\,0 & 0\,0\,0\,0 \\
0\,0\,0\,0 & 0\,0\,0 & 0\,0\,X & 0\,0\,0 & 0\,0\,0\,0 \\
0\,0\,0\,0 & 0\,0\,0 & 0\,0\,0 & X\,0\,0 & 0\,0\,0\,0 \\
0\,0\,0\,0 & 0\,0\,0 & 0\,0\,0 & 0\,X\,0 & 0\,0\,0\,0 \\
0\,0\,0\,0 & 0\,0\,0 & 0\,0\,0 & 0\,0\,X & 0\,0\,0\,0 \\
0\,0\,0\,0 & 0\,0\,0 & 0\,0\,0 & 0\,0\,0 & X\,0\,0\,0 \\
0\,0\,0\,0 & 0\,0\,0 & 0\,0\,0 & 0\,0\,0 & 0\,X\,0\,0 \\
0\,0\,0\,0 & 0\,0\,0 & 0\,0\,0 & 0\,0\,0 & 0\,0\,X\,0 \\
0\,0\,0\,0 & 0\,0\,0 & 0\,0\,0 & 0\,0\,0 & 0\,0\,0\,X
\end{vmatrix} \times \begin{Bmatrix} DP1 \\ DP2 \\ DP3 \\ DP4 \\ DP5 \\ DP6 \\ DP7 \\ DP8 \\ DP9 \\ DP10 \\ DP11 \\ DP12 \\ DP13 \\ DP14 \\ DP15 \\ DP16 \\ DP17 \end{Bmatrix} \quad (5)
$$

Enterprise parameters from the process domain justify the results of the process, as new growth companies are dependent on the latest technology to create new products. In terms of production and service maintenance, the popular processes tend to suit the customers' requirements, which is convenient for research and development, increasing employment numbers in organizations with a desire for growth and orienting them toward an incremental innovation perspective. Success can be seen in the investment targets of large companies, which reserve competitiveness for the market, as new types of systems are required. The simplicity of the manufacturing organization can lead to slightly more complex (applied [15]) systems. A design practice that achieves the goal of AD can be found in indices [16]:

Building an organization to resolve challenges while supporting sustainable development is not the task of one company. The company's best strategy depends on the customers' requirements for innovation. Innovations are based on capital flow, increasing sales and controlling labor costs, (for example, [6, 7]). Strategically, product design and manufacturing can be decentralized among stakeholders and can influence the development of new corporations in terms of coupled design.

3.1 Optimizing Total Sustainability

As a result, organizations can achieve greater efficiency and effectiveness in their sustainability endeavors by considering few aspects. Quality control measures, for example, help preventing unwanted working culture from product contamination while promoting process innovation to maximize quality [7]. A human resource vision that fosters a culture of competence and adaptability supports system and operational control, ultimately enhancing product delivery to market [17]. This approach to human resource management, emphasizing competence training, cultivates new cultural advantages and contributes to the effective management of products reaching the market [17]. The resulting solution is a decoupled sustainability approach optimized for an organization's economic design enhances workforce performance, contributing to the long-term success of sustainable organizations [18]. This matrix-based method isolates the interdependencies of various sustainability tasks, allowing each to prioritize maximizing revenue, minimizing costs, or supporting operations.

Table 3. Original matrix for production management or control to maximize return. The solution governs an organization to correspond to a variable process domain for solutions (adapted [19]), organizing production (OP) is primarily focused on maximizing revenue, production management/control (PMC) is centered on minimizing cost, and training & competence development (TCD) is geared toward supporting operations.

Sustainability tasks	Maximizing revenue	Minimizing cost	Operations
OP	X		
PMC		X	
TCD			X

The solution becomes as the decoupled version of the original integration matrix. The matrix focuses on decoupling, this arrangement ensures that each sustainability task contributes to a specific aspect of sustainable manufacturing, allowing for more targeted and efficient efforts in achieving long-term success and growth while emphasizing sustainability, is adjustable with different weights.

4 Discussion

In this research, we have explored the application of AD principles in business operations to enhance system-level sustainability and reduce unnecessary procedures. The primary focus is on improving organizational efficiency and long-term sustainability through the optimization of design matrices and organizational development [7]. This chapter discusses the key findings and implications of our research.

4.1 Key Findings

Our research has revealed several important insights into the application of AD principles in organizational development:

a) Early-stage prototype matrices for strategic options can help organizations focus on growth, performance, and customer value, enabling them to achieve their goals with minimal complexity [7].
b) Effective communication, quality control, and supplier integration are crucial factors for achieving customer-centered demand in organizational development.

c) Maximizing profits, turnover, and quality while minimizing product contamination involves incorporating process innovation and engaging competent employees in technology usage [7, 19].
d) Competency training is essential for developing a human resource vision that contributes to system and operations control [17].
e) Sustainability is a key consideration in refining complex systems and maximizing organizational performance [18].

4.2 Implications

The findings of our research have several implications for organizations and their approach to sustainability and efficiency. By building and managing an expert organization suitable for future Horizon financing and agile business support for continuous development. This implies:

A) AD can be applied across multiple domains, highlighting the potential for a more integrated and collaborative approach to organizational development [4].
B) AD optimizes organizational design parameters and acts as enabler for organizations to adapt and evolve existing systems more effectively, promoting the integration of organizational culture into business processes [5].
C) AD principles can facilitate the redesign of existing organizations or the implementation of new advancements in a systematic and goal-oriented manner [7].

4.3 Future Research Directions

Our research has provided valuable insights into the application of AD principles in organizational development. However, further research is needed to ensure the horizon management and applicant selection criteria among the agile organization development. We need to investigate:

a) the practical implementation of AD principles in product development contexts.
b) the potential limitations and challenges associated with the application of AD principles in organizations design processes.
c) the relationship between AD principles and other organizational development frameworks and methodologies.

4.4 Future Research

Axiomatically designed, innovative organizations follow multifaceted design domains. Product-to-process design may be further developed in systems engineering, and efficiency practices require additional research into technology strategies. The organization's services within the customer interfaces are ecological practices in systems engineering. A supply chain simulation of the manufacturing operations of deterministic models matches those of sustainable operations.

4.5 Conclusions

In conclusion, sustainable organizations can achieve long-term success and growth by adopting a strategic approach that encompasses various aspects, such as production control, management, and competency development [7, 17, 18]. By utilizing a decoupled version of the sustainability matrix, organizations can effectively focus on specific sustainability tasks to maximize revenue, minimize costs, and support operations [18]. This matrix-based method facilitates a targeted approach to sustainable manufacturing, allowing for increased efficiency and adaptability in a competitive market environment. Regularly supplementing competitiveness and promoting continuous improvement further ensures the organization's sustainable growth and success.

In summary, our research has highlighted the potential benefits of applying AD principles in organizational development to improve system-level sustainability and efficiency. Future research should build upon our findings to further explore the practical implications and applications of these principles in various contexts.

References

1. Rauch, E., Matt, D.T., Dallasega, P.: Application of axiomatic design in manufacturing system design: a literature review. Procedia CIRP **53**, 1–7 (2016). https://doi.org/10.1016/j.procir.2016.04.207
2. Smith, A., Todd, W.B.: An Inquiry into the Nature and Causes of the Wealth of Nations. In: Smith, A., Todd, W.B. (eds.) The Glasgow Edition of the Works and Correspondence of Adam Smith, Vol. 2: An Inquiry into the Nature and Causes of the Wealth of Nations, Vol. 1. Oxford University Press (1975). https://doi.org/10.1093/oseo/instance.00043218
3. Clauer, D., Fottner, J., Rauch, E., Prüglmeier, M.: Usage of autonomous mobile robots outdoors - an axiomatic design approach. Procedia CIRP **96**, 242–247 (2021). https://doi.org/10.1016/j.procir.2021.01.081
4. Matt, D.T., Modrák, V., Zsifkovits, H. (eds.): Industry 4.0 for SMEs. Springer, Cham (2020). https://doi.org/10.1007/978-3-030-25425-4
5. Malik, A., Napitupulu, H., Ginting, R. Comparison and Integration of AD with Quality Function Deployment as a Design Method: a Literature Review. IOP Conference Series: Materials Science and Engineering, Volume 1003, 2nd International Conference on Industrial and Manufacturing Engineering (ICI&ME 2020) 3–4 September, Medan, Indonesia. https://iopscience.iop.org/article/10.1088/1757-899X/1003/1/012017/meta. Accessed 21 Dec 2022 (2020)
6. Cochran, D.S., Eversheim, W., Kubin, G., Sesterhenn, M.L.: The application of axiomatic design and lean management principles in the scope of production system segmentation. Int. J. Prod. Res. **38**(6), 1377–1396 (2000)
7. Franzellin, Vittorio & Matt, Dominik & Rauch, Erwin. The (future) customer value is the focus. An AD method combined with a Delphi approach to improving the success rate of new strategies, products, or services (2010)
8. Own study. Referred as of knowledge extracted, referred, and investigated through European manufacturing survey 2019–2021 during 2022 data-acquisition, https://sites.utu.fi/ems/en/, Accessed 20 Dec 2022 (2022)
9. Paulson, L. The quaternions—and type classes. Machine logic. At the junction of computation, logic, and mathematics. https://lawrencecpaulson.github.io/2022/03/09/Quaternions.html, Accessed 13 Dec 2022 (2022)
10. Suh. N. Ergonomics, AD, and complexity theory. https://doi.org/10.1080/14639220601092509, Accessed 13 Dec 2022 (2007)

11. Suh, N.: Designing-in of Quality Through AD. Reliability, IEEE Transactions on. **44**, 256–264 (1995). https://doi.org/10.1109/24.387380
12. Albano, L. An axiomat approach to performance-based design. Doctoral dissertation at the Massachusetts Institute of Technology. https://www.axiomaticdesign.com/wp-content/uploads/thesis_sweden96.pdf. Accessed 20 Dec 2022 (1992)
13. Farid, & Suh, N. P. AD in Large Systems: Complex Products, Buildings and Manufacturing Systems. Springer International Publishing AG (2016)
14. Nordlund, M., Kim, S.-G., Tate, D., Lee, T., Oh, H.: 'AD: Making the Abstract Concrete.' Procedia CIRP 50, 216–221 (2016)
15. Podolny, J., Hansen, M.: How Apple Is Organized for Innovation. https://hbr.org/2020/11/how-apple-is-organized-for-innovation, Accessed 17 Dec 2022 (2020)
16. Suh, N. Chapter 1 Introduction to AD. Web http://web.mit.edu/2.882/www/chapter1/chapter1.htm (1996)
17. Cavique, M., Dodun, O., Brown, C.: Axiomatic design as an innovation framework. IOP Conf. Ser.: Mater. Sci. Eng. **1037**(1), 012015 (2021). https://doi.org/10.1088/1757-899X/1037/1/012015
18. Perera, S. Digital maturity characteristics in New South Wales, Keynote VII: Wednesday, December 21, 2022, 11:40 am – 12:20 pm. IEOM Sydney Conference (2022)
19. Engelseth, P., Salman, M., Mushtaq, W., Awaleh, F., Kozlowski, R., Glavee-Geo, R.: Additive manufacturing: currently a disruptive supply chain innovation? Operations and Supply Chain Manage.: Int. J., 410-422 (2021).https://doi.org/10.31387/oscm0470312
20. Swanston, W. J. & Biggar, G. T. Developing a framework for establishing cross-functional integration within a product development project. Paper presented at Project Management Institute Annual Seminars & Symposium, Houston, TX. Newtown Square, PA: Project Manage. Inst. (2000)

Axiomatic Design and Sustainability

Education Design Requirements Towards a Sustainable and Autonomous Europe

Miguel Cavique[1,2](✉), João Fradinho[2], António Gabriel-Santos[2,3], and António Mourão[2]

[1] Naval Academy, Base Naval de Lisboa, Alfeite, 2810-001 Almada, Portugal
cavique.santos@escolanaval.pt

[2] UNIDEMI, Department of Mechanical and Industrial Engineering, Faculty of Science and Technology, Universidade NOVA de Lisboa, Campus de Caparica, 2829-516 Caparica, Portugal

[3] MANO – Equipamentos de movimentação e automação, SA, Rua D. Luís de Noronha, 14 – Apartado 2, 2635-380 Rio de Mouro, Portugal

Abstract. EU lacks chains in its industrialization model. In a new multipolar world, the absence of autonomy may cause disruptions. The new model of reindustrialization should be sustainable. Sustainability has the well-known three vectors of environment, enterprises, and society. All changes will be possible with the willingness of people. It is through education redesign that sustainability will be achieved, focusing on persons – "Make EU humanist again" could be the motto for the next generation. The break of product chains in the EU removed a middle class that worked on production and services. Salaries and social recognition go to extremes. Nonsuccess persons with middle Education turned disposable. However, all parts of a value chain need different skilled persons. The new reindustrialization process can include all types of persons. The requirement is to "shape the education system to create a sustainable EU defined in the environmental, economic, and societal" to achieve it. It asks for developing group skills and individual acceptance of the other. It demands a change in Education to focus on secondary Education, other than the tertiary level. Countries in the EU are making a huge effort to raise young people to high Education, making such persons work in sub-employment. The ability to work with all levels of Education is the lever for self-esteem, personal recognition, and social cohesion.

Keywords: Sustainability · Societal Cohesion · Reindustrialization · Education

1 Introduction

Physical products imported per capita by the EU in 2021 were 3.6 tons against 1.6 tons of physical exports [1]. Regarding carbon import, the CO_2 footprint in 2018 represents one-third of the total EU carbon footprint, more significant than the 23% from the EU manufacturing sector [2]. These data give a rough idea of the EU imports of manufacturing products. EU needs a new reindustrialization process that should take place in the following decade.

E. Puik et al. (Eds.): ICAD 2023, LNNS 849, pp. 91–100, 2024.
https://doi.org/10.1007/978-3-031-49920-3_6

The reindustrialization of the EU must not repeat the past industry tasks and processes. It must focus on producing goods with current technologies framed by environmental, economic, and societal sustainability. The industry will likely change to digitalization, with a sustainable policy of shorter reuse, remanufacturing, and recycling cycles.

Sustainable policies should include environmental, economic, and social activities. The focus on environmental sustainability is leaving apart social sustainability. Social sustainability interacts with the economic value chain and environment; its front end is mutual respect in the work environment. A significant part of self-respect regards each person's contribution as a social being. Therefore, the cut of the value chains in the EU has as a counterpart a social disrespect.

These facts show that a reorientation of Europe's strategic position is imperative: it must be self-sufficient to the greatest possible degree. In addition, recent times are already showing the great danger of a lack of social cohesion with immense unemployment and very low wages. It is necessary to act as soon as possible and take energetic measures, breaking with the industrial organization and the global positioning of Europe since the middle of the 20th century.

The idea of reindustrializing Europe restarted during the pandemic because of EU's lack of goods. EU people experienced mixed feelings regarding societal behavior, seeing some countries suffering from dilation processes. At the same time, Europeans experienced mixed feelings regarding societal behavior concerning garbage collection, package, and food delivery.

Reindustrialization fosters the perception of an EU recast on its global strategic positioning. This perception comes from the shortage of product availability, including assembly components and those with low technological content.

This paper aims to show the requirements of an Education Redesign, on the secondary and tertiary level, to allow a sustainable future regarding the environment, economy, and society.

The following section presents the challenges of the EU in the aforementioned context. Then, Sect. 3 defines the problems, or, otherwise, the Customer Needs (CN). The mapping of Functional Requirements (FRs) to Design Parameters (DP) is described in Sect. 4. Finally, the main conclusions are presented and discussed.

FRs are the crucial elements to define an EU policy, each country or region policy, and the type of Education. Essential solutions need to be addressed, enabling people with problem-solving skills. Moreover, people need to be able to work in groups with different education levels to enhance merit dependence and foster self-esteem. This work ends by showing the FRs at the second level of decomposition.

2 EU Challenges

The education role will take decisive importance regarding both EU strategic position and social cohesion challenges. In the long term, education activities will define the possible policies. As defined by Stephanie Spencer, "we take it as axiomatic that the history of politics, government, religion, economics, ethics and so on cannot correctly be studied without reference to the nature of the educational activity, in both its formal and informal settings" [3].

Social networks and artificial intelligence (AI) are increasing their influence on informal settings of educational activity—moreover, a similar trend has surged in formal settings of the education process. Social networks and AI are technical issues that can foster Education and relationships but can create personal segregation due to individual, intellectual, or professional specialization. Many of the traditional cohesion forces are wearing. The societal reshape, Education included, needs institutional redesign to reinforce links in between. Institutions need to remain whatever governments are, ensuring reliable and democratic states.

Self-segregation grows in all social and economic personal status raising cultural niches or even nihilistic niches, which are the opposite of culture.

Widespread culture and economic wealth are outcomes of XIX-century liberalism. Still, taking it for granted is a colossal error that can prevent societies from solving their problems. Ultraliberalism of the end of the XX-century turned economic efficacy to rule over justice and ethics. The multipolar World creates a neo-realism world focusing on autonomy and State strength. The authors' opinion about autonomy and reduction of EU external dependency regards avoiding disruption of institutions and governance. EU should maintain a strong relationship with all world players, spreading its historic assets, ethics, and humanism. To do it, redesigning institutions can help solve current and future problems.

Problems will be more and more multidisciplinary and collaborative. The ability to solve problems and cooperate in groups are educational contributions to social health.

Solving problems is a way of creating something new. It can be solved by algorithms when the solution is foreseen. Otherwise, a new problem needs a creative new key, or in other words, requires a customer needs' definition [4]. Creative thinking should have a solid scientific basis paramount for developing any new design. Creativity depends on a "certain minimum amount of scientific and technical knowledge" [5]. However, it requires an expansive view of engineering, which contrasts with the specialization of science. Culture is a way to define societal needs and can be a branch of the engineering process to find solutions for problems.

Additionally, the previous personal examples of success and intuition can strongly support finding new solutions [6]. Education must address new paradigms on the shift from analysis to synthesis and from competition to cooperation. 'Engineering is design', stated Goran Putnik [7]. All processes in engineering need solutions. They may have an algorithmic solution or need an utterly new method. It happens in all engineering fields, and at all levels of the engineering process, from technical to high graduate levels.

Design theories, like Axiomatic Design (AD), help the inventive process [8] by fostering the design and identifying research needs. AD provides a structure of axioms, theorems, and a methodology for reaching a good design solution. The AD theorems are hints of knowledge to fulfill a design.

Research is of paramount importance in the innovation process. However, enterprises and societies must link the three innovation vectors to keep continuous innovation. First, scientific research develops new concepts and analyzes solutions. Plus, the capability of the enterprises to put things working and make them happen. Both need financial support and risk of capital. Working simultaneously with the three vectors can make vague ideas into an innovation process [8]. Innovation is possible with enterprises where they exist and with the support of the value chains as a whole. If part of the value chain disappears, skills to allow innovation also lack. So, a long-term redesign of all systems is needed, which is a tricky task while maintaining the system working.

Design and design science can give some help. Project learning and AD can assist in developing ideas using a constructivist approach [9]. Groups foresee new scientific fields while solving new problems. It warrants the researcher or the developer a broader view of science and asks for knowledge in various areas.

Design theories belong to the science of the virtual [10] and need development in the scientific and applications fields. Unfortunately, the research on the theoretical foundations of Design theories has been low since the beginning of this century [11]. Design theories and applications need a comprehensive view of the problems, from identifying the social context to solving the manufacturing issues. Reducing engineering to technique can be a huge error.

In order to face the great challenges identified for the next generations in the EU (namely, strategic autonomy and social coexistence), action in the educational process is considered decisive to lead Europe to a new path. The unparalleled potential of AD framework is used to conceptually design the entire European educational system, with the ultimate aim of reducing Europe's external dependency and simultaneously creating a society based on the personal and professional fulfillment of its citizens, always framed by the Sustainability concept.

3 Customer Needs

Customer Needs (CN) domain expresses the needs of all design stakeholders. The needs of the stakeholder have the counterpart view of the problem.

So the question is, "What the problem is?".

Europe neglected the medium/long-term effect of sending the production of products to other regions. It caused a reduction in the amount of technically specialized personnel and qualified workers in production. Similarly, the design of the industrial process variables has been neglected.

The lack of production made EU fragile regarding economic sustainability, although creating a dystopic vision of helping the environment. The absence of production makes EU lack political action in the event of disruptions. Moreover, the industry deficiency creates a more severe problem of lack of social sustainability. As mentioned by Prabhu Kandachar [12], social sustainability is the neglected component of sustainability.

The reduction of workers and skills targets the middle of the value chain of the product creation life cycle. Europe focused on activities at the beginning of the product value chain and distribution. The first activity needs high-level academic qualifications, while distribution tasks require low-qualified persons. As a consequence, jobs with qualified technical knowledge get reduced. The design of technical systems and means of production needs fewer workers. It diminishes the attractiveness of technical working functions. Therefore, many young people with high Education work in sub-employment or find jobs in other regions. It may turn the EU high Education dystopic. The continuous innovation system in EU should keep all parts of the value chains engaged.

EU cannot run a reindustrialization process by denying non-graduated workers, which in turn creates a lack of social cohesion. It creates unfavorable social cohesion in Europe due to the dismissal of ranges of society, part of it with tertiary Education. Moreover, lack of creative work, loss of work dignity, and reduced personal autonomy degrade self-esteem.

According to the United Nations [13], "Social sustainability is about identifying and managing business impacts, both positive and negative, on people. .... Directly or indirectly, companies affect what happens to employees, workers in the value chain, customers, and local communities, and it is important to manage impacts proactively".

Thus, reindustrialization in Europe must be framed in a vision of sustainability that promotes social cohesion. The value chains of the industry must consider the entire product development cycle as a way to generate valuable and recognized jobs, thus promoting self-esteem.

Satuf. C. et al. [14] define self-esteem as the personal feeling regarding the individual's opinion. As mentioned by Samantha Krauss [15], individuals with high self-esteem are more likely to succeed at work by building and maintaining positive social relationships and receiving more social support from coworkers and supervisors. In the macro view, social cohesion is a source of wealth and economic growth [16].

The breakdown of value chains in the development of products and the respective production chains is a way of destroying social cohesion. The break of supply chains redistributes the workforce to the margin neighborhoods of the social spectrum. Unfortunately, these redistributions are away from any strategic social policy.

Any product needs a massive number of components whose conception, project, and manufacture constitute the bulk of the work inherent to the industry. Thus, it is necessary to redesign the education model and implement the entire training chain.

"What the problem is?". According to AD, it is the CN0 that can be described as follows:

CN0 - Implementing sustainability in the EU through environmental, economic, and societal pillars. The implementation is a design defined by a set of functional requirements.

Unfortunately, EU does not shape but react to events, geostrategic, pandemic diseases, or environmental.

EU needs a vision for the enhancement of its position in the global context. This vision must anchor on the three pillars of sustainability, environment, economy, and society (see Fig. 1). The CN0 can be decomposed as follows:

CN1 – regarding the environment – there is a need to develop a well-being way of life within stable integration with natural conditions;

CN2 – concerning the economy – there is a need for autonomy in all value chains;

CN3 – on society – there is a need for social cohesion by making all people feel useful.

According to AD, CNs map into the FR domain. Next section shows the FRs and the corresponding Design Parameters (DPs) regarding Education for a sustainable EU.

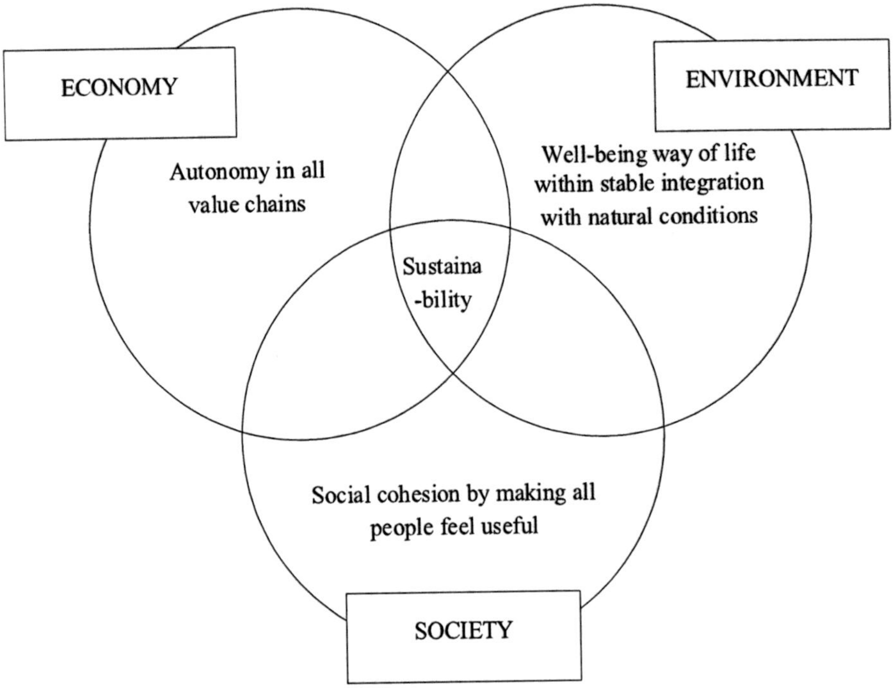

Fig. 1. The common vision of sustainability

4 Functional Requirements and Design Parameters

"No design is better than its FRs" is a statement always said at each AD conference. This section defines the FRs to fulfill the CNs above regarding Education. This work focuses on the ontological problem of Education by defining the high-level FRs. This work purposely narrows to Education as a contribution to redesigning a society.

Education politics may feel odd not to mention in this work requirements as education affordability, facilities, namely, libraries and labs, career opportunities, or personnel certification. Moreover, curriculae discussion is out of the frame of this work.

CN0 can map to FR0, as below.

FR0 - Shape the education system to create a sustainable EU defined in the environmental, economic, and societal fields.

CN1, CN2, and CN3 map to the following FRs regarding the framework of design education:

FR1 – Understand the World and the dependency of the technical systems on the natural ones;

FR2 – Grow students personally and professionally while developing the skills to solve problems with an economic impact in a new industrialized era;

FR3 – Create self-esteem and respect for professionals of all levels of the new industrial process, emphasizing the value of their work.

In the physical domain, one can define the following DPs at the same level 1:

DP1 – Societal activities in all levels of Education;

DP2 – Interaction with economic activities in all levels of Education;

DP3 – Professional output in all levels of the value chain of the industrial economy.

Equation 1 shows the relations between FRs and DPs. It reveals a decoupled design, beginning with understanding the World, then personal and professional development, and ending with promoting self-esteem.

$$\begin{Bmatrix} FR1 \\ FR2 \\ FR3 \end{Bmatrix} = \begin{bmatrix} \times & 0 & 0 \\ \times & \times & 0 \\ \times & \times & \times \end{bmatrix} \begin{Bmatrix} DP1 \\ DP2 \\ DP3 \end{Bmatrix} \tag{1}$$

This equation is a global approach for Education as a whole with expression at all levels of Education.

The decomposition of FRi for each of the three education levels is below. The FRi.j regards the decomposition of DPi into the educational level j. Therefore, as an example, tertiary Education is defined by FR1.3, FR2.3, and FR3.3.

FR1.1 – Interact and live in connection with nature;

FR1.2 – Promote internships to safeguard nature;

FR1.3 – Design and develop systems and utilities in the social context that follows eco-design criteria;

FR2.1 – Cooperate with economic activities;

FR2.2 – Link with enterprises to know the technics of some sectors of a value chain;

FR2.3 – Understand and design enterprise systems regarding management and technical systems;

FR3.1 – Promote the usefulness of the different professions for life in society;

FR3.2 –Know how-to-do production and maintenance of systems that ensure the functioning of society;

FR3.3 – Create scientific knowledge and the creative capacity for design and manufacturing.

Each of the FRs from the second level of decomposition has physical concretization and can be decomposed according to the specificities of each country or region. Next, some hints help create the DPs at the second level and decompose the FRs.

The International Standard Classification of Education (ISCED) [17] is the reference classification for organizing formal education programs and related qualifications by education levels and fields into internationally agreed categories. The term "tertiary education" refers to ISCED levels 5–8. UNESCO adopted the most recent version of the classification in November 2011.

Educational structure is divided into three groups: 1) primary, level 1; 2) secondary, levels 2 to 4; and 3) tertiary, levels 5–8.

According to ISCED, secondary Education gives the ability to a student to begin a professional life through skills of responsibility and autonomy [18], which make these people able to work [18]. However, in 2021, more than 40% of the 25–34-year-old persons in the EU decided to finish tertiary Education (see Fig. 2). Almost 85% of the 20–24 completed at least an upper-secondary level of Education [19].

Most manufacturing activities do not need tertiary Education. Secondary professional skills and lifelong learning can create excellent and motivated workers in the

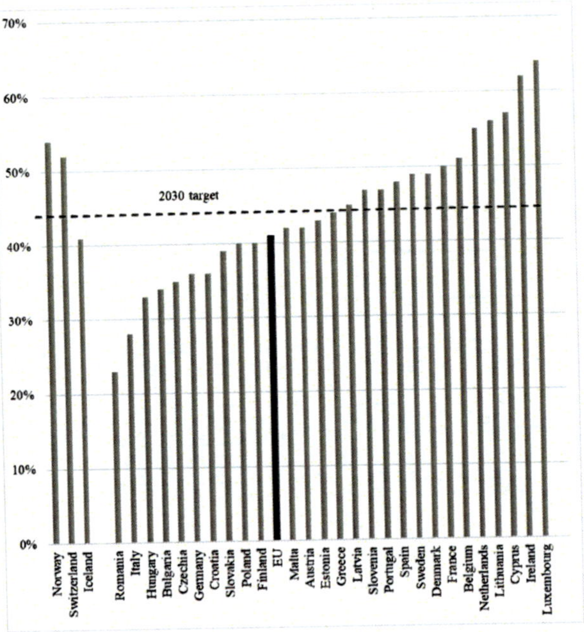

Fig. 2. Percentage of European population aged 25–34 with tertiary educational attainment (ISCED 5–8), 2021, adapted from [19]

manufacturing field. In the EU (2020) 16% of the workforce devoted to manufacturing, 23% to Distribution and Transport, and 26% to Non-Market, including Education, health care, and defense [20]. In the EU in 2020, elementary workers represented 26% of the unemployed persons, followed by 17% of service and sales workers, 12% of operators and assemblers, and 11% of clerks. However, the scarcity of manufacturing highly skilled workers may avoid launching a reindustrialization process. This fact implies to foresee the need for Education and corresponding information processes.

Despite their natural skills, all people can help in a healthy society. Work is a way of personal and societal fulfillment.

5 Discussion and Conclusions

"Europe has forgotten its role in the World since the beginning of the XX century. The World denied their rules without replacing them with others", said Ortega y Gasset in 1930. If it had then been right, now, at the beginning of the XXI century, the EU has no political purpose. This work gives a high-level framework for the EU educational system.

Education for future generations needs to ensure social cohesion links and spiritual values. We need to insert human beings into the environmental problems to avoid social asymmetries in the World. Finally, the economy needs to have a social role by putting politics to decide the role of markets and not the opposite. In a word – the EU needs to redirect to humanism. Humanism in the new economic and environmental ambiance might be the strength of the EU.

Each individual needs to be autonomous through Education. Education gives contribution to a self-reliant and interdependent society. The same idea can encompass the role of each country in the EU or the EU's role in the World.

Axiomatic Design is a way to "do things right". We must focus on "how to do the right things". The main conclusions of this work are as follows.

- EU needs an education system for sustainable environmental, economic, and societal development. The social field has been overlooked in favor of the environment.
- During the education period, people need to understand the World and the dependency of the technical systems on the natural ones; grow personally and professionally and, develop skills to solve the problems with economic impact; create self-esteem and respect for professionals of all levels.

The authors focused this paper on the working environment, although they believe humanism is a wider concept than working. However, personal work is a way to achieve self-fulfilment, self-esteem and contribute to a healthy society.

At lower decomposition levels, the FRs should also accomplish the personal desire for pure knowledge despite the envisaged use of the subjects.

Acknowledgment. Authors acknowledge Fundação para a Ciência e a Tecnologia (FCT - MCTES) for its financial support via the project UIDB/00667/2020 and UIDP/00667/2020 (UNIDEMI).

References

1. Eurostat, Statistics explained. https://ec.europa.eu/eurostat/statistics-explained/index.php?title=Physical_imports_and_exports#Physical_trade_by_stage_of_manufacturing. Accessed 18 Mar 2023
2. Institut national de la statistique et des études économiques. https://www.insee.fr/fr/statistiques/6474294. Accessed 18 Mar 2023
3. Spencer, S.: Out of the classroom: 'informal' education and histories of education. Hist. Educ. **4**(50), 468–484 (2021). https://doi.org/10.1080/0046760X.2021.1900407
4. Suh, N.P.: Introduction to design. In: Suh, N.P., Cavique, M., Foley, J.T. (eds.) Design Engineering and Science, pp. 1–33. Springer, Cham (2021). https://doi.org/10.1007/978-3-030-49232-8_1
5. Slatineanu, L., Coteata, M., Hrituc, A., Besliu-Bancescu, I., Dodun, O.: Information-seeking in education for creative solving of design and engineering problems. Designs **7**, 22 (2023). https://doi.org/10.3390/designs7010022
6. Stam, L., Verbeek, P.P., Heylighen, A.: Between specificity and openness: how architects deal with design-use complexities. Des. Stud. **66**, 54–81 (2020). https://doi.org/10.1016/j.destud.2019.11.010
7. Putnik, G.D., Putnik, Z., Pinheiro, P., Alves, C.: Engineering is design and only design - Part I: the value of making a distinctive sign. Procedia CIRP **109**, 678–683 (2022). https://doi.org/10.1016/j.procir.2022.05.313
8. Suh, N.P.: A theory of innovation and case study. Int. J. Innov. Manag. **14**(5), 893–913 (2010). https://doi.org/10.1142/S1363919610002921

9. Arcidiacono, G., Yang, K., Trewn, J., Bucciarelli, L.: Application of axiomatic design for project-based learning methodology. Procedia CIRP **53**, 166–172 (2016). https://doi.org/10.1016/j.procir.2016.08.003
10. Simon, H.: The Sciences of the Artificial. 3rd ed. MIT Press, Cambridge, MA, USA (1996)
11. Cash, P.: Where next for design research? Understanding research impact and theory building. Des. Stud. **68**, 113–141 (2020). https://doi.org/10.1016/j.destud.2020.03.001
12. Kandachar, P.: Chapter 7 - Materials and social sustainability. In: Karana, E., Pedgley, O., Rognoli, V. (eds.) Materials Experience. Butterworth-Heinemann, (2014). ISBN 9780080993591. https://doi.org/10.1016/B978-0-08-099359-1.00007-2
13. United Nations Global Compact. https://unglobalcompact.org/what-is-gc/our-work/social. Accessed 25 Mar 2023
14. Satuf, C., Monteiro, S., Pereira, H., Esgalhado, G., Afonso, R.M., Loureiro, M.: The protective effect of job satisfaction in health, happiness, well-being and self-esteem. Int. J. Occup. Saf. Ergon. **24**(2), 181–189 (2018). https://doi.org/10.1080/10803548.2016.1216365
15. Krauss, S., Orth, U.: Work experiences and self-esteem development: a meta-analysis of longitudinal studies. Eur. J. Pers. **36**(6), 849–869 (2022). https://doi.org/10.1177/08902070211027142
16. Berger-Schmitt, R.: Considering social cohesion in quality of life assessments: concept and measurement. Soc. Indic. Res. **58**, 403–428 (2002). https://doi.org/10.1023/A:1015752320935
17. UNESCO Institute for Statistics. International Standard Classification of Education - ISCED 2011. ISBN 978-92-9189-123-8
18. European Union Europass. https://europa.eu/europass/en/description-eight-eqf-levels. Accessed 25 Mar 2023
19. Educational attainment statistics - Statistics Explained (europa.eu). https://ec.europa.eu/eurostat/statistics-explained/index.php?title=Educational_attainment_statistics. Accessed 25 Mar 2023
20. Employed population by occupation and sector_CEDEFOP. https://www.cedefop.europa.eu/en/tools/skills-intelligence/employed-population-occupation-and-sector?year=2020&country=EU. Accessed 25 Mar 2023

Axiomatic Design Facilitating Integrated Building Design and Operation

Clarice Bleil de Souza[1(✉)], Camilla Pezzica[1], Marianna Marchesi[1], and Christopher A. Brown[2]

[1] Welsh School of Architecture, Cardiff University, Bute Building, King Edward VII Avenue, Cardiff CF10 3NB, UK
bleildesouzac@cardiff.ac.uk

[2] Mechanical and Materials Engineering Department, Worcester Polytechnic Institute, Worcester MA01609, USA

Abstract. The paper presents an initial reflection on the possibility of adopting Suh's Axiomatic Design (AD) as a framework to integrate and coordinate the different methods used by the Architecture Engineering and Construction (AEC) industry to evidence-base decision-making for sustainability. A taxonomy for aggregating the different methods used by the AEC industry is proposed so they can be inserted into the AD framework and used to support and formulate evidence-based design decisions in a way that is systematic, informed and, among other things, traceable so true records of the design process can be retrieved. The taxonomy is based on the new EN ISO standards and the literature on design research and decision-making. It organizes the discussion about the feasibility of having AD as a facilitator for integrated building design and operation in the AEC industry. Three potential issues are identified from this proposition, which call for further research and academic debate in the AEC and AD communities, laying the conditions for fruitful future interactions between them. On the one hand, the paper proposes that the AEC community can consider AD as a catalyzer to promote integrated design for sustainability. On the other hand, it pushes AD to be adapted to accommodate the needs of a fragmented design industry which often produces one-off design solutions.

Keywords: AEC industry · Design for sustainability · Integrated design · Axiomatic Design

1 The Fragmented AEC Design for Sustainability Process

This paper examines a proposition for integrating Suh's Axiomatic Design (AD) with common methods used by the Architecture, Engineering and Construction (AEC) industry, providing a methodological and theoretical framework to support integrated design for enhancing sustainability-oriented design decision-making processes. Integrated design, in this context, refers to a reconciliation of methods, decisions and solutions from different specialisms towards providing a single and comprehensive material

E. Puik et al. (Eds.): ICAD 2023, LNNS 849, pp. 101–117, 2024.
https://doi.org/10.1007/978-3-031-49920-3_7

response to complex, potentially conflicting, and interwoven requirements, considering the whole life cycle of a building – from conception to operation, re-use, demolition and recycling.

Design for sustainability in the AEC industry has to address multiple types of stakeholders' needs, deals with large amounts of multi-domain information, and should be evidence-based to show solutions proposed work effectively (prior to and after an asset is built) while being heavily regulated and permeated by liabilities. This is particularly the case in projects that wish to apply for sustainability certification by, for instance, the Living Building Challenge (LBC) [1], which has clear requirements to achieve living within planetary boundaries. These standards have requirements which, to be fulfilled, need cross-disciplinary interactions and concerted action throughout the design process; e.g., relying solely on solar power for energy supply balancing demand accordingly; integrating renewable energy systems with electric vehicles on-site; maintaining a balance between water supply and demand through rainwater harvesting; using materials that are renewable, recyclable and do not release volatile organic components which can compromise the health of building occupants. In these types of projects, successfully coordinating the AEC design process for integrated design to achieve sustainability-related design requirements is not a trivial task. It involves coordinating the design delivery process from multiple aspects including stage outcomes, core tasks, core statutory processes, procurement routes and information exchanges among the design teams [2] to achieve design solutions that outperform current building industry standards; while protecting professionals from unforeseen and uncommon liabilities. This is particularly relevant because over the last few decades, technology and specialization have created a disconnect in design decision-making, reducing opportunities for integrated and innovative propositions.

In general, architects adopt solution-focused approaches to ill-defined design problems, and the process follows a spiral, cyclic structure. On the other hand, engineers apply problem-oriented strategies to well-defined design problems and the design process is implemented through a linear sequence of activities [3]. Architects normally rely on precedence and repertoire to make complex decisions whereas experts, and engineers tend to work in silos with domain-specific computational models designed to perform specific activities, but not integrated into a coherent procedural framework [4]. Precedence, repertoire, and modelling approaches of the physical world, which share common design parameters but are developed to achieve different performance objectives, do not enable decisions to be integrated. Rather, many times they push conflicts to be reconciled through Decision Support Systems which are highly deterministic (e.g., multicriteria analysis, optimization, etc.).

In practice, approaches such as the Integrated Design Process (IDP), attempt to support sustainable building design and construction with strategies for project teams to share a vision of sustainability and work collaboratively to implement goals at appropriate design phases during the project development process [5, 6]. IDP seems to successfully describe generic management procedures outlining guidance on roles, tasks, and critical activities during each stage of the process [7]. It complements conventional project management approaches but falls short in support of coordinating integrated decision-making within multidisciplinary design teams [7].

Some authors in the literature [8, 9] suggested the re-integration of the architects' and engineers' models into a common procedural framework suitable for both disciplines to support design teams in transdisciplinary collaboration. According to [8], a common model of the design process should reproduce the process of going to-and-from problem and solution, and sub-problems and sub-solutions. In this process, problem definition should depend upon solution conjectures which, in turn, help clarify the design problem. The latter should be hierarchically decomposed into sub-problems, while the overall design solution should be developed by generating, combining, evaluating, and choosing sub-solutions which respond to different sub-problems. On the basis of these premises, Suh's Axiomatic Design (AD) has been proposed as an appropriate common approach for supporting architects and engineers in performing decision-making in conceptual building design and modular design [10–12], but a common model able to integrate, into the design process for sustainability, the multiple methods used by different AEC design disciplines to make decisions is still missing.

The lack of an integrated design decision-making framework in both design practice and research poses a challenge in producing truly sustainable solutions. Key performance indicators (KPIs) alone are not sufficient to produce sustainable design, especially when they need to be achieved through manipulating design parameters that are common to different knowledge domains. KPIs, in which liabilities are higher or KPIs which are connected to higher profits, tend to be achieved, many times to the detriment of KPIs which promote health, wellbeing and/or other environmental gains. Relationships between KPIs and design parameters common to many knowledge domains have to be coordinated through concerted action so design solutions can achieve multiple requirements [13]. The AEC industry does not have a framework to specifically support this coordination; it does not have a framework to support the generation of design solutions which respond to requirements from multiple domains, particularly those which are difficult to measure and/or to cost but promote health, wellbeing and/or other environmental gains.

This paper conjectures that AD could help coordinate the different stakeholders' needs, design, decision-making, and project control methods used by the AEC industry to achieve integrated building design and operation. In this way solutions may better address the different sustainability goals of the 21st century.

The paper starts by proposing a place for AD in the AEC industry. It then groups common methods used by the AEC industry to extract stakeholders' needs, make design proposals, decide upon and test design alternatives, including methods which control the performance of the end product when designing for sustainability. It finishes by discussing the position of these methods within the AD approach highlighting fits and misfits with AD components (e.g., applied principles, axioms etc.), outlining areas for deeper investigation and future joint AEC and AD development.

2 Proposing a Place for AD in the AEC Industry

Disagreement in approach between architects and engineers in practice can complicate collaboration particularly in relation to "how design decisions are balanced to achieve overarching project targets, negotiated among project team members, propagated into the

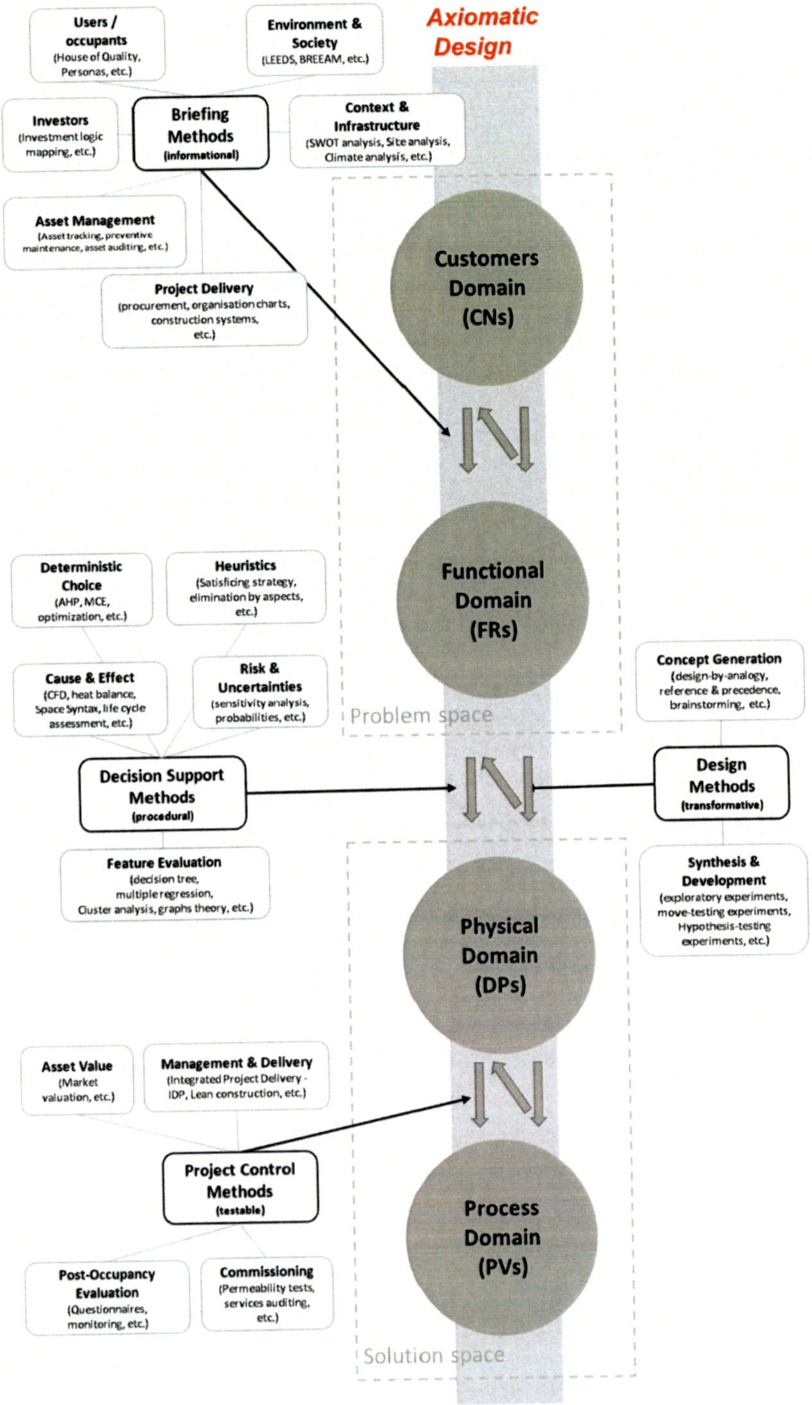

Fig. 1. Placing methods to design for sustainability used by the AEC industry in AD.

information flow of the design process, and subsequently revised as the project develops" [14]. Information management systems and the increased specialisation and automation of the construction industry call for project coordination to happen in a systematic way with a clear push for the entire process to be traced, with true records to be put in place so the diversity of liabilities behind them can be monitored.

AD enables the co-evolution between problem and solution to be traceable, and the decision-making informed, while facilitating knowledge and information transfer, storage, and retrieval as well as enabling the engagement and coordination of multiple stakeholders. Moreover, AD provides a sequence of stages and activities to progress the project (AD domains) and a sequenced creation process based on going to-and-from problem and solution, plus to-and-from sub-problems and sub-solutions (zigzagging). In AD, problem and solution are systematically and consistently specified in parallel, moving down a hierarchy, and design decisions are made in an explicit way, maintaining data. AD is supported by general decision-making principles (Suh's axioms, corollaries, and theorems) which help define effective designs with respect to specified requirements, evaluate the synthesized ideas, and select the most feasible solution among valuable alternatives [15, 16].

AD has been applied to sustainability issues in designing manufacturing systems using constraints to avoid undesirable outcomes while enhancing creativity to enlarge solution spaces [17]. Suh's axioms guide the selection between candidate design solutions. However, AD is somewhat silent on the generation of candidate solutions. Moreover, a constraint-based approach to design for sustainability can result in diminished solution spaces and over constrained problems, but this can be addressed fostering creativity enhancement in design processes [17].

While the AD approach may be suitable to support an integrated design process, it is important to assess how it can accommodate current AEC methods used in design. These AEC methods include identifying and mapping stakeholders needs, supporting decision-making by proving designs attend to these multiple needs, as well as controlling the delivery process towards fulfilling them as best as possible. To the best of the authors' knowledge, there are no records of how these different methods can be integrated through AD, neither is there a taxonomy that enables their integration within an overarching framework to be properly coordinated and assessed as the design process progresses.

Figure 1 shows where the most common methods used by the AEC industry to design and how they can potentially be integrated via AD, together with a taxonomy used to aggregate these methods into four different groups namely 'Briefing Methods', 'Design Methods', 'Decision Support Methods' and 'Project Control Methods'. The taxonomy was put together combining recommendations from the EN ISO 19650 series, which refer to information and asset management, together with the literatures in design research and decision-making in engineering. The rationale behind each group is presented in Sect. 3, whereas a discussion about how each group of methods can potentially fit within the AD framework is presented in Sect. 4.

3 Methods Used to Design for Sustainability in the AEC Industry

Methods and information management systems are operational for designers to design. They are part of the 'systems of knowing in practice' [18] and are important elements of design practice. They are more informative than design inputs and outputs (as proposed by [19]) to understand decision-making and enable decision chains to be recorded, while at the same time inferring the potential decision-makers behind them. Records of this type aid project coordination and provide evidence of correct attribution of responsibilities and liabilities, which permeate the AEC industry, while also enabling knowledge transfer within and across the different disciplines involved in the design process.

Briefing methods, design methods and project control methods reflect the employment of tacit knowledge in solving design problems, while decision support methods and information management systems make designers' ontologies and epistemologies explicit, facilitating scrutiny when prioritizing, coordinating and reconciling decisions.

Briefing methods are informational and therefore used to collect information to specify design requirements and constraints. Design methods are transformative, moving from what a situation is to what a situation will be, and used in the co-generation of problems and solutions. Project control methods are testable and used to keep in check the different aspects related to product development process and performance in use. Decision support methods are procedural as they contain clear procedures to aid in decision-making. Transversal to all these methods, and hence falling out of the scope of the discussion on AEC methods' integration in AD, are information management systems: ontological and relational, enabling different types of project information to be tracked throughout the whole design process.

3.1 Briefing Methods

Briefing methods were classified by type of information needed to formulate design problems based on the different information management perspectives presented by EN ISO 19650-1 [20]. They depend on the needs and aspirations of different stakeholders who are part of the design process as well as on the needs, opportunities and constraints imposed by the context in which the design will be inserted, from site to society.

The sub-category 'Users/occupants' groups methods is used to identify needs and aspirations of building users/occupants to ensure the design solution satisfactorily responds to them. This sub-category is well known to the Axiomatic Design community, and it contains methods commonly used in marketing (e.g., House of Quality) and in human-computer interaction (e.g., personas), to cite a few.

The sub-category 'Investors' groups methods used to identify the needs and aspirations of project clients, who might not necessarily be the occupants or users of an asset but have clear financial targets for it. Methods commonly used to map investors' needs and aspirations come from the business domain (e.g., investment logic mapping).

The sub-category 'Asset management' is a particular category in the AEC industry with specific needs for the operational phase of an asset (the longest phase in any asset life cycle). Methods commonly used to map asset management needs come from maintenance engineering, building controls and operation (e.g., preventive maintenance, asset auditing, asset tracking) and are potentially alien to the AD community which mainly deals with the asset up to the end of its production life.

The sub-category 'Project delivery' groups methods predominantly used to extract needs related to the coordination of the different parts of a project supply chain, so they are satisfactorily completed and delivered to clients. They include project management, construction, and procurement methods (e.g., procurement routes, organization charts, construction assemblage systems) and deal with specific needs affecting project requirements from the beginning, many of which cannot be changed after planning application or analogous project milestones.

The sub-category 'Context & infrastructure' is particular to the AEC industry as it focuses on methods to extract needs, opportunities, and constraints of the context an asset will be inserted in, more specifically its site, climate, neighborhood, and its social and environmental ecosystems. Methods used in this subcategory come from architecture (e.g., site analysis), planning (e.g., SWOT analysis) and building physics (e.g., climate analysis) and provide usually unique information to design a 'prototype of one' as every building is a one-off custom job with little economy of scale or customized information to design modular solutions.

The sub-category 'Environment & society' focuses on methods used to extract wider societal and environmental needs of a project which are normally prescribed by, for instance, building sustainability standards (e.g., LEED, BREEAM). These methods ensure needs are set based on collective interests, rather than individual ones from clients and users/occupants alone.

3.2 Design Methods

Design methods were classified based on the type of transformation they enable by merging the reflective practice approach proposed by Schon [18] with the disintegrated design process proposed by Jones [19].

Therefore, 'Concept generation' methods resemble what Jones [19] describes as "methods of searching for ideas". They reflect the more intuitive part of the design process in which searches for potential design solutions are undertaken to select a subset, or one of them, to be further tested and developed. Methods used in this sub-category come from engineering design (e.g., design-by-analogy), architecture (e.g., reference or precedence search) or both (e.g., brainstorming).

On the other hand, 'Synthesis & development' methods resemble what Schon [18] describes as "design experiments in a wider sense", i.e., not only including Schon's experiments but any other potential types of experiments which enable design ideas to be synthesized and further developed. Design experiments proposed by Schon are fundamentally different from Jones's transformation methods. The former expresses what designers want to achieve out of the experiments they propose, whereas the latter is a collection of different procedures to connect problem and solution spaces. Thus, methods in this sub-category comprise the three classic experiments proposed by Schon [18] (exploratory experiments, move-testing experiments and hypothesis-test experiments) but can be extended to include digitally assisted design experiments, which connect synthesis and development with decision support systems (e.g., parametric design methods, digital fabrication methods, etc.).

3.3 Decision Support Methods

Decision support methods were grouped based on the type of evaluation they enable designers to use when making decisions. These can vary from rational decision-making [21] to decision analysis [22] up to heuristic [22–24] methods.

Rational decision-making methods assume there are optimal design solutions and/or the best choice among design alternatives [21]. Methods of this sort provide value judgement about the desirability of a design and were grouped under the sub-category 'Deterministic choice' methods. They are commonly applied to detailed building design stages (optimization, multi-criteria evaluation, etc.) to, for instance, fine-tune design decisions about building materials, service components, etc. They have gradually been pushed to be implemented in early design stages to optimize building energy performance as a means to rationalize decisions related to, e.g., façade components and construction systems [25].

'Decision analysis methods' are decision support methods which enable designers to identify, represent and assess decisions to be made [22]. They are tools for decision analysis and can be categorized under four different sub-groups; 'Cause & effect', 'Feature evaluation' and 'Risks & Uncertainties'.

The sub-category 'Cause & Effect' is widely used to inform performance-based building design through the application of, for instance, building physics models and simulations of different sorts (e.g., heat balance, computational fluid dynamics, pollution dispersion, etc.). Methods of this sort are used to predict the behavior and performance of different design alternatives and can be used in isolation, to describe the consequences of different design decisions, or in combination with other decision analysis and/or rational decision-making methods when judgments need to be made.

The sub-category 'Feature Evaluation' groups methods predominantly used to extract information from data through machine learning algorithms of different types (e.g., decision trees, multiple regression, cluster analysis, etc.). Methods in this category are used to identify characteristics between different design variables such as window and balcony size, which influence daylight performance [26] by post-processing building simulation results.

The sub-category 'Risk & Uncertainties' groups methods used to undertake systematic data analysis based on mathematical models developed to assess how variations in design parameters affect design solutions. Methods of this type are widely used with building performance simulation (e.g., sensitivity analysis, risk assessment, robustness, etc.), to assess for instance, how uncertainty in relation to material properties affects building performance [27]. Recent experimental research can also be found in [28] who proposes a methodology which integrates robustness and risk assessment examining decisions made at the early design stages considering reversal in ranks, delayed discovery and insufficient gain or loss of performance gains.

The sub-category 'Heuristics' comprises groups of methods applied when decisions need to be made under uncertainty [22], when intuitive judgement is needed [23], when multiple alternatives are available [24], etc. basically when a choice needs to be made in the absence of a deterministic method. This is a commonly used method in the AEC industry. It can be found, for instance, in early design stages when deciding to proceed with a specific design hypothesis if it satisfies a basic aspiration level (e.g., satisficing strategy). And it can also be found during design development when the number of candidate solutions is reduced by eliminating one-by-one alternatives that do not meet certain aspirational levels (e.g., elimination heuristics).

3.4 Project Control Methods

Project control methods, providing feedback on whether the purpose of the different stakeholders' needs and aspirations for the product are met, were classified by their testing objective. To this end, they reflect the purposes for an asset listed according to different stakeholders' perspectives in ISO 19650-1 [20]. However, their sub-categories were mainly defined based on the Soft-Landings Approach [29], which focuses on asset operational performance and meeting of client's expectations.

The sub-category 'Asset value' groups methods related to assessing the value of the asset to its investors and/or owners. It is a particular sub-category of the building sector as "buildings [are] financial assets that figure in forms of market exchange…" [30] and therefore need to fulfill specific investors/owners needs related to strategic business cases for ownership and operation [20]. Methods in this category come from business finance and operation (e.g., Market valuation, etc.) and are used to gauge the value of the asset to investors throughout project development up to buildings in operation.

Since all stakeholders have aspirations for the asset behavior and performance, the sub-category 'Post-Occupancy Evaluation' groups methods that deal with asset performance in operation. It groups the methods used to assess building performance in use, i.e., while the building is already being occupied [31]. It includes user/occupant satisfaction, building and energy use, providing feedback on how well the asset is fulfilling users/occupants needs while in use as well as how well the asset is responding to contextual, infrastructural, societal, and environmental requirements. Post-occupancy evaluation methods come from Psychology, Social Sciences and Economics (e.g., questionnaires, interviews, etc.) when referring to user/occupants' satisfaction, and from Engineering (e.g., monitoring, etc.) when referring to asset and energy use.

The sub-category 'Commissioning', on the other hand, deals with building response and functioning right after construction, when a series of procedures are undertaken to test, check and ensure the building and its services are operating as designed. Methods in this category come from different Engineering domains (e.g., permeability tests, services' auditing, etc.) and form part of a mandatory building delivery stage in many countries [2].

The sub-category 'Management & Delivery' refers to methods employed to control the whole project organization and delivery, from design to manufacturing and construction of an asset up to its handover to the client. Specific needs for this category are included in the project brief and followed throughout the life of a project using different types of project management methods. Examples of project management methods applied to sustainability include Integrated Project Delivery (IPD) and Lean Construction, which respectively focus on the development of sustainable integrated project solutions and construction waste reduction.

3.5 Information Management Systems

Contrarily to methods, information management systems were not classified and inserted into the classic AD framework [15, 16]. They are a standalone group mentioned in this section only to highlight that the AEC industry uses a collection of complementary models, databases, and schemas to represent assets and record associated information. These models and schemas collect information throughout the design process using different ontologies and epistemologies, not always reconciled through interoperable software features. For instance, Building Information Management Systems [32] are structured to represent asset construction properties and the relationships between them, whereas metadata schemas such as Brick or Haystack are structured to represent buildings in operation, mainly the operation of their services and controls [33].

4 Can AD Coordinate Integrated AEC Sustainability Projects?

An outline on how different AEC methods to design for sustainability can be integrated via AD has been proposed in Fig. 1. This outline acknowledges that problem and solution are progressively specified, starting from an analysis of needs, and moving to the generation of possible solutions through an iterative process of zigzagging between the problem (what) and solution (how) spaces, including Process Domain [34]. Prior to any empirical testing it is already possible to highlight some conceptual issues in the proposed framework, which emerge from misalignments between how AD was developed and is supposed to be applied in product design, and the current AEC design practice.

The first issue refers to how far one can go with zigzagging in the AEC design practice. Delivery methods in the AEC industry are structured based on a complex system in which contracts, procurement and core statutory processes are interwoven. Clear design stages are put in place for projects to be developed so professional services, information exchange and contracts are prepared accordingly (e.g., [2]). These stages contain not only milestones for client approval but also milestones for core statutory process approvals (e.g., for the UK are specifically planning, building regulations and health and safety approvals). Statutory processes of this type imply freezing solutions as submitted since only minor changes can happen after approval. This means zigzagging is *de facto* restricted between core statutory process approval points throughout the design process and design delivery stages, contractually used to set up milestones for client's approval.

Acknowledging this limitation and attempting to better bridge issues appearing in the early design stages with issues related to construction and operation, delivery processes have been amended to include Soft-Landing principles [2, 29]. However, much is still to be done with regards to how Soft-Landing principles can be made operational as the implementation of these principles mostly depends on the existence of a consistent framework for their integration throughout the design process. AD can be extremely helpful in this front if it integrates AEC briefing methods in the zigzag happening between the Customers and Functional Domains. This is particularly the case if 'Asset management' and 'Project delivery' methods are brought to the early design stages and integrated with 'Investors' and 'users/occupants' methods to better inform the definition of Functional Requirements (FRs). FRs are the functions that must be fulfilled by the physical elements, Design Parameters (DPs), in order to satisfy customer and stakeholder needs [15]. AD can also be extremely helpful if it integrates 'Project delivery' methods in the zigzag happening between the Functional, Physical and Process Domains filling a particular gap essential in designing for sustainability [2]; the one of considering Process Variables (PVs), the variables involved in producing the specified DPs [15], in form generation. According to Frampton [35], the architectural form/shape is the result of "the constantly evolving interplay of three converging vectors, the *topos*, the *typos*, and the *tectonic*" [35] where the term "tectonics" encompasses the construction process from the materials up to the finished building [35, 36].

Interestingly, integrating these methods to the AD approach addresses some of the flaws highlighted by [37] in the AD literature. Specifically, 'Briefing methods', as defined in this paper, address issues with regards to identifying the key stakeholders involved in the design process, which for the AEC industry are clearly listed in the [20]. These same briefing methods are also powerful to identify different stakeholders' needs enabling them to be mapped and specified separately, to ensure that all aspects of the problem are properly defined and addressed [38] as the project progresses.

On the other hand, the mapping of stakeholders' needs to functional requirements and constraints as described in AD is not a straightforward task. AD – in its purest form – does not offer adequate instruments to capture the variety of requirements that the AEC design process has to master, with consequent problems in the design specification phase and in the application of axioms 1 and 2. To this end, the classification proposed by Thompson [37, 38] may supplement the framework proposed in this paper by providing clear strategies to identify constraints and non-FRs: both common elements in the AEC design process and requiring special consideration with regards to decision-making methods used to address and assess them.

Because AD is not prescriptive with regards to design methods to be used throughout the design process, the variety of methods employed by the AEC industry to deal with 'Concept generation' can be seamlessly integrated in the presented integrating framework. 'Synthesis & development' methods proposed in Sect. 3 can be used to augment AD design matrixes, relating DPs and FRs, with matrices relating DPs to DPs. Design matrices alone do not support specification of interactions between physical components (DPs) for the physical integration of system elements into a whole-design solution. This is particularly the case because 'Synthesis & development' methods are centered in design experiments having holistic assessment goals. Therefore, easily admitting the introduction of, for instance, Design Structure Matrix (DSM) [39, 40] or Interaction Matrixes [19] which represent how each element in the overall system relates to every other element in the system. Such matrices relating DPs to each other are used to assure that physical integration does not violate Suh's Axiom 1, maintaining independence of the FRs [41]. In this context, combining AD and DSM, for instance, can well be used to assess the implementation of sustainability requirements such as reusing, repairing, and remanufacturing towards resource circularity. Recent applications show an initial effort to use AD and DSM in construction projects for better control on changes [42].

A second conceptual issue, however, can be identified when attempting to integrate 'Decision support methods' to AD. On one side, the authors acknowledge that the AD approach already provides designers with two principles, Suh's Axioms 1 and 2, independence and information axioms, to support decision making in order to define effective designs with respect to specified requirements, to evaluate the synthetized ideas and to select the most feasible solution among valuable alternatives. One the other side, it could be said that Suh's Axiom 2, minimize the information content, does not admit 'Deterministic choice' methods because value judgement should never be deterministic. This, in principle, prevents such methods from being implemented in any design stage, despite these clearly gaining traction in the AEC industry. Whilst 'Deterministic choice' methods can be unsuitable if used in the early design stages as they freeze solutions rather early in the process, their potential to assist decision-making in detailed design stages can accelerate choice (e.g., use of multi-criteria evaluation or optimization routines in façade design to integrate construction and energy performance). Thus, the case for using Suh's Axiom 2 in AEC projects should be further examined.

Moreover, Axiom 2 prescribes the use of a specific decision analysis method, namely boundary searching [19], in which limits to acceptable solutions are specified based on probabilities of DPs fulfilling FRs. This prescription leaves room for multiple 'Cause & effect' methods to be applied to assess the success of manipulating different DPs towards achieving specified FRs. However, it excludes the application of some 'Risk & uncertainties' as well as 'Feature evaluation' methods. Despite not being prescriptive about how probabilities are calculated, boundary searching determines how probability results should be assessed, ruling out methods such as decision trees (part of the 'Feature evaluation' group), and expected relative performance losses (part of the 'Risk & uncertainties' group), to cite a few.

Axiom 2 also limits the use of 'Heuristics', including formal methods of heuristics, by not admitting, among others, the use of Satisficing Strategy and Recognition Heuristics, whilst promoting Elimination by Aspect [24]. Limitations in the use of 'Heuristics' can be a problem when assessing non-FRs and constraints as these many times do not have an associated probability function and therefore require 'softer' methods of assessment, such as for instance Simon's Satisficing Strategy. Relaxing the use of Axiom 2 would potentially increase the range of admissible decision analysis methods. However, more work is needed to understand, in detail, how each different decision-making method can be used if AD becomes the main decision-making framework used by the AEC industry to promote integrated design. Also, more work is needed to determine how this collection of methods complements the AD decision-making framework so it can better respond to the particularities of different design domains.

As part of this examination consider that Axiom 2, on minimizing information, should be applied after Axiom 1. That is, the best design solution (DP) is the one among those candidates that maintains the independence of the FRs equally well (Axiom 1), that has the least information content [14]. Axiom 2 ranks the candidate solutions that satisfy Axiom 1. Information content is defined as the log of the reciprocal of the probability of success in fulfilling FRs and avoiding constraints, therefore minimizing the information content (Axiom 2) is equivalent to maximizing probabilities of success. There can be important uncertainties in determining probabilities of success, hence uncertainty in these Axiom 2 based rankings. Considering Axiom 2 secondarily, however, limits the need for its applications and the associated difficulties.

The third conceptual issue arises from a set of particularities of the AEC design process which make assessing the success of a design solution a substantially difficult task. Every building is unique, not possible to be prototyped, and has to respond to the needs of multiple stakeholders with different goals and involved in different stages of the process.

Buildings have to respond to a specific site, climate, client, and occupant needs. Whereas standard solutions can be deployed in different parts of the design process (e.g., construction pre-fabrication, etc.), the combined response is always an idiosyncratic, large, and expensive intervention which cannot be tested through prototyping. There is no possibility for zigzagging to be implemented until the best version of a product can be developed. Therefore, testing the response of a building in its fullness can only be done after the building is built. This means only Project Control methods related to 'asset value' and 'management & delivery' can be used in the design stage, but these primarily respond to the needs of the investors and the project team. Methods such as 'Commissioning' can only be applied at the end of the construction phase as they specifically check if the building and its services are operating as designed. 'Post Occupancy Evaluation' methods, can only be assessed for building already in operation as they depend on how user/occupants interact with the building while it is managed.

The absence of prototyping makes the predictability of success highly dependent on decision-making methods used throughout the design process. These methods have to factor in uncertainties related to use and operation combined with climate related uncertainties. After all, building performance will depend on how the occupants interact with the building as well as how the building responds to climatic variations. As a result, there are large investments in research and practice towards developing decision-making methods related to predicting these uncertainties (IEA annex 79).

Initial attempts to record occupant-centric design patterns to inform design have been made in [14]. These patterns contain records of the application of different 'Cause & effect' methods to assess buildings' environmental performance together with 'Risk & uncertainty' methods related to occupant behavior, for facilitating the use of both in coordination when assessing design proposals. Whereas this proposal does not fully cover for uncertainties in relation to building usage in general, but mainly energy usage, it enables user behavior to be directly factored in the EAC design process, thus enabling performance to be assessed by applying Axiom 2. Occupant centric design patterns can be used also to simulate design robustness to different types of occupancy behavior, pushing the use of control methods to the zigzagging between the Functional and Physical domains. However, these are yet to be tested in practice so they can be expanded to include further aspects of performance testing, among which user/occupants' satisfaction. These call for further research, potentially between the Annex 79 and AD communities.

5 Conclusions: Adapting AD to the AEC Community Needs

This paper presents the initial results of an attempt to use AD to coordinate and integrate, in orderly fashion, many methods used to support evidence-based design in the AEC industry to produce design solutions which respond to current sustainability challenges. In theory, using AD to this end would not necessarily clash with the way the AEC industry currently operates. On the contrary, AD could promote the integration of many of these methods throughout the design process in a coordinated, traceable, and informed way, promoting transparency. However, three conceptual issues emerged from this study, calling for caution in the use of off-the-shelf AD notions in a field which presents some critical differences with the product design field.

The first issue identified by this research concerns limitations to the possibility to apply zigzagging in the AEC design practice beyond the boundaries of each of the prescribed project development stages. As a matter of fact, these are highly regulated by core statutory process approval points and formally stated in contracts, meaning choices cannot be changed after these points unless external conditions allow. Nonetheless, future research could clarify if AD can play a role in responding to this challenge by facilitating the coordinated implementation of Soft-Landing principles throughout the design process. One area of interest is the use of AD to coordinate how the different briefing methods can be used to map stakeholders' needs in the zigzagging between the Customer and Functional domains as well as between the Functional, Physical and Process Domains. From an AEC design perspective, exploring this issue provides opportunities to build a sufficiently complete problem framing for staggered project development. From an AD development perspective, this opens a debate on the opportunity to have domain-specific guidance to capture a complete set of requirements and constraints (possibly considering FRs and non-FRs) to respond to the specific challenges involved in designs which cannot be prototyped.

The second issue identified by this research shows the critical points in bringing AD to the AEC industry when coordinating the application of decision support methods while zigzagging from the Functional to the Physical domains. This calls for further empirical research and practice-based investigations to verify in more detail the compatibility between axioms and each of the decision support methods used by the AEC industry. Potential starting points could be to further investigate the admissibility and complementarity of: (i) 'Deterministic choice' in detailed design stages, in relation to Axiom 1; (ii) different heuristic methods in assessing non-FRs and constraints; (iii) each of the different decision analysis methods ('Cause & effect', 'Feature evaluation' and 'Risk & uncertainties'), one by one.

The third issue, however, is the most difficult one to address. It refers to the fact that the AEC deals with a prototype of one, meaning it has very limited means, if at all, to enable full zigzagging between all domains (which hinders the assessment of user/occupant satisfaction, among other things). Although the AEC industry has attempted to put in place mechanisms to transfer this assessment to the design stages by producing more sophisticated decision-making methods, much is still needed in relation to testing and deploying these methods in practice. The methods are mainly limited to assessing occupant behavior in relation to building energy consumption and need to be expanded to account for other aspects of building usage such as, for instance occupancy satisfaction. Challenges remain in relation to how this can be done so that predictability can be increased to enable the application of Axiom 2.

In a nutshell, AD seems promising to support the coordination and integration of the different methods used by the AEC industry to achieve evidence-based designs which are able to respond to current sustainability challenges. It can be useful to coordinate the different stakeholders' needs, promote the translation of design problems into sustainable design solutions, test design alternatives and control projects. The taxonomy produced to group and organize different stakeholders' needs, design, decision-making and project control methods supports new AD applications and opens new avenues for design research as well as new topics for AD development.

Acknowledgements. This paper was produced as part of the work of the Chair, Vice-Chair, and members of the International Association for Axiomatic Design (IAAD) Inter Society Board related to deploying AD to the AEC industry. To this end the researchers acknowledges support from: the UK Engineering and Physical Science Research Council (grant number EP/S03126X/1) to disseminate the work of the IEA Annex 79 via Prof Clarice Bleil de Souza; the Axiomatic Design Community for productive peer-review of Dr Marianna Marchesi contributions during her PhD research project implementation; the European Commission for the support in the development of the MSCA IF CircuBED project (Grant Agreement n. 793021) which has allowed Dr Marianna Marchesi to identify new potential applications of AD approach for the implementation of a circular economy in the built environment; colleagues in the Department of Energy, Systems, Territory and Construction Engineering (DESTEC, University of Pisa) for the fruitful discussions had with Prof Clarice Bleil de Souza and Dr. Camilla Pezzica during two international exchanges sponsored by the EU ERASMUS + program focused on applying AD to the built environment.

References

1. International Living Future and Institute: Living Building Challenge 3.1: a visionary path to a regenerative future. Living Futur. Inst. **31**, 1–82 (2016). https://living-future.org/sites/default/files/16-0504LBC3_1_v03-web.pdf
2. RIBA: Plan of Work 2020 Overview by Royal Institute of British Architects. Contract Adm. (2020). www.ribaplanofwork.com
3. Cross, N., Roozenburg, N.: Modelling the design process in engineering and in architecture. J. Eng. Des. **3**(4), 325–337 (1992). https://doi.org/10.1080/09544829208914765
4. Thompson, M., Beck, M.B.: Coping with change: urban resilience, sustainability, adaptability and path dependence. In: Future of Cities: Working Paper, Foresight, Government Office for Science, London, UK (2015). Accessed 11 June 2021. http://www.gov.uk/government/publications/future-of-cities-coping-with-change
5. Larsson, N.: The integrated design process; history and analysis. Int. Initiat. Sustain. Bult. Environ., 1–16 (2009)
6. Zimmerman, A.: Integrated Design Process Guide. Canada Mortgage and Housing Corporation, Ottawa (2006). http://www.smartgrowth.org
7. Ikudayisi, A.E., Chan, A.P.C., Darko, A., Adegun, O.B.: Integrated design process of green building projects: a review towards assessment metrics and conceptual framework. J. Build. Eng. **50**, 104180 (2022). https://doi.org/10.1016/j.jobe.2022.104180
8. Roozenburg, N., Cross, N.: Models of the design process: integrating across the disciplines. Des. Stud. **12**(4), 215–220 (1991)
9. MacMillan, S., Steele, J., Austin, S., Kirby, P., Spence, R.: Development and verification of a generic framework for conceptual design. Des. Stud. **22**(2), 169–191 (2001). https://doi.org/10.1016/S0142-694X(00)00025-9
10. Marchesi, M., Matt, D.T.: Application of axiomatic design to the design of the built environment: a literature review. In: Farid, A.M., Suh, N.P. (eds.) Axiomatic Design in Large Systems, pp. 151–174. Springer, Cham (2016). https://doi.org/10.1007/978-3-319-32388-6_6

11. Marchesi, M., Fern, J.E., Matt, D.T.: Axiomatic design in large systems. Axiomat. Des. Large Syst., 175–200 (2016). https://doi.org/10.1007/978-3-319-32388-6
12. Marchesi, M., Kim, S., Matt, D.T.: IMECE2015-50517, pp. 1–10 (2015)
13. Bleil de Souza, C., Dunichkin, I.V.: Axiomatic design in regenerative urban climate adaptation. In: Andreucci, M.B., Marvuglia, A., Baltov, M., Hansen, P. (eds.) Rethinking Sustainability Towards a Regenerative Economy. FC, vol. 15, pp. 5–21. Springer, Cham (2021). https://doi.org/10.1007/978-3-030-71819-0_1
14. Bleil de Souza, C., Tucker, S., Deme Belafi, S., Reith, A., Hellwig, R.: Occupants in building design decision-making. In: O'Brien, W., Tamasebi, F. (eds.) Simulation-Aided Occupant-Centric Building Design: Theory, Methods, and Detailed Case Studies. Taylor & Francis (2023)
15. Suh, N.P.: The Principles of Design. Oxford University Press (1990)
16. Suh, N.P.: Axiomatic Design: Advances and Applications. Oxford University Press (2001)
17. Brown, C.A., Rauch, E.: Axiomatic design for creativity, sustainability, and industry 4.0. MATEC Web Conf. **301**, 00016 (2019). https://doi.org/10.1051/matecconf/201930100016
18. Schon, D.: The Reflective Practitioner: How Professionals Think in Action. Ashgate Publishing Limited (1991)
19. Jones, J.C.: Design Methods: Seeds of Human Futures. John Wiley & Sons (1980)
20. EN ISO 19650-1: Organization and Digitization of Information about Buildings and Civil Engineering Works, Including Building Information Modelling (BIM) - Information Management Using Building Information Modelling. Part 1: Concepts and Principles, pp. 1–46 (2018)
21. Hazelrigg, G.A.: Fundamentals of Decision Making: for Engineering Design. Neils Corp (2012)
22. Simon, H.A.: The Sciences of the Artificial, 3rd Editio. MIT Press, Cambridge, MA (1996)
23. Hammond, K.: Judgement and decision making in dynamic tasks. Inf. Decis. Technol. **14**, 3–14 (1988)
24. Tversky, A.: Elimination by aspects: a theory of choice. Psychol. Rev. **79**(4), 281–299 (1972)
25. Moghtadernejad, S., Chouinard, L.E., Mirza, M.S.: Multi-criteria decision-making methods for preliminary design of sustainable facades. J. Build. Eng. **19**, 181–190 (2018). https://doi.org/10.1016/J.JOBE.2018.05.006
26. Loche, I., Bleil de Souza, C., Spaeth, A.B., Neves, L.O.: Decision-making pathways to daylight efficiency for office buildings with balconies in the tropics. J. Build. Eng. **43**, 102596 (2021). https://doi.org/10.1016/j.jobe.2021.102596
27. Hopfe, C.J., Hensen, J.L.M.: Uncertainty analysis in building performance simulation for design support. Energy Build. **43**(10), 2798–2805 (2011). https://doi.org/10.1016/j.enbuild.2011.06.034
28. Agarwal, M.: Solar potential in early neighborhood design (2016)
29. Building Services Research and Information Association – BSRIA: Soft Landings Core Principles. 2nd Edn. Constr. Res. Innov. (2018)
30. Cass, N., Shove, E.: Standards? Whose standards? Archit. Sci. Rev. **61**(5), 272–279 (2018). https://doi.org/10.1080/00038628.2018.1502158
31. RIBA: Post occupancy evaluation: an essential tool to improve the built environment, London, UK (2008). https://doi.org/10.4324/9780080518251
32. Wong, J.S., Yang, J.: Research and application of building information modelling (BIM) in the architecture, engineering and construction (AEC) industry: a review and direction for future research. In: Proceedings of the 6th International Conference on Innovation in Architecture, Engineering and Construction (AEC), pp. 356–365 (2010)
33. Brickschema (2022). https://brickschema.org/
34. Brown, C.A.: Axiomatic design of manufacturing processes. In: Eighth International Conference on Axiomatic Design - ICAD 2014 (2014)

35. Frampton, K.: Studies in Tectonic Culture. Harvard University, Graduate School of Design (1985)
36. Deplazes, A.: Constructing Architecture: Materials, Processes, Structures. A Handbook, 4th edn. Birkhauser (2018)
37. Thompson, M.K.: Improving the requirements process in axiomatic design theory. CIRP Ann. **62**(1), 115–118 (2013). https://doi.org/10.1016/J.CIRP.2013.03.114
38. Thompson, M.K.: A classification of procedural errors in the definition of functional requirements in axiomatic design theory. In: Proceedings ICAD2013, Seventh International Conference on Axiomatic Design (2013). http://www.mkthompson.net/wp-content/uploads/2013/06/ICAD2013-16-Thompson-FR-Error-Classification.pdf
39. Eppinger, S.D., Browning, T.R.: Design Structure Matrix Methods and Applications. MIT Press (2012)
40. Ulrich, K.T., Eppinger, S.D., Yang, M.C.: Product Design and Development. McGraw Hill Education (India) Private Limited (2019)
41. Suh, N.P.: Design systems. CIRP Ann. **46**(1), 75–80 (1997). https://doi.org/10.1145/3380851.3416743
42. Padala, S.P.S.: Application of axiomatic design and design structure matrix for early identification of changes in construction projects. J. Inst. Eng. Ser. A **103**(2), 647–661 (2022). https://doi.org/10.1007/s40030-021-00612-2

Designing an Ergonomic Geothermally Heated Pinecone Seed Extractor

Eric Wolf Ruge(✉), Maxim Rudkov(✉), and Joseph Timothy Foley(✉)

Department of Engineering, Reykjavik University,
Menntavegur 1, Reykjavík 102, Iceland
{Eric22,Maxim17,Foley}@ru.is

Abstract. Reforestation is one key element to counteract the global climate catastrophe. The Icelandic Forest Service (IFS) provides pine tree seeds in Iceland, using a labor-intensive, time-consuming, and unergonomic process with two various machines for drying and separating the seeds from the cones. To optimize this procedure, the two machines were combined into one machine (SeedEx) following Axiomatic Design and Product Design principles. Aligning Customer Needs (CNs) to desired functionality allowed the team to realize the project in twelve weeks by minimizing design iterations through careful modularity. Changing the main rotation axis of the extractor enables operators to load cones and receive seeds ergonomically. The SeedEx prototype processes 260% of the former daily capacity while reducing labor by 92% and eliminating time delays between processes.

Keywords: Seed Extraction · Pinus Contorta · Convective drying · Axiomatic Design

1 Introduction

Deforestation is a major concern that is highly prevalent throughout the world, not only accelerating global warming but also declining biodiversity [1]. The country of Iceland has been hit particularly hard by this human-made catastrophe. A nation that once had 40% of its countryside covered by forests began to lose its tree cover in the 9th century. Today, only 2% of Iceland is forested [2]. The IFS is reforesting Iceland, doubling the total amount of woodland and forests since 1950 by planting tree species such as Russian Larch, Alaskan poplar, Sitka spruce, and especially the Lodgepole pine tree.

Pine tree seeds are located in the pinecone, protected from weather impacts and predators during winter, by remaining closed and sealed with resin. In nature, Lodgepole pinecones open up during summer due to the increase in temperature and evaporation of water with resin [3], as shown in Fig. 1. However, not all pinecones open up through this natural process. Studies show that a significant amount remains closed until the resin bond is melted by temperatures above 52 °C [4,5]. In fact, these serotinous cones enable the seeds to survive

E. Puik et al. (Eds.): ICAD 2023, LNNS 849, pp. 118–131, 2024.
https://doi.org/10.1007/978-3-031-49920-3_8

Fig. 1. Sealed and open Lodgepole pinecone with released seeds

wildfires and spread after the wildfire is over. Additionally, the IFS states, if the seeds are exposed to temperatures over 55 °C, the chance of the seeds drying out increases. Once a seed is dried there is no chance to germinate and develop into a tree.

From October to March, the IFS collects Lodgepole pinecones to plant new trees (Pinus Contorta) and sell seeds to other companies for reforestation purposes. The process the IFS is currently using to extract the seeds takes over 24 h. The IFS starts by soaking the cones to dissolve the resin. Afterward, the cones are dried in one machine (Fig. 2a) and then shaken by a different machine (Fig. 2b) to bring out the seeds. Both machines require manual loading and unloading, which takes up to 60 min for a total capacity of 90 L. On average, the IFS states that 30% of the pinecones do not open up during the first process cycle due to serotinous cones. This process of extracting the seeds is labor-intensive, time-consuming, and unergonomic. To fix these issues, the Seed Extractor (SeedEx) was designed with the goal of both drying soaked pinecones and extracting the seeds in one process, by enabling less time-consuming and ergonomic loading and unloading capabilities. Following the concepts of Axiomatic Design theory in a product design context [6], a top-level need was phrased: "**CN_0**: Dry pinecones and separate the seeds in one convenient process by the worker."

Based on this, the decomposed CNs can be identified as follows:

- **CN_1.** Dry the pinecones to open up at least 70% of the loaded pinecones while serotinous cones can remain closed.
- **CN_2.** Separate the seeds from the pinecones.
- **CN_3.** Enable one person to load and unload the machine.
- **CN_4.** Cheaper than the project's budget of 500 000 ISK.
- **CN_5.** Fit through standard doors (width less than 85 cm).
- **CN_6.** Reliable in unheated greenhouse conditions, since it will be stored in an unheated greenhouse, all year round.
- **CN_7.** Minimum loading capacity of 50 L.
- **CN_8.** Pinecone temperature not exceeding 55 °C.

The name SeedEx was chosen because the machine extracts seeds from pinecones. The target customer is the IFS as well as other planting companies and the companies buying seeds from the IFS. Future projects like the

(a) Dryer (b) Shaker

Fig. 2. IFS's current seed extraction machines.

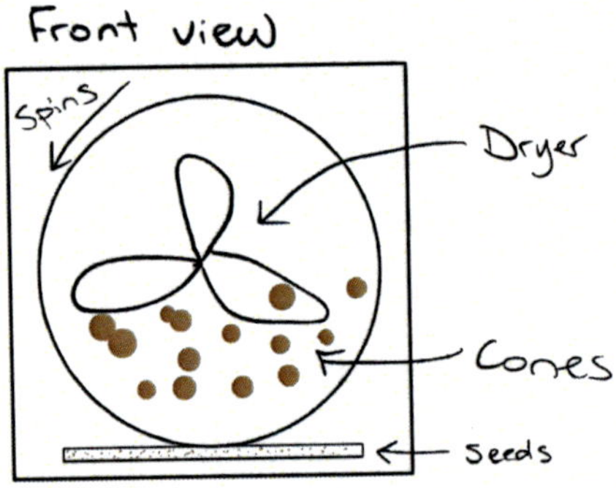

Fig. 3. Front loading concept with single bearing

Reforest'Action will increase the need for seeds, which emphasizes the potential need for an additional machine for the IFS. Besides the IFS in Iceland, other governments or companies around the globe are considered to be interested in the prevalent design. The IFS was involved in the design process and provided financial resources. Therefore, the final machine design will not be turned into a business but due to the rising global demand, the potential is estimated to be 100 machines a year.

2 Prior Art

As stated in Sect. 1, the IFS separates seeds from Lodgepole pinecones. Considering a total variety of 115 different pine species with different physical characteristics, the prior art on pine seed extraction varies substantially. In the following chapter, recent concepts are analyzed and evaluated.

2.1 Machines for Pinecones

Two current available designs on the market are the dehuller Extractor 1300 and the smaller dehuller Extractor 800 from BCC AB. These machines are designed to extract seeds from the same species of pine trees and are therefore comparable to SeedEx.

The BCC AB Seed Extractor 1300 and 800 are large capacity and multipurpose units performing pre-cleaning and drying of pinecones and seed extraction. Pre-cleaning removes small impurities, e.g. needles, scales, and twigs before drying. This takes place due to the friction of the loaded pinecone mass inside the machine and is therefore a feature that SeedEx would also perform when applying motion to the cones.

Both machines use convectional drying and rotational moving of the cones. During the drying process, the cone gradually opens for the seed to be extracted [7]. Applying this process for SeedEx appears promising, since the current machines of the IFS work with the same principle. The inside of Extractor 1300 and the front of Extractor 800 are similar to the first design ideas of SeedEx, shown in Fig. 3.

However, the BCC AB units do not fulfill all the CNs of the IFS, for example, reliability (CN_6). BCC AB machines are not built from stainless steel. In an unheated greenhouse environment, it is possible for a small scratch in the paint to cause severe consequences over time.

Additionally, the units consume 1.5 kW of electricity for the heating element [8]. Electric heating elements burn up over time and require an advanced controller. SeedEx aims to outperform the BCC AB units by using hot water for heating, considering the fact that geothermal hot water eliminates fire hazards and is comparatively cheap in Iceland. Next, the two units do not fit through doors. Even though the capacity of the Extractor 1300 with 200 L fulfills the CN of the IFS, the design was not chosen to fit through doors. Hence, Seedex must rely on two bearings and a long-shaped drum.

Finally, the machines of BCC AB are not ergonomically beneficial for the IFS. Like the current machine, it is essential for one person to be able to manually load and unload the pinecones with small bags. SeedEx can outperform this process by utilizing gravity for unloading and allowing heavy machinery to load the machine. In conclusion, the overall drying and separating process of the Extractors 1300 and 800 applies to SeedEx, but has to be modified to accomplish the CNs of the IFS.

Another available design on the market comes from Jiaozuo Zhoufeng Machinery Co. Ltd. The pinecone and seeds separating sheller processing machine differs from other designs. It is built on a trailer and is powered by a 20 hp diesel engine, enabling a flexible field of operation. The machine consists of a feeding hopper and a conveyor belt, allowing constant feeding of the machine. The capacity of the machine is 500 kg. The main shaft speed is 800 RPM, the fan speed is 800 RPM, and its dimensions are 3250 mm by 750 mm by 1280 mm [9].

The high flexibility of the machine comes with both high operating and fixed costs, as well as a notable environmental impact due to the greenhouse gas emission of the diesel engine. The IFS does not require a self-sufficient design. Consequently, this design is solving distinct CNs.

However, designing SeedEx on wheels enables one person to handle the machine and is therefore an important takeaway from this design.

2.2 Machines for Pine Nuts

The variety of different pine species resolves in a range of various pine seeds. Pine nuts are larger than Lodgepole pine seeds and are edible. The seed extraction process is therefore different but still worth considering since both have the same goal of extracting the seed undamaged.

The pine nut shelling machine from Zhengzhou Tonde Machinery Co. Ltd cracks pine nuts without causing any damage to the kernels. The electrical power of this machine is 9.68 kW with a capacity is 300 L. The machine is 3 m tall and the gross weight is 4000 kg. The machine consists of a feeding hopper and a laminated spring plate conveyor belt allowing constant feeding. The machine processes the maximum capacity in one hour. After the pinecone is cracked open the pine nuts are shaken by sizing decks and the kernels flow to gravity tables [10].

Applying this process to Lodgepole pinecones comes with the risk of not succeeding in the given project time of twelve weeks, considering breaking these species open without damaging the seed was not proven to work before. Moreover, filtering out the seeds from impurities would resolve in a higher complex machine which is not suitable for the budget of 500.000 ISK and project time of twelve weeks.

2.3 Functional Requirements and Constrains

The CNs in Sect. 1 result in a list of Functional Requirements (FRs) and Constraints (Cs) by using Axiomatic Design Theory. FR satisfies a CN by adding a required function, whereas Cs are boundaries on acceptable solutions [6].

The following list shows the Functional Requirements regarding the CN for the SeedEx machine, starting with the overall: "$\mathbf{FR}_0$ Dry and separate the seeds from the pinecones by requiring one operator.

Add mechanical advantage for unloading." which is then decomposed into further FRs.

$\mathbf{FR}_1$. Dry the soaked pinecones to open up at least 70% of the loaded cones.
$\mathbf{FR}_2$. Add kinetic energy to pinecones to separate the seeds.
$\mathbf{FR}_{2.1}$. Add kinetic energy to the pinecone to get the seeds released.
$\mathbf{FR}_{2.2}$. Strain seeds from pinecones.
$\mathbf{FR}_3$. Allow one person to load and unload.
$\mathbf{FR}_{3.1}$. Allow one person loading.
$\mathbf{FR}_{3.2}$. Allow one person unloading.

The following list shows the Constraints regarding the CNs for the SeedEx machine:

C1. The project cost must not exceed 500.000 ISK.
C2. The machine has to fit through doors with a maximum width of less than 85 cm.
C3. Corrosion resistance, particularly against moisture.
C4. Minimum loading capacity of 50 L.
C5. Maximum pinecone temperature of 55 °C.

3 Design

In this chapter, the development of the design is explained. The design process was conducted with Axiomatic Design to determine the most simple and independent solution [6]. Brainstorming ideas, drawing sketches, and crafting prototypes assisted in the process to identify design issues and to be creative in decomposing the FRs into Design Parameters (DPs) while satisfying the Cs.

3.1 Design Process

The design process starts with analyzing the CNs and examining the prior art as described in Sect. 2. Figure 3 shows the first idea of SeedEx being influenced by the current designs from BCC AB. As the brainstorming and analysis continued, more changes were developed. As shown in Fig. 3 the idea consists of one bearing holding the drum. To satisfy the minimum capacity and maximum width, the design was developed in a longer drum. To then support the increased stress, two bearings are necessary, and therefore the loading and unloading needed to be reconsidered. This prototype was then produced in cardboard and is shown in Fig. 4.

Autodesk Inventor Professional 2021 was used for the CAD drawings shown in Fig. 5a and 5b. At this point, the loading height was designed to be at 1.125 m, so the loading is comfortable and ergonomic [11]. The length was set to 1.5 m which was calculated for a maximum loading capacity of 200 L. For loading and unloading, the top half of the hexagon is opened and one-third of the drum slides open. The bottom part of the hexagon was designed to be open, so the seeds can fall out during the process and the empty pinecones can be dumped after the process is over.

During the design process, the movement of the pinecones was estimated to behave like a liquid. This way the load on the bearings, motor, and frame was calculated. Soaked Lodgepole pinecones have half the density of water. When spinning the drum, the pinecone mass will flow inside the drum. To provide an even drying process on every cone, SeedEx needs to guarantee a consistent concoction of the loaded mass inside the drum. Therefore, fins were added inside the drum.

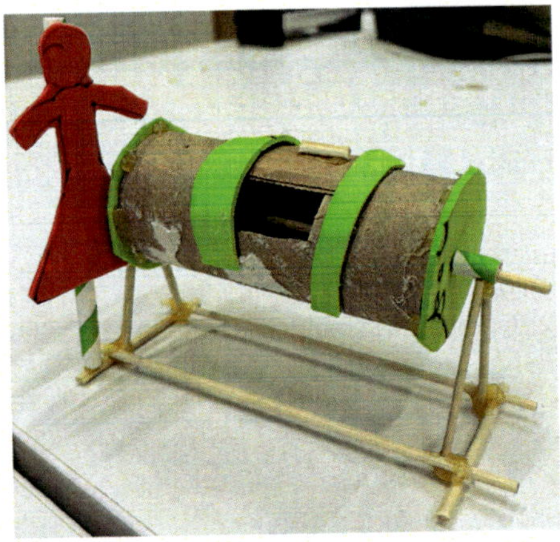

Fig. 4. Prototype with two bearings enables relative sizing and ergonomics

In this design the motor, radiator, fan, and, piping for the airflow direction were added. The opening section of the drum was then redesigned before manufacturing, simplifying the handling and increasing the clearance of the drum inside the hexagon. After the first conducted experiment of the system, bottom rails for standard 1/1 Gastronorm stainless steel containers were installed to catch the seeds and reduce heat loss.

3.2 Design Parameters

From the Functional Requirements listed in Sect. 2, a list of Design Parameters is developed while following the Independence Axiom and the Information Axiom [6], as shown in Table 1.

Starting with the top-level Design Parameter "$\mathbf{DP}_0$ Fan with a radiator to blow hot air and dry the pinecones. Motor for rotating the drum. Mesh to separate the empty pine cones from the seeds. The opening section of the drum is at a height of 1.125 m above the ground." and decomposing into further DPs, as summarized in Table 2:

$\mathbf{DP}_1$. Water radiator and electric fan
Hot geothermal water at 80°C with a flow rate of $0.05 \mathrm{Ls}^{-1}$ flows through the 2.5 kW radiator and exchanges heat with the passing air from the $240\,\mathrm{m}^3\mathrm{h}^{-1}$ fan. This heated air is at 48°C fulfills CON_5 and dries the pinecones.

$\mathbf{DP}_{2.1}$. Electric motor and gearbox
A 1.1 kW electric motor with a maximum rpm of 1445 and a gearbox with a 1:67 ratio. Bring the pinecones in motion by spinning the drum at 10 rpm, so the seeds fall out. The powertrain provides 500 Nm of torque which was chosen by the following equation:

$$T_{\text{requirend}} = m_{\max} x_{\max} g \tag{1}$$

(a) Closed while in operation and dispensing seeds.

(b) Open for loading of cones.

Fig. 5. CAD 3D Models of final design show cover positions of extractor.

Table 1. Top level FR-DP mapping.

	Functional Requirement	Design Parameter
1	Drying pinecones	Water radiator and electric fan
2	Add kinetic energy to pinecones to separate the seeds	Electric motor, Gearbox and Mesh with 8 × 8 mm holes
3	Allow one person to load and unload	Machine opening 1.125 m above ground
		and unloading through gravity

where $T_{\text{requirend}}$ is the maximum torque required, $m_{\max}$ the highest possible loading weight with a safety factor of 2, $x_{\max}$ is the maximum lever arm as shown in Fig. 6, and g is the gravitation constant. The approximated values result in:

$$T_{\text{requirend}} = (250\,\text{kg})(0.18\,\text{m}(9.8\,\text{m s}^{-2}) = 441\,\text{Nm} \quad (2)$$

As shown in Fig. 6, the lever arm reaches its maximum when the pinecones accumulate on one side of the drum.

DP$_{2.2}$. Stainless steel mesh
Stainless steel mesh made out of 1 mm thick wire with square holes of 8 × 8 mm. This mesh is wrapped around the drum, holds the pinecones and lets seeds fall through to the bottom. The size of 8 × 8 mm was recommended by the IFS as the ideal size to let the seeds fall through, but stop even the smallest pinecones.

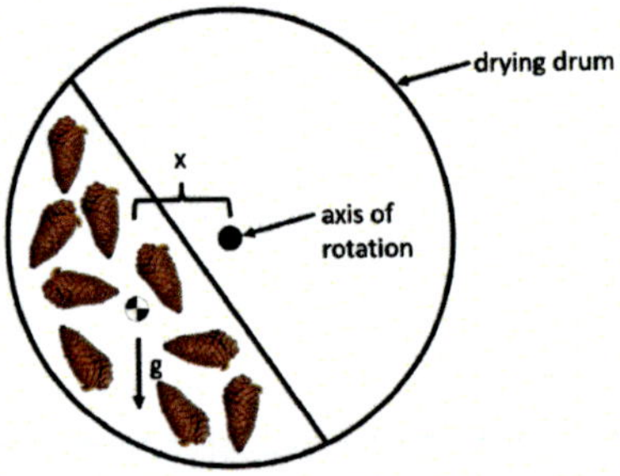

Fig. 6. Estimated behavior of the pinecones inside the drum with resulting torque

$DP_{3.1}$. Machine opening for loading at 1.125 m above ground.
The height of the frame with wheels is designed to be 0.9 m above ground. The loading height ends at 1.125 m. so the loading is comfortable and ergonomic [11].

$DP_{3.2}$. Unloading through gravity.
The machine has an open bottom. By turning the drum 180° from the loading position with the removable section removed, the pinecones fall through the bottom into a collection box.

3.3 Design Matrix

The top and second level design matrices [12] were developed as shown in Eqs. 3 & 4. This matrix is uncoupled, i.e. diagonal matrix, meaning by varying one DP each FR can be changed without affecting other FRs [12]. Therefore, it is possible to work on individual DPs simultaneously. This simplifies the optimization of the system (Fig. 7).

$$\begin{Bmatrix} FR_1 \\ FR_2 \\ FR_3 \end{Bmatrix} = \begin{bmatrix} X & 0 & 0 \\ 0 & X & 0 \\ 0 & 0 & X \end{bmatrix} \begin{Bmatrix} DP_1 \\ DP_2 \\ DP_3 \end{Bmatrix} \tag{3}$$

Table 2. Second level FR-DP mapping.

	Functional Requirement	Design Parameter
1	Drying pinecones	Water radiator and electric fan
2.1	Add motion to the pinecones	Electric motor and gearbox
2.2	Strain seeds from pinecones	Mesh with 8 × 8 mm holes
3.1	Allow one person loading	Machine opening at 1.125 m above ground
3.2	Allow one person unloading	Unloading through gravity

Fig. 7. First Test of SeedEx at Reykjavík University's power and energy lab.

$$\begin{Bmatrix} \mathrm{FR}_1 \\ \mathrm{FR}_{2.1} \\ \mathrm{FR}_{2.2} \\ \mathrm{FR}_{3.1} \\ \mathrm{FR}_{3.2} \end{Bmatrix} = \begin{bmatrix} X & 0 & 0 & 0 & 0 \\ 0 & X & 0 & 0 & 0 \\ 0 & 0 & X & 0 & 0 \\ 0 & 0 & 0 & X & 0 \\ 0 & 0 & 0 & 0 & X \end{bmatrix} \begin{Bmatrix} \mathrm{DP}_1 \\ \mathrm{DP}_{2.1} \\ \mathrm{DP}_{2.2} \\ \mathrm{DP}_{3.1} \\ \mathrm{DP}_{3.2} \end{Bmatrix} \tag{4}$$

4 Results and Discussion

In this chapter, the results of the different tests are shown and discussed. This was conducted starting with the FEM simulation of the frame, followed by the testing of the drying system, and concluding with the testing of the whole machine in operation.

4.1 Frame FEM Simulation

The frame was simulated in Autodesk Inventor 2021 bearing a load of 6.25 kN split between the pillow block-bearing seats. This is equal to 2.5 times the expected loading. The expected loading was calculated from the weight of the drum 50 kg with the maximum allowed amount of pinecones at 200 kg. This results in a force of 2.5 kN acting on the bearing seats. As can be seen in Fig. 8 the max von Mises stress did not exceed 75 MPa on the bearing seats. Given that the yield strength of 304 stainless steel is 205 MPa, the frame experiences no fatigue under the expected operation. The load is estimated to remain below half the yield strength. The deformation with the expected loading is 0.0372 mm as can be seen in Fig. 9. The observed deformation in the real unit was too small to measure with the available tools, as it was less than 1 mm.

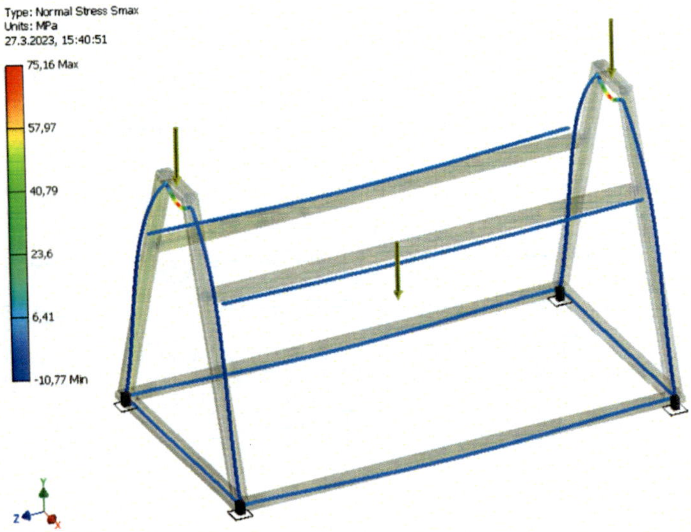

Fig. 8. FEM simulation of the frame with loading equal to 2.5 times the expected loading

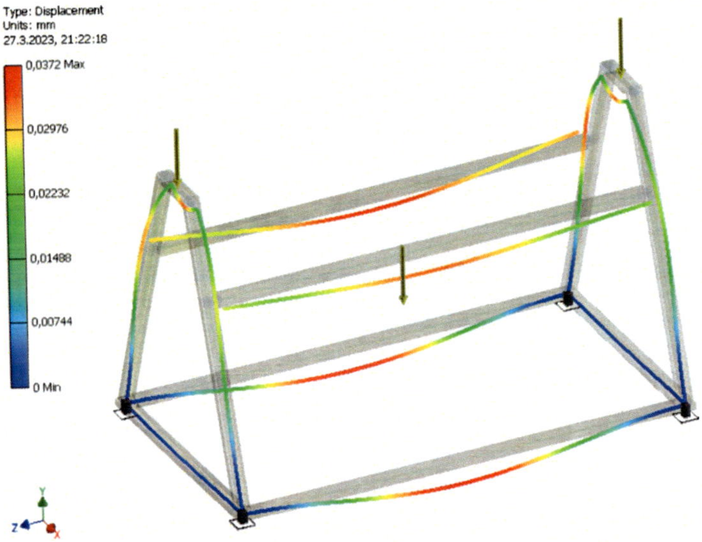

Fig. 9. FEM simulation of the deformation in the frame with the expected loading

4.2 FR$_1$ Test: Airflow and Air Temperature (Radiator and Fan)

The fan and the radiator were tested together to evaluate the range of temperature and airflow expected from this module. The steady temperature at the final location will be 10 °C with a relative humidity of 80%. The relative humidity

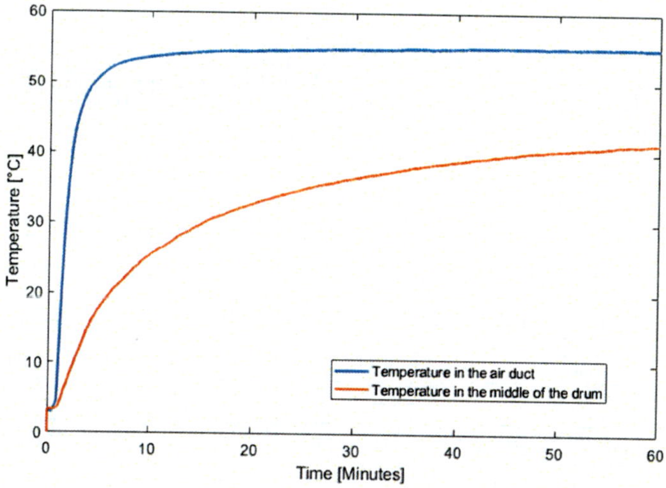

Fig. 10. SeedEx heating up at the 3°C greenhouse with 80°C water

will drop to 14% after the air is heated to 40 °C according to the psychrometric chart. Therefore, the relative humidity should not affect the process. The radiator was connected to the water source with a water flow of 0.16 Ls^{-1} at 52 °C, and high water flow was required to compensate for sub 70 °C water at the university. The fan was set to a max setting of 230 m^3h^{-1}. The fan and the radiator were connected with a pipe so that the fan blows air through the radiator. The in- and outflow temperatures of air were measured with a DTP6 thermometer, and the air velocity was measured with AN310 Anemometer. The temperature of the discharging water was 50 °C. The inflow air temperature was at 20°C and outflow at 42 °C. The airflow dropped to 150 m^3h^{-1} due to the pressure drop in the radiator. The air temperature for pinecone drying must not exceed 50 °C. Given access to hotter water, the desired temperature will be easily reached even with ambient temperatures of 10 °C according to the data from this test.

4.3 Functional Testing of the System with 100 L/50 Kg of Pinecones

To begin system testing, the pine cones were soaked for 24 h. Afterward, they were loaded into the machine and dried with a continuous rotation of 10 RPM at 40.3 °C. After 20 h, 70% of the pine cones got dry and opened up. The data log of the first 60 min can be seen in Fig. 10. The temperature in the air duct reached a horizontal asymptote after 16 min. The measured data was:

- The air temperature inside the drum was 41.8 °C.
- The ambient temperature was 3 °C.
- The water temperature going in was 75 °C.
- The water temperature going out was 59 °C.
- The airflow was 155 m^3h^{-1}.
- The water flow was 0.04 Ls^{-1}.

The remaining unopened 30% of the pinecones are due to the cones level of serotiny. These pine cones will be reentering the process, as they require multiple pre-soaks before they open up during the drying process. The unloading through gravity worked as desired and the machine was unloaded in under 10 s. Adding 4 m and 50 s for the loading process, the labor work was reduced by 92% in comparison to the current IFS process (60 m).

This testing evaluated all the FRs, from loading the machine in the beginning to unloading the machine after the drying and seed separation was done.

5 Conclusion

Replanting trees is increasing in importance due to global warming and preservation of biodiversity. The IFS is responsible for the required Russian Larch, Alaskan poplar, Sitka spruce, and Lodgepole pine tree seeds in Iceland. The government agency is extracting seeds with two machines from pinecones in a labor-intensive process. By using Axiomatic Design the CNs of the IFS were identified and entirely accomplished in a simplified and independent design. This enabled SeedEx modules to develop over the design process without affecting others. The Seed Extractor (SeedEx) dries pinecones (CN_1) and separates the seeds (CN_2) in one process, while additionally enabling one person to unload and load the machine (CN_3). SeedEx is performing a convection drying process with a water heat exchanger and an industrial fan. The pinecones are secured in a cylindrical drying drum, which consists of a 8×8 mm square hole mesh and a removable section. The drum is brought into circular motion by an electric motor. This motion is forcing the small seeds to be released from the pinecone and fall through the mesh and out of the machine. The loading and unloading mechanism of the drum consists of one removable section of the drum. The section enables one average-height person to remove it for loading and unloading at a height of 1.125 m, while the unloading process is executed by gravity by turning the open drum 180°, reducing the labor work by 92%. The implementation of the cylindrical drying drum results in the maximum width to fit through 85 cm doors (CN_5), while at the same time providing five times the stated minimum loading capacity of 50 L (CN_7). The customer budget of 500 000 ISK was not exceeded, as the cost of building the machine resulted in 490 000 ISK (CN_4). At the same time, reliability in unheated greenhouse conditions was accomplished by choosing 304 stainless steel for the components (CN_6). The major challenge was combining the process into one machine. SeedEx accomplished all CNs in the conducted experiment. 70% of the pinecones opened up and released the seeds over a time of 20 h which fulfills (CN_1).

Overall, in addition to the fulfillment of customer needs, the designed SeedEx contributes to the acceleration of reforestation efforts and strengthens the conservation of biodiversity.

References

1. IPBES-IPCC: Ipbes-ipcc co-sponsored workshop biodiversity and climate change workshop report (2021). https://doi.org/10.5281/zenodo.4782538
2. Government of Iceland: Forestry (2022). https://www.government.is/topics/environment-climate-and-nature-protection/forestry. Accessed 11 Nov 2022
3. Song, K., Chang, S.-S., Lee, S.J.: How the pine seeds attach to/detach from the pine cone scale? Front. Life Sci. **10** (2017). https://doi.org/10.1080/21553769.2017.1287777
4. Lamont, B.B., Le Maitre, D.C., Cowling, R.M., Enright, N.J.: Canopy seed storage in woody plants. Botanical Rev. **57**(4) (1991). https://doi.org/10.1007/BF02858770
5. Nathan, R., Ne'eman, G.: Serotiny, Seed dispersal and seed predation in Pinus Halepensis (2000). https://skat.ihmc.us/rid=1GQDJLNHZ-236XZSC-CK/Nathan%26Neeman2000MedPine1.pdf. Accessed 27 Mar 2023
6. Suh, N.P., Cavique, M., Foley, J.T. (eds.): Design engineering and science (2021). https://doi.org/10.1007/978-3-030-49232-8
7. Agriexpo: Pine cone dehuller Extractor 1300. https://www.agriexpo.online/prod/bcc-ab/product-181245-49928.html. Accessed 30 Aug 2022
8. BCC: Extractor 1300. https://web.archive.org/web/20201031071922/https://www.bccab.com/products-planting/cone_and_seed_handling/wet-dewinger-seed-extractor-1300/. Accessed 30 Aug 2022
9. Pine Nuts Shelling Machine. https://www.alibaba.com/product-detail/Pine-cone-and-seeds-separating-sheller_60559869090.html. Accessed 15 Nov 2022
10. Pine Nuts Shelling Machine. https://www.shellingmachine.com/cracking-shelling-machine/pine-nut-shelling-machine.html. Accessed 02 Oct 2022
11. International Organization for Standardization: EN ISO 6385, Ergonomic principles in the design of work systems (2016)
12. Suh, N.P.: Axiomatic Design - Advances and Applications. Oxford University Press, Oxford (2001)

Axiomatic Design and Product Design

Axiomatic Design Using Multi-criteria Decision Making for Material Selection in Mechanical Design: Application in Different Scenarios

Alessandro Giorgetti[1](✉), Fernando Rolli[2], Vincenzo La Battaglia[1], Stefano Marini[1], and Gabriele Arcidiacono[2]

[1] Department of Industrial, Electronic and Mechanical Engineering, Roma Tre University, 00146 Rome, Italy
alessandro.giorgetti@uniroma3.it

[2] Department of Engineering Science, Guglielmo Marconi University, 00193 Rome, Italy

Abstract. In mechanical design, the selection of material alternatives has become a pressing issue due to the progressive growth in the complexity of mechanical systems in search of a continuous increase in performance and the presence of a wide range of possible materials. Moreover, there are many requests for projects, which makes the choice of material a decisive activity for the success or failure of the project itself. Multi-criteria decision-making (MCDM) describes the systematic approaches developed to evaluate alternatives in terms of multiple and often conflicting objectives and identify the cluster of optimal choices. Several methodological approaches combining axiomatic design and MCDM methods have been proposed in the literature. However, it is only in recent years that this methodological combination has found fertile ground as a decision-support tool with interesting applications in the field of material selection. This paper aims to analyze the current state of the art in integrating of axiomatic design and MCDM methods considering different scenarios of the available information in material selection in the field of mechanical design.

Keywords: Material Selection · Multi-Criteria Decision Making · Axiomatic Design

1 Introduction

1.1 Materials Selection Problem

Material selection has been considered one of the critical elements of sustainable development as the process motivates the selection of materials that aid in following cleaner production, saving resources and energy, and bringing economic efficiency to any manufacturing enterprise. Materials have a key role throughout the manufacturing as well as the design process. However, selecting the best possible material alternative is a challenging task [1–3] owing to the increasing availability of a large number of materials

E. Puik et al. (Eds.): ICAD 2023, LNNS 849, pp. 135–148, 2024.
https://doi.org/10.1007/978-3-031-49920-3_9

[4, 5]. During the selection process, many attributes of the materials need to be considered, e.g. the mechanical properties, physical properties, thermal properties, magnetic properties, wear, oxidation, and corrosion behavior. Moreover, a sustainable lifestyle has become necessary due to environmental constraints [4]. Therefore, sustainability adds another criterion that should be considered when selecting a suitable material. In short, the material selection process is a multiple-criteria decision-making problem. In order to achieve the best solution, the researchers have proposed different procedural steps to arrive at the optimal decisions for material selection strategy [4, 5]. The findings suggest the following stages of a typical material selection process: 1) creating a group of alternative solutions based on the performance requirements, which constitute the selection criteria; 2) screening of the initial solution; 3) ranking and comparing the set of alternatives and 4) identifying an optimal solution. The findings of the past research have enunciated that regardless of the relation between material and process selection, the two main critical aspects for an appropriate material selection are screening and ranking [4].

1.2 Scope

The introduction of axiomatic design as a design methodology in the industrial field is based on two axioms [6]:

- The independence axiom consists of maintaining the independence of functional requirements (FRs), where FRs are defined as the minimum number of independent requirements that characterize the project objectives.
- The information axiom allows us to select the least complex design solution from a finite set of independent solutions. This second axiom states that the design with the highest probability of meeting the requirements is the best design choice, i.e., the one with the least information content.

Several methodological approaches have been proposed in the literature that combines axiomatic design and multi-criteria decision making (MCDM) methods [7–17]. However, only in recent years, this methodological combination has found fertile space as a decision-making tool with interesting applications in the field of material selection [7, 8]. The present paper aim is to analyze how these two methodologies can be integrated in this context.

The proposed methodologies use the concept of information content as a discriminator in the selection of mutually alternative solutions. Then, each approach follows an independent development, as authors often re-interpret the information axiom based on MCDM methodologies. We believe no comprehensive theoretical study still defines the conditions of applying axiomatic design as a material selection tool in mechanical design. The risk is that the full potential of axiomatic design will not be exploited or, even worse, only formally optimal solutions will be obtained. Therefore, it is necessary to present a complete and comprehensive overview of how axiomatic design can be used as a decision-making tool in material selection. This goal can be pursued by first defining scenarios for applying the method. Based on each of them, specific conditions of applicability can be detected. In this study, we have identified three basic scenarios in the highest possible generalization. The first scenario corresponds to a situation of complete information on material selection criteria, which coincide with the functional requirements of the material to be selected. The second scenario, on the other hand, is more complex and corresponds to incomplete information on selection criteria corresponding to the functional requirements of the problem. The third and last scenario analyzed is called the partial information scenario. It corresponds to the partial correspondence between selection criteria and functional requirements. Some selection criteria are nonfunctional requirements (NFRs).

2 Multi-attribute Selection: Methodological Background

MCDM methods are methodologies for selecting a solution from a set of different possible alternatives on the basis of a set of criteria, which may even be conflicting [5]. These methods can essentially be divided into two categories. Multiple objective decision making (MODM) and multiple attribute decision making (MADM). The main difference between the two approaches is that MODM methods perform comparison analysis on a very large set of solutions, potentially even in infinite numbers. In contrast, MADM methodologies aim to select the best solution from a predefined and limited number of alternatives [5]. Usually, MODM methods are based on decision variables that are continuous functions or integers, whereas in MADM methods, the decision variables are discrete values. In this paper, we refer only to MADM approaches because they are directly compatible with the axiomatic design framework. This compatibility stems from the fact that both methodological approaches perform comparative evaluations on a finite set of alternatives. However, at the same time, this finding presents us with the first significant difference between the two methodologies. While axiomatic design allows the generation of alternative solutions based on the application of the independence axiom, MADM methods do not provide any rational mechanism for pre-selecting alternatives (A_i) to be candidates for final evaluation. This pre-selection consists of formal verification of the candidate material's compliance with the properties it is to possess and the simultaneous exclusion of any incompatibilities. This reduces the number of materials to be submitted for final evaluation, facilitating the selective process. In Subsect. 1.1, we have introduced that the materials selection process consists of 4 stages. Whereas axiomatic design performs the entire four planned stages, from the identification of materials to be evaluated to the selection of the robust product, MADM methods are designed to perform only the last two stages, i.e., the comparison activities and the determination of the best solution. The last two steps are accomplished by constructing an appropriate dimension n x m matrix called the decision matrix [4] (Table 1). This matrix turns out to be characterized by four essential elements:

- n rows corresponding to the finite set of materials (A_i) subject to selection;
- m columns representing the selection criteria, which in the terminology of MADM methods are called *attributes* (C_j);
- m weighting coefficients (W_j) defining the relative importance of the selection criteria, where $\sum_{j=1}^{m} W_j = 1$;
- $n \times m$ elements a_{ij} internal to the decision matrix that constitute the evaluation attributed to alternative A_i with respect to the evaluation criterion C_j.

Table 1. Decision matrix

	C_1 (W_1)	C_2 (W_1)	-(-)	-(-)	C_m (W_m)	Score
A_1	a_{11}	a_{12}	-	-	a_{1m}	$\sum_{j=1}^{m} W_j a_{1j}$
A_2	a_{21}	a_{22}	-	-	a_{2m}	$\sum_{j=1}^{m} W_j a_{2j}$
-	-	-	-	-	-	-
-	-	-	-	-	-	-
A_n	a_{n1}	a_{n2}	-	-	a_{nm}	$\sum_{j=1}^{m} W_j a_{nj}$

The decision matrix, as formulated in Table 1, provides a deterministic solution to the material selection problem, although some evaluation criteria may conflict with each other. This solution consists of providing an ordering to the predefined set of alternatives (A_i) based on the weighting coefficients (W_j) [4, 5]. Nevertheless, to achieve this, we have to resort to a process called normalization, representing a specific material selection problem in a corresponding decision matrix. This process varies depending on the particular MADM technique being adopted. In general, it aims to make a_{ij} evaluation elements comparable and define the weighting coefficients of the selection criteria. In this regard, we must consider that the comparison criteria can be heterogeneous. They may be the physical, chemical and mechanical properties of materials, but also economic considerations, environmental sustainability assessments, or cultural and aesthetic aspects. Therefore, MADM methods make comparisons of a multidimensional nature [18], the final results of which may not coincide, as each method has its own particular specificities. In fact, each method proposes a different model for representing selection preferences.

3 Methodological Compatibilities

In this section, we analyze under what conditions axiomatic design can be a viable alternative to MADM methods, in what cases, on the contrary, the two approaches can be combined, and finally, what are the conditions of incompatibility. Before continuing the discussion, let us assume that the identification of candidate materials for final selection is always made through the formal application of axiomatic design. In this way, as anticipated in Sect. 2, a restricted set of materials is pre-selected based on a formal verification of the characteristics that the mechanical component to be designed must possess. Axiomatic design allows these characteristics to be translated into terms of neutral functional requirements, which constitute the criteria for the final selection. Based on this preliminary hypothesis, we can introduce at least three different operational scenarios.

Scenario 1. We can define a *complete information scenario* as a material selection problem for which the evaluation criteria are exclusively the functional requirements that led to the identification of a predefined set of alternatives. In addition, we know the quantitative data needed to apply the information axiom.

Scenario 2. We can define an *incomplete information scenario* as a material selection problem for which the evaluation criteria continue to be the functional requirements, but we do not have all the quantitative data needed to apply the information axiom. Some criteria may have an evaluation in qualitative terms based on subjective expert judgments.

Scenario 3. We can define a *partial information scenario* as a material selection problem for which the evaluation criteria are only partly the functional requirements, which have guided activities to identify the set of candidate materials. In this case, the final selection also takes place based on nonfunctional criteria.

3.1 Materials Selection Under Conditions of Complete Information

In a complete-information scenario, applying the information axiom allows us to identify the most suitable material for our objective. In this case, the accuracy of selection is related to the designer's ability to exhaustively represent the specifications of the material to be selected in terms of functional requirements and design constraints. Axiomatic design can directly carry out all four steps involved in the selection process. It is not necessary to implement a normalization process to obtain a decision matrix such as the one introduced in Sect. 2 [19]. Therefore, the application of the information axiom is an alternative tool to traditional MADM methods in selecting a material based on a finite set of candidates. The application of this axiom is to identify the material with the least information content [20]. By definition, the information content associated with a specific functional requirement FR_i is defined as follows:

$$I_i = log\left(\frac{1}{P_i}\right) \tag{1}$$

In this case, P_i is the probability that the material under evaluation meets the *i-th* functional requirement. To extend this concept to a complete system, we have to resort to the mathematical properties of logarithms and algebraic properties of square matrices. Preliminary application of the independence axiom allows us to submit candidate materials to a functional verification, which, in essence, establishes the existence of the requirements and the absence of incompatibilities. This preliminary verification allows us to represent the mapping between the problem's intended functional requirements (FR_i) and candidate material properties (DP_i) in terms of a design matrix (Fig. 1).

In axiomatic design, design matrices that meet the independence axiom can only be diagonal (uncoupled) or triangular (decoupled). For diagonal matrices, the total information content (I_{tot}) is equal to the sum of the information content of all functional requirements (I_i) since they are, by definition, independent [19].

$$I_{tot} = \sum_{i=1}^{n} I_i = \sum_{i=1}^{n} log\left(\frac{1}{P_i}\right) = -log\left(\prod_{i=1}^{n} P_i\right) \tag{2}$$

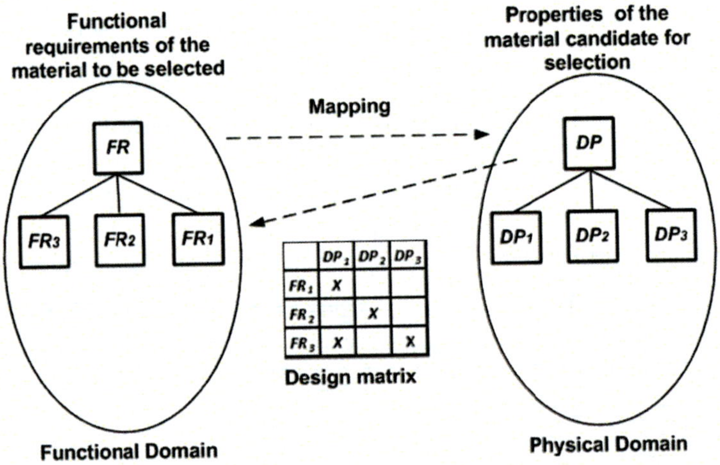

Fig. 1. Axiomatic design as a material selection tool for mechanical component design

Instead, there are functional coupling situations for decoupled representations [21]. In these cases, the FR_{i+1} requirement depends on the occurrence of the FR_i requirement [22]. This means that the probability that the FR_{i+1} requirement is satisfied by the DP_{i+1} property of a given candidate material is conditional on the probability that the previous FR_i requirement is satisfied by the DP_i property of the same material. However, it is always possible to identify a sequence of conditional probabilities that guarantee functional independence in the corresponding design matrix. Therefore, Eq. 2 is also valid in the decoupled matrix case, albeit using conditional probabilities in the functional coupling relations [19]. Under these conditions, axiomatic design allows the selection of a robust material with respect to the functional requirements that have been formalized. Unfortunately, in several cases, the application of the information axiom in its standard formulation has limitations. First, in complex selection problems, for example consisting of the presence of many functional requirements, the use of the information axiom can be complicated [21]. For this reason, there is sometimes a tendency to replace the application of the information axiom with the adoption of MADM methods. As we saw in Sect. 2, these methods are designed to allow us to provide a finite set ordering of alternative solutions.

Nevertheless, the solution identified may not be robust because the use of weighting coefficients determines functional dependence among the selection criteria [19]. Second, the available data may not be quantitative or there may be numerous nonfunctional aspects to consider. In these cases, the information axiom can no longer be applied in its standard approach. The following subsections elaborate on these situations.

3.2 Materials Selection Under Conditions of Incomplete Information

In an incomplete information scenario, some criteria admit only subjective judgments. For example, in mechanical design, sometimes we can only provide a subjective assessment based on linguistic terms (low, medium, high) to assess the corrosion level of a material. In these cases, incomplete information depends on the vagueness of attributable judgments [4, 5]. Therefore, many authors have proposed the use of fuzzy theory so that the information axiom can be applied to numerical data [7, 12].

Fuzzy Approach. Fuzzy AD (FAD) methodology is based on conventional axiomatic design. However, crisp ranges are replaced by fuzzy numbers representing linguistic terms (Fig. 2). In Fig. 2, triangular fuzzy numbers (TFNs) are shown. The intersection of TFNs representing design and system ranges presents the common area [9, 23–26]. Firstly, the information content is calculated in a non-fuzzy environment. Then information content in a fuzzy environment is calculated as follows:

- $I_i = \infty$ if the intersection between two adjacent triangles is an empty set;
- $I_i = \log\left(\frac{\textit{Area of system range}}{\textit{common area}}\right)$ if, on the other hand, the common area is not an empty set.

Even in this case, the best solution is the one with the least information content.

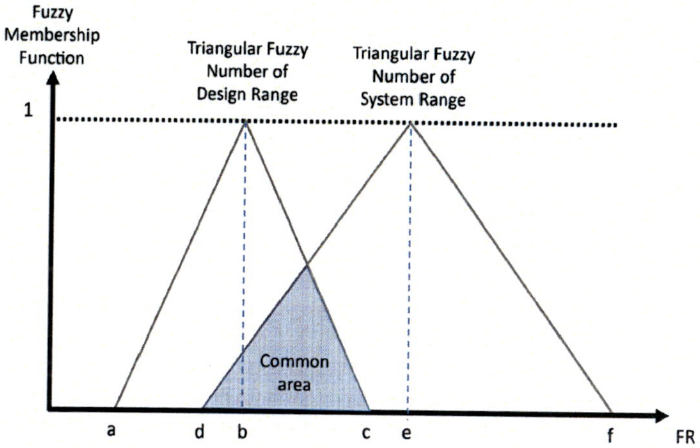

Fig. 2. System-design ranges and common area in fuzzy environment.

Fuzzy axiomatic design presents two fundamental limitations. First, this methodology relies on expert judgment to determine the degree of fuzziness in the design parameters. This can introduce subjectivity into the design process, which may lead to inconsistencies or biases in the design. Recently, advanced approaches derived from fuzzy theory have been proposed, such as the intuitionistic fuzzy set (IFS) and the neutrosophic (NS) method. In particular, the latter approach seeks to overcome this limitation of fuzzy theory by introducing evaluations of "truth", "indeterminacy" and "falsity" into the model. Abdel-Basset et al. [13] used this approach in conjunction with axiomatic design in the selection of medical instrumentation. Another approach that has been proposed in recent years is based on Z-numbers [14, 15]. In all these cases, adopting these methodologies aims to reduce uncertainty in applying axiomatic design in incomplete information application scenarios. Uncertainty in a selection problem can also arise from the risk of adopting a particular material. For example, overheating may result in undesirable effects on some mechanical components. Hafezalkotob et al. report in [27] a real-life material selection case for the construction of gas turbine blades. Temperature variation was considered as a risk factor. In this case, the FAD method was modified to include the risk variable associated with blade overheating. Thus, combining elements of risk and the FAD approach, the Risk Fuzzy Axiomatic Design (RFAD) method is obtained. The information content for the RFAD technique is calculated as follows [28, 29]:

$$I_{ij}^{r} = log\left(\frac{1}{P_{ij}(1 - r_{ij})}\right) \tag{3}$$

In this case, r_{ij} is a risk factor with a value in the range of zero and one. Comparing the information contents of FAD and RFAD approaches, it is explicit that each I_{ij}^{r} is greater than its corresponding I_{ij}. Greater risk factor r_{ij} leads to a higher value of information content I_{ij}^{r}. In addition, FAD can be a complex and time-consuming process, especially when dealing with systems with many functional requirements or design parameters. In these cases, the information axiom can be reformulated in terms of MADM methods. Generally, this methodological re-interpretation consists of introducing weighting coefficients to be assigned to the selection criteria.

Selection Based on Weighted Attributes. The importance of criteria in decision-making problems is often not similar. Consequently, the relative importance of criteria should be considered to achieve a realistic solution. In general, the significance coefficients can be computed using objective, subjective, or integrated techniques [27]. Subjective significance coefficients are achieved from experts' opinions while objective significance coefficients are obtained using the decision matrix's values without utilizing experts' judgments. The two types of significance coefficients may be combined. Different techniques are borrowed from MADM methods for calculating the significance coefficients of criteria [29–32]. From a methodological point of view, the use of these weighting techniques consists of combining FAD and RFAD approaches with MADM methods. In this paper, we briefly introduce the three main MADM approaches used to determining weighting coefficients in situations where information is incomplete: the information entropy method, the analytic hierarchy process and the best-worst method.

Information Entropy Method. Entropy is based on the classical measures of Boltzmann and the second law of thermodynamics [27, 34]. The idea of entropy in information science, initially suggested by Shannon [34], is a tool for specifying the uncertainty of a variable. The general concept of Shannon's entropy is to evaluate the significance coefficient of each criterion from the distribution of data over variables. The Shannon entropy has been utilized with combinations of many MADM techniques for various applications in material selection problems [4, 5, 27, 31, 33]. Hafezalkotob et al. [31] developed the RFAD method with the integrated Shannon entropy significance coefficients to generate an entropy-weighted risk-based fuzzy axiomatic design (WRFAD) approach. The information content of the WRFAD technique is based on the integrated Shannon significance coefficients. However, this approach has a fundamental disadvantage. It requires a significant amount of input data, which can sometimes be difficult to obtain. The accuracy of the results depends on the quality of the data, and inaccurate or incomplete data can lead to erroneous decisions.

Analytic Hierarchy Process. The analytic hierarchy process (AHP) is a decision support method developed to complete problems by breaking the solution problems, grouping them, and arranging them into a hierarchical structure [4, 5]. This method uses a comparison of criteria paired with a measurement scale that has been determined to obtain priority criteria. The main input of the AHP method is experts' perception, so there is a factor of subjectivity in retrieval decisions [35–37]. This aspect is both a strength and a weakness of this method. It is a strength of the method because it is a powerful tool for modeling complex decision-making situations. However, considerable uncertainty and doubts in the evaluation affect the accuracy of the data and results obtained. Based on this consideration, another theory was developed, namely Fuzzy 40.

Analytic Hierarchy Process. The fuzzy AHP is a method of AHP developed with fuzzy logic theory [11, 38]. The fuzzy AHP method is used similarly to the method of AHP. It is just that the fuzzy AHP method sets the AHP scale into the fuzzy triangle scale to be accessed priority.

Best-Worst Method. Rezaei [32] developed the best-worst method (BWM) based on a consistency comparison system. The method has been further extended by Guo and Zhao [39] by integrating the fuzzy set data into the approach (called fuzzy BWM or FBWM). In BWM, the pairwise comparison between the best and worst criteria is defined as a reference comparison in which the best and worst criteria are computed. Additionally, the secondary comparison occurs when neither of the selected criteria is defined as the best or the worst element. In real-world problems where uncertainty and ambiguity of decision-maker exist, it is tough to evaluate the accurate weights of the criteria. Based on the BWM approach, the hybrid hierarchical best-worst fuzzy axiomatic design (HB-WFAD) selection method, which combines axiomatic design and FBWM, has been proposed [38]. In this case, criteria weights are calculated by exploiting the BWM technique, which has the advantage of requiring a limited amount of information. However, subjective value judgments always remain a critical issue in the selection process.

3.3 Materials Selection Under Conditions of Partial Information

In some situations, selection procedures may involve elements that difficultly can be formalized in terms of functional requirements. Design constraints most often belong to this category. Design constraints are limitations to design, which may depend on the material's physical, chemical, or mechanical properties to be selected or on economic reasons, product availability, environmental and social sustainability of the production process, or even aesthetic and cultural motivations [4, 40–44]. In terms of axiomatic design, these constraints can be classified into two categories [45]: input constraints if they relate to the fulfillment of a specific functional requirement that a mechanical component must possess, for example, a certain threshold of maximum allowable heat transmittance. In contrast, system constraints do not relate to a specific material property. For example, materials from different countries with the same properties may have been produced through processes with different environmental and social impacts. Many companies require their suppliers to ensure high levels of environmental and social sustainability [28]. Axiomatic design does not provide formal rules for treating these elements, as is the case with functional requirements. To overcome this limitation, Mabrok et al. [36] proposed to consider these elements as nonfunctional requirements (NFRs) and to replace the functional domain in Fig. 2 with a new domain, called the requirement domain, which includes both types of requirements (Fig. 3).

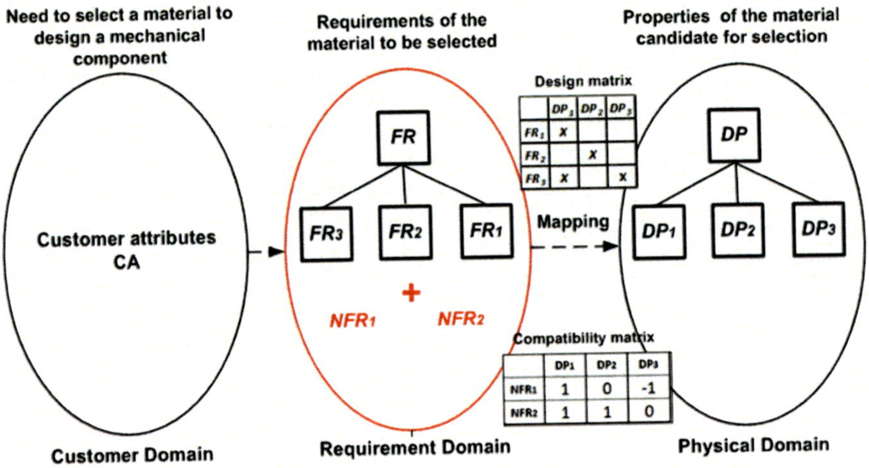

Fig. 3. Selecting materials for mechanical component design by including nonfunctional requirements in axiomatic design.

This intuition allows the formal inclusion of nonfunctional requirements in axiomatic design. At the operational level, this idea can be accomplished through the definition of a new matrix called the extended design matrix. This new matrix includes two blocks, the design matrix (*n* x *n*) related to the *n* functional requirements of the selection problem and the compatibility matrix of size *k* x n for the *k* associated nonfunctional requirements. The latter matrix relates the nonfunctional requirements to the properties of the material submitted for verification. In the example shown in Fig. 3, the compatibility matrix can be constructed based on three values:

- $a_{ij} = 1$ if design parameter *j-th* satisfies nonfunctional requirement *i-th*;
- $a_{ij} = 0$ if design parameter *j-th* is indifferent to nonfunctional requirement *i-th*;
- $a_{ij} = -1$ indicates, on the other hand, that design parameter *j-th* violates the nonfunctional requirement *i-th*.

If we refer to the example in Fig. 3, the relationship between the requirement and physical domains can be represented by Eq. 4.

$$\begin{bmatrix} \begin{pmatrix} FR_1 \\ FR_2 \\ FR_3 \end{pmatrix} \\ \begin{pmatrix} NFR_1 \\ NFR_2 \end{pmatrix} \end{bmatrix} = \begin{bmatrix} \begin{pmatrix} X & 0 & 0 \\ 0 & X & 0 \\ X & 0 & X \end{pmatrix} \\ \begin{pmatrix} 1 & 0 & -1 \\ 1 & 1 & 0 \end{pmatrix} \end{bmatrix} \begin{bmatrix} DP_1 \\ DP_2 \\ DP_3 \end{bmatrix} \tag{4}$$

In practical terms, the design matrix represents the traditional mapping between functional requirements and selected material characteristics (design parameters). In contrast, the compatibility matrix measures the level of compliance with respect to the nonfunctional requirements provided by the selection. This way, Eq. 4 introduces two pre-selection operations on candidate materials for final selection. An initial pre-selection is made to identify candidates that meet the functional requirements as in the traditional axiomatic design approach. This pre-selection allows a finite set of candidate materials to be identified. Then, the compatibility matrix allows us to evaluate with respect to this first set of materials, which ones meet the nonfunctional requirements posed by our problem. This second intervention allows us to restrict the set of candidate materials even further for final selection. Finally, we must proceed to the final selection.

On the other hand, as far as the final selection is concerned, we can no longer resort to the information axiom, or at least as defined in Sect. 2, since we also must consider evaluation criteria that derive from nonfunctional requirements. In this case, the various MADM methods provide a powerful tool for making the final selection [36]. Recently, some studies have proposed the AHP methodology, which has the advantage of grouping the selection criteria hierarchically-mindedly into groups and subgroups [35, 36]. In this sense, the selected material constitutes the best solution with respect to modeling the operational context. However, it may not coincide with the robust solution. Therefore, we may have found a suboptimal solution with respect to the functional requirements due to the simplifications introduced by assuming the nonfunctional requirements and then applying the MADM methods.

4 Conclusions

In mechanical design, the selection of materials to be used is becoming an increasingly complex problem because of the wide availability of alternative materials and the progressive emergence of new constraints. Currently, it is common practise to consider elements and performance related to environmental and social sustainability in the design process, in addition to the usual selection criteria such as the physical, chemical and mechanical properties and the cost of the component. The component must be manufactured to meet specific technical and economic requirements and minimize the social and environmental impacts throughout its entire life cycle, from raw material extraction to end-user use and dismantling. In this context, axiomatic design allows these even mutually conflicting requirements to be articulated in formal terms of functional requirements and design constraints. This specificity of axiomatic design allows these elements to be included as selection criteria in a unified framework. However, the increasing complexity of problems in scenarios with incomplete or partial information makes the final choice very difficult. Therefore, scholars resort to the extension of the information axiom to simplify its applicability. In chronological order, the use of fuzzy theory was the first step in this process of adapting axiomatic design in complex application contexts. Then, other methodologies were presented that further simplify the information axiom by applying MADM methods (AHP, information entropy, BWM). In this case, a robust solution is given up for suboptimal solutions. The latter has the advantage of providing a solution even to very complex selection problems, unsolvable with the traditional approach. The partial-information scenario is emblematic of this trade-off situation. The presence of several system constraints can make it complicated to formulate the selection problem solely in terms of functional requirements. Instead, the interpretation of system constraints as nonfunctional requirements simplifies the determination of a solution. However, the final choice may not coincide with the robust solution with respect to axiomatic design. It is certainly a better solution than the simplified model based on nonfunctional requirements. In this sense, research is increasingly focused on studying the compatibility between axiomatic design and MADM methods in order to reduce the gap between robust solutions and suboptimal choices.

References

1. Giorgetti, A., Cavallini, C., Arcidiacono, G., Citti, P.: A mixed C-Vikor fuzzy approach for material selection during design phase: a case study in valve seats for high performance engine. Int. J. Appl. Eng. Res. **12**(12), 3117–3129 (2017)
2. Arcidiacono, G.: Development of a FTA versus parts count method model: comparative FTA. Q. Reliab. Eng. Int. J. **19**, 411–424 (2003)
3. Ceccanti, F., Giorgetti, A., Cavallini, C., Arcidiacono, G., Citti, P.: Comparative evaluation of fuzzy axiomatic design and IAMS comprehensive VIKOR approaches for material selection in mechanical design. J. Eng. Des. Technol. **13**(1), 80–87 (2020)
4. Jahan, A., Edwards, K., Bahraminasab, M.: Multi-criteria Decision Analysis for Supporting the Selection of Engineering Materials in Product Design, 2nd edn. Butterworth-Heinemann, Oxford (UK) and Cambridge (USA) (2016)
5. Rao, R.V.: Decision Making in the Manufacturing Environment. Using Graph Theory and Fuzzy Multiple Attribute Decision Making Methods. Series in Advanced Manufacturing. 1st edn. Springer, London (2007). https://doi.org/10.1007/978-1-84628-819-7
6. Suh, N.P.: Axiomatic Design - Advances and Applications, 1st edn. Oxford University Press, New York (2001)
7. Dhivya, J., Maheswari, K., Saroja, M.N.: A fuzzy logic approach for material selection problem. In: International Conference on Materials Research in Science and Engineering (KMRSE 2021). AIP Conference Proceedings 2446, Coimbatore, India (2022)
8. Gul, M., Celik, E., Gumus, T.A., Guneri, A.F.: A fuzzy logic based PROMETHEE method for material selection problems. Beni-Suef Univ. J. Basic Appl. Sci. **7**, 68–79 (2018)
9. Cicek, K., Celik, M.: Multiple attribute decision-making solution to material selection problem based on modified fuzzy axiomatic design-model selection interface algorithm. Mater. Des. **31**(4), 2129–2133 (2010)
10. Candan, G., Kir, S., Yazgan, H.R.: Solution of material selection problem using fuzzy axiomatic design and DEMATEL methods. Acta Physica Polonica Series **131**(1), 24–27 (2017)
11. Kahraman, C., Cebi, S.: A new multi-attribute decision making method: hierarchical fuzzy axiomatic design. Expert Syst. Appl. **36**(3), 4848–4861 (2009)
12. Maghsoodi, A.I., Hafezalkotob, A., Azizi, I., Maghsoodi, S.I., Hafezalkotob, A.: Selection of waste lubricant oil regenerative technology using entropy-weighted risk-based fuzzy axiomatic design approach. Informatica **29**, 41–74 (2018)
13. Abdel-Basset, M., Mohamed, M., Mostafa, N.N., El-Henawy I.M., Abouhawwash, M.: New multi-criteria decision-making technique based on neutrosophic axiomatic design. Sci. Rep. **12**(1) (2022)
14. Aydoğan, S., Günay, E.E., Akay, D., Kremer, G.E.O.: Concept design evaluation by using Z-axiomatic design. Comput. Ind. **122** (2020)

15. Liu, Q., Chen, J., Wu, Y., Yang, K.: Linguistic Z-numbers and cloud model weighted ranking technology and its application in concept evaluation of information axiom. J. Supercomput. **78**, 6061–6089 (2022)
16. Rolli, F., Parretti, C., Giorgetti, A., Arcidiacono, G., Citti, P.: A mixed axiomatic design/MADM approach for the sustainability representation of an offshore hydrocarbon extraction facility. Sustainability (forthcoming)
17. Arcidiacono, G., Berni, R., Cantone, L., Nikiforova, N.D., Placidoli, P.: Fast method to evaluate payload effect on in-train forces of freight trains. Open Transp. J. **12**, 77–87 (2018)
18. Zheng, P., Wang, Y., Xu, X., Xie, S.Q.: A weighted rough set based fuzzy axiomatic design approach for the selection of AM processes. Int. J. Adv. Manuf. Technol. **91**(5–8), 1977–1990 (2017)
19. Gonçalves-Coelho, A., Fradinho, J.M.V., Gabriel-Santos, A., Cavique, M., Mourão, A.J.F.: How to handle the design preferences with axiomatic design. IOP Conf. Ser. Mater. Sci. Eng. **1235** (2022)
20. Fradinho, J., Gonçalves-Coelho, A.: The information axiom and robust design. In: Suh, N.P., Cavique, M., Foley, J.T. (eds.) Design Engineering and Science, pp. 307–325. Springer, Cham (2021). https://doi.org/10.1007/978-3-030-49232-8_11
21. Slătineanu, L., et al.: Selection of a solution when using axiomatic design. In: International Conference on Axiomatic Design ICAD 2017. MATEC Web of Conferences, vol. 127, p. 01019 (2017)
22. Brown, C.A.: Kinds of coupling and approaches to deal with them. In: Proceedings of 4th ICAD2006, The Fourth 612 International Conference on Axiomatic Design, Firenze (2006)
23. Vinodh, S., Kamala, V., Jayakrishna, K.: Application of fuzzy axiomatic design methodology for selection of design alternatives. J. Eng. Des. Technol. **13**(1), 2–22 (2015)
24. Karatas, M.: Multiattribute decision making using multiperiod probabilistic weighted fuzzy axiomatic design. Syst. Eng. **20**(4), 318–334 (2017)
25. Kannan, D., Govindan, K., Rajendran, S.: Fuzzy axiomatic design approach based green supplier selection: a case study from Singapore. J. Clean. Prod. **96**, 194–208 (2015)
26. Celik, M., Cebi, S., Kahraman, C., Er, I.D.: Application of axiomatic design and TOPSIS methodologies under fuzzy environment for proposing competitive strategies on Turkish container ports in maritime transportation network. Expert Syst. Appl. **36**(3), 4541–4557 (2009)
27. Hafezalkotob, A., Hafezalkotob, A.: Risk-based material selection process supported on information theory: a case study on industrial gas turbine. Appl. Soft Comput. **52**(1), 1116–1129 (2017)
28. Gormus, E., Tasci, D.Z.: A novel approach for green supplier selection problem: fuzzy axiomatic design with risk factors. J. Manag. Econ. Res. **19**(2), 1–16 (2021)
29. Kulak, O., Goren, H.G., Supciller, A.A.: A new multi criteria decision making approach for medical imaging systems considering risk factors. Appl. Soft Comput. **35**, 931–941 (2015)
30. Kulak, O., Kahraman, C.: Fuzzy multi-attribute selection among transportation companies using axiomatic design and analytic hierarchy process. Inf. Sci. **170**(2–4), 191–210 (2005)
31. Hafezalkotob, A., Hafezalkotob, A.: Fuzzy entropy-weighted MULTIMOORA method for materials selection. J. Intell. Fuzzy Syst. **31**(3), 1211–1226 (2016)
32. Rezaei, J.: Best-worst multi-criteria decision-making method. Omega **53**, 49–57 (2015)
33. Zhang, G., Marvel, S., Truong, L., Tanguay, R.L., Reif, D.M.: Aggregate entropy scoring for quantifying activity across endpoints with irregular correlation structure. Reprod. Toxicol. **62**, 92–99 (2016)
34. Shannon, C.E.: A mathematical theory of communication. Bell Syst. Tech. J. **27**, 379–423 (1948)
35. Pourabbas, E., Parretti, C., Rolli, F., Pecoraro, F.: Entropy-based assessment of nonfunctional requirements in axiomatic design. IEEE Access **9**, 157831–156845 (2021)

36. Mabrok, M., Ryan, M., Efatmaneshnik, M.: Integrating nonfunctional requirements into axiomatic design methodology. IEEE Syst. J. **11**(4), 2204–2214 (2015)
37. Saaty, T.L.: How to make a decision: the analytic hierarchy process. Eur. J. Oper. Res. **48**(1), 9–26 (1990)
38. Maghsoodi, A.I., Mosavat, M., Hafezalkotob, A., Hafezalkotob, A.: Hybrid hierarchical fuzzy group decision-making based on information axioms and BWM: prototype design selection. Comput. Ind. Eng. **127**, 788–804 (2019)
39. Guo, S., Zhao, H.: Fuzzy best-worst multi-criteria decision-making method and its applications. Knowl.-Based Syst. **121**, 1–9 (2017)
40. Citti, P., Giorgetti, A., Millefanti, U.: Current challenges in material choice for high-performance engine crankshaft. Procedia Struct. Integrity **8**, 486–500 (2018)
41. Giorgetti, A., Monti, C., Tognarelli, L., Mastromatteo, F.: Microstructural evolution of René N4 during high temperature creep and aging. Results Phys. **7**, 1608–1615 (2017)
42. Baldi, N., Giorgetti, A., Palladino, M., Giovannetti, I., Arcidiacono, G., Citti, P.: Study on the effect of inter-layer cooling time on porosity and melt pool in Inconel 718 components processed by laser powder bed fusion. Materials **16**, 3920 (2023)
43. Vezzù, S., Cavallini, C., Rech, S., Vedelago, E., Giorgetti, A.: Development of high strength, high thermal conductivity cold sprayed coatings to improve thermal management in hybrid motorcycles. SAE Int. J. Mater. Manf. **8**(1) (2015)
44. Giorgetti, A., Millefanti, U., La Battaglia, V., Citti, P.: Investigations of fatigue damage in a nitriding low-carbon bainitic steel for high-performance crankshaft. Metals **12**(12), 2052 (2022)
45. Liu, A.: Problem definition. In: Suh, N.P., Cavique, M., Foley, J.T. (eds.) Design Engineering and Science, pp. 167–189. Springer, Cham (2021). https://doi.org/10.1007/978-3-030-49232-8_5

Design of Bio-inspired Gripper Arm for Mars Sample Retrieval

Aron I. Sverrisson[1], Bjarki Laxdal[1], Ignas Urbonas[1], and Joseph Timothy Foley[2](✉)

[1] Department of Engineering, Reykjavik University, Menntavegur 1, Reykjavík 102, Iceland
[2] Robotics and IOT (RIOT) Lab, Department of Engineering, Reykjavik University, Menntavegur 1, Reykjavík 102, Iceland
foley@ru.is

Abstract. In the last decades, the idea of traveling to Mars has become increasingly popular as technology progresses and the journey is becoming more than just fiction. As a result, existing rovers on Mars are preparing drilling samples of Martian soil for eventual pickup. These soil samples can be located on rough terrain in all orientations and angles and need to be retrieved. The geometry and mass of these samples are predefined, it is cylindrical with a length of 15.2 cm, a width of 2.3 cm, and a mass of 57 g. Traditional gripping approaches focus on opposed fingers, industrial-style scoops, or vacuum adhesion which are very orientation specific and not secure enough for a drone. Our MARS-DOG design acquires sample tubes securely without precision positioning or orientation in a sandy environment. The biomechanical-inspired design is a spring-loaded claw inside a net that closes the gaps between each claw. The net's aperture is closed with a string on the claw tips. A 3D-printed prototype is capable of picking up, carrying, and dropping a sample tube analog that is at least 57 g, regardless of the object's pitch and yaw in a dusty environment. Stress-testing the unit revealed a lifting capability of 6 kg which is sufficient for additional sample collection tasks in addition to the original Mars sample tube goals.

Keywords: Claw · Axiomatic Design · gripper Arm · Design

1 Introduction

The problem we are trying to solve is the retrieval of rock and dust samples on Mars. Currently, a rover or drone digs up samples, puts them in tubes, drops them on the ground, and moves on. The proposed way to retrieve these samples is to have a drone fly to them, pick them up, and transport them back to the base. This poses problems, since the samples can be oriented in any way, i.e. roll and yaw, and could be sitting at an angle, i.e. pitch. The idea and goal for the project are to create a gripping mechanism that can easily grab the sample,

E. Puik et al. (Eds.): ICAD 2023, LNNS 849, pp. 149–163, 2024.
https://doi.org/10.1007/978-3-031-49920-3_10

which has a predefined geometry and mass. The sample is cylindrical with a length of 15.2 cm, a width of 2.3 cm, and a mass of 57 g.

The gripping mechanism needs to be able to pick up the samples regardless of the samples' orientation and angle. The name of the gripping mechanism is MARS-DOG, Mars Drone Omniorientational Gripper because it's a gripping mechanism that goes on a drone that will be on Mars and will be omniorientational, i.e. it can pick up items, regardless of their orientation as long as they fit in the net. The gripping mechanism is going to be created for the RIOT lab at Reykjavik University, the customer and stakeholder of this project. The product can also be used by other potential customers, for example, other space agencies, such as Space X or NASA, that would have an interest in picking up samples on planets, and even be applied to other fields such as garbage companies, cities, municipalities, and anyone that needs to pick something up easily and reliably.

1.1 The Sample

The sample tube upon which all FRs and CNs were based and tested was made using a 3D printed replica of the Perseverance Sample Tube 266 from NASA [1]. This sample was chosen as it was the newest sample tube that could be found and is used by the Perseverance rover on Mars. The length of this sample is 15.2 cm, the width is 2.3 cm and the mass is 57 g. An STL drawing of the sample tube, created by the Jet Propulsion Laboratory, California Institute of Technology [2] was used to create the 3D printed sample. The 3D-printed sample tube can be seen in Fig. 1.

1.2 Customer Needs

A customer need is a way of noting down what the customer needs the product to be able to do. Our customer needs were acquired by brainstorming what the customer might need from this project, and from what information Joseph T. Foley, a professor and a member of the RIOT lab at Reykjavik University, told the team about it. Using that information, we developed a top-level need (CN_0), which was then decomposed into smaller, more manageable pieces (CN_1-CN_4).

The team's main priority is stated in CN_0 from our Axiomatic Design [3].

CN_0. The gripping mechanism needs to be able to grab, carry and drop the sample no matter its pitch and yaw in a dusty environment.

After the top-level need is identified, it is broken down into the following smaller customer needs CN_1–CN_4 in our Axiomatic Design.

CN_1. The gripping mechanism needs to be able to pick up the sample regardless of the sample's pitch.

CN_2. The gripping mechanism needs to be able to pick up the sample regardless of the sample's yaw.

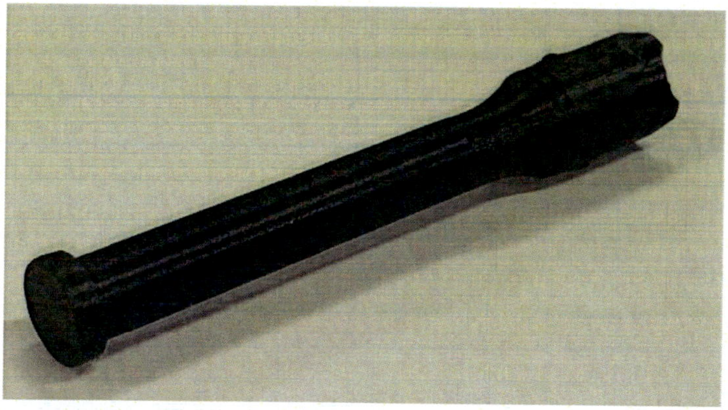

Fig. 1. FDM printed model of NASA's sample tube to test gripper compatibility

CN_3. The gripping mechanism needs to be able to pick up, carry and drop a sample that is 57 g [4].

CN_4. The gripping mechanism needs to be able to pick up, carry and drop a sample that is covered in dust.

2 Background and Prior Art

There are many applications for which a drone or a robot can be used to grab stuff. Cities and municipalities could use a drone with a gripper to pick up trash around the city or to transport small objects between offices quickly. Production lines could transport material between different warehouses, conveyor belts, or production lines.

Picking up objects has been a goal for many companies for various reasons. As the usage of drones for commercial and personal use has increased in the last decade a link between them has emerged. Grabbing objects with little complexity in any orientation of the object has been the goal of many companies and these companies have achieved this goal in various ways, such as using a four-armed gripped like the Small Hammer Robot Arm Gripper by AlexNLD [5], a drone with 5-axis robotic arms by ProDrone [6] and a vacuum gripper by Robotiq [7]. There exist many other ways to pick up objects that don't use claws or a vacuum to do so, such as a purse seines net, which is often used by fishing boats to catch fish [7], or the Ogre-Eyed Spider, which spins web between its legs to securely hold on to its prey [8]. These methods will be shown and talked about in this Sect. 2.

2.1 Small Hammer SNM2500 Robot Arm Gripper

The Small Hammer Robot Arm Gripper is your typical four-armed gripped and can be seen in Fig. 2a. The problem with having only the claws is that it will

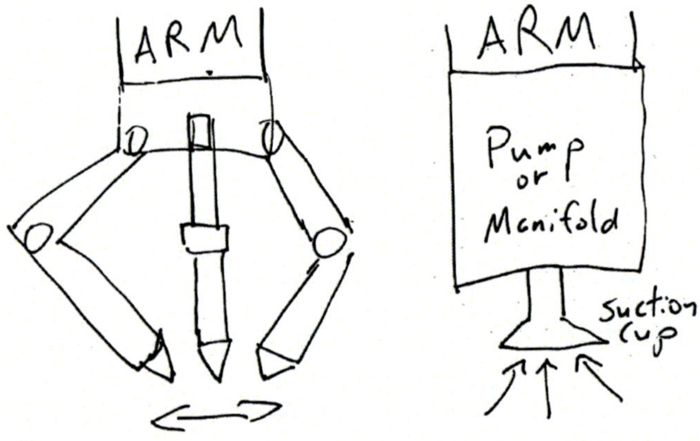

(a) A four arm gripper closes 4 "claws" to scoop up items

(b) A vacuum gripper creates a pressure differential on a compliant suction cup.

Fig. 2. A four arm gripper [5]. Robotiq's vacuum gripper [9].

have problems holding onto the sample during flight, and the sample will fall out of the claws, this design, therefore, does not fulfill our requirements.

2.2 Robotiq's Vacuum Gripper

The Robotiq Vacuum Gripper picks up objects by creating a vacuum between the vacuum cup and the object [9]. This works when the difference between atmospheric pressure and the negative pressure, is enough to provide the ability to lift, hold and move the object. This would not work on Mars, since the atmospheric pressure there is extremely low and therefore the suction needed to pick up an object would be gigantic in comparison to the atmospheric pressure. Another problem with vacuum grippers is that they are extremely sensitive to dust and require a lot of electricity to power a vacuum or air pump, as well as having the requirement that the object it is grabbing has to be compatible with a vacuum gripper, which is not suitable for this project.

2.3 ProDrone PD6B-AW-ARM Commercial-Use Drone

The ProDrone is a large-format drone equipped with two internally-developed high-performance, completely original 5-axis robotic arms [6] used to pick up objects, as can be seen in Fig. 3a. This on the other hand has the problem that it needs both grippers to secure the sample and each arm has to be rotated based on the rotation of the sample on the ground. It also runs the risk of the sample falling out of the grippers, if it is not held on precisely by the ends of the claws, and since the sample is cylindrical it is likely to slip and result in the gripper losing grip and dropping it during flight.

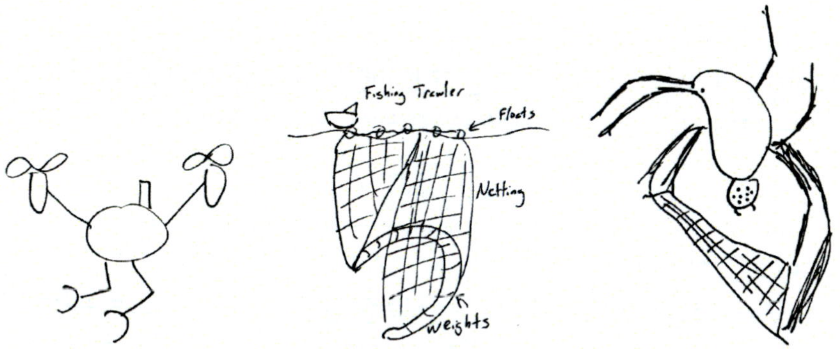

(a) VTOL drones such as the ProDrone often use simple closing claws

(b) A Purse Seines net defines a vertical boundary in the water

(c) The Ogre Eyed Spider's limbs guide its net.

Fig. 3. ProDrone PD6B-AW-ARM drone [6]. Purse Seines net caching fish [7]. Ogre Eyed Spider using net to catch prey [8].

2.4 Purse Seines

A purse seine is a net that is deployed like a wall encircling an area and any fish inside it. The seine floats along the top line with a lead line threaded through rings along the bottom. Once a group of fish is located, a vessel encircles the group with the net. The lead line is then pulled in, "pursing" the net closed on the bottom, preventing fish from escaping by swimming downward. The net is then either hauled aboard or brought alongside the vessel. This method can be seen in Fig. 3b. The problem with using a purse seines net is that since one of the sides is open until the line is pulled, the sample may slide out of that hole if it is oriented in that particular way, resulting in the gripping mechanism not being able to pick up the sample.

2.5 The Ogre-Eyed Spider

The team decided to look outside of standard technology and viewed how nature solves the problem of picking objects up. One of the best animals using "claw"-like arms to grab things were spiders. Thus the ideal spider was found, The Ogre-eyed Spider. The spider is faced with the same problems that we are, it has to somehow grab, secure, and bring its prey back to itself, despite the prey having different orientations, sizes, and terrain. How the Ogre-eyed spider solves these problems is by not simply relying on its four legs to grab its prey, but also spinning a net between its legs to secure the prey [8]. As seen in Fig. 3c.

Our idea is bio-mechanically inspired, to create a gripper with claws similar to the design in Fig. 2a. We want to add a net that connects to each claw, inspired by Fig. 3c and 3b. By adding a net to the claws we think the sample will be more secure, i.e. the net will help the claws to hold onto the sample no matter the orientation of the sample, and prevent it from falling out of the gripper.

2.6 Functional Requirements

Using Axiomatic design the project was broken down into the following Functional Requirements(FR) and Constraints(CON) [3].

The team's main requirement is stated in FR_0 from our Axiomatic Design:

FR_0. Pick up, carry, and drop the sample tube regardless of its yaw and pitch.

From the Customer Needs, we built a list of Functional Requirements and Constraints.

FR_1. Grab the sample geometry regardless of its yaw.

FR_2. Grab the sample geometry regardless of its pitch.

FR_3. Release the tube when commanded.

CON_1. Grab the sample in the presence of granular contamination (sand).

CON_2. Hold 57g while the structure is moving $22\,\mathrm{km\,h^{-1}}$ (Approximate speed of existing Mars drones.)

3 Design

The only assumption the team has in this project is that the maximum angle the sample will be laying at is 15° since another drone has to land and mine the sample so we can retrieve it, and we assume that the drone can not land and mine it out at angles exceeding 15°.

3.1 Design Methodology

This project is developed and implemented with the methods and guidelines found in the books *Design Engineering and Science* [10], *FUNdaMENTALS of Design* [11] and Axiomatic Design [3].

Axiomatic design is a systems design methodology developed by Nam P. Suh. The method gets its name from its use of Axioms, which control the analysis and decision-making process in product design. The two axioms are The Independence Axiom (i.e. maintain the independence of the Functional Requirements) and the Information Axiom (i.e. minimize the information content). The Independence Axiom provides adjustability and control while avoiding unintended consequences. It focuses on modularity, which in turn ensures that teams can progress on any part without delays. The Information Axiom focuses on robust designs, compensating for errors, and minimizing the effect of wear and tear. Axiomatic Design analyses the process of transforming customer needs into functional requirements, design parameters, and process variables. The coupling between FRs and DPs is extremely important and can be mathematically demonstrated using the design matrix, which can be seen in Eq. 1. In the design matrix, a non-zero entity represents a connection between the DP and FR. When it has a diagonal line of non-zero values, with all other values as zero, the design is said to be uncoupled. This is said to be ideal, as changes in a DP only alter its respective FR [3].

Table 1. First level FR-DP mapping.

ID	Functional Requirement	Design Parameter
1	Grab the sample regardless of its yaw	Net on the end of grippers
2	Grab the sample regardless of its pitch	Suspended by wire
3	Release the sample when commanded	Retractable gripper claws

3.2 FR-DP Mapping

After building a list of the Functional Requirements, the next step is to develop lists of Design Parameters, keeping in mind the Independence Axiom (i.e. maintain the independence of the Functional Requirements) and the Information Axiom (i.e. minimize the information content of the design) [3].

The team's main Design Parameter is as stated in DP0 from our Axiomatic Design.

DP_0. Retractable four-fanged gripper claw with a net.

From the Functional Requirements, the following Design Parameters were built.

DP_1. Net on the end of grippers
The net on the gripper will encompass the item being picked up, wrapping around it, ensuring that it can be picked up regardless of orientation.

DP_2. Suspended by wire
Using a suspension fixture will allow the claw to angle and align itself and the ends of the claws to the surface of the ground as the drone lands, enabling it to grab the sample regardless of its angle.

DP_3. Retractable gripper claws
Using retractable gripper claws allows for closing and opening the claw, allowing the claw to hold on and release grabbed objects.

We continue a "zig-zag" procedure to decompose and map the Functional Requirements to the Design Parameters as shown in Table 1.

From this mapping, we develop an uncoupled design matrix as shown in Eq. 1.

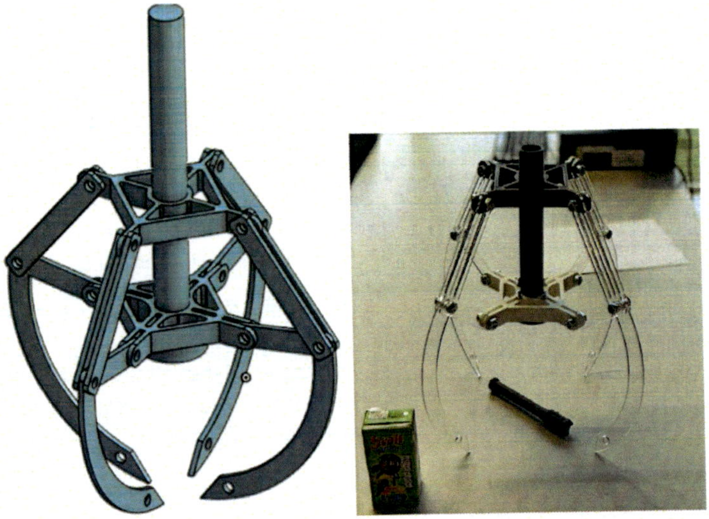

(a) CAD model of early design driven along central shaft.

(b) Early design prototype with sample tube for size comparison.

Fig. 4. MARS-DOG First Iteration

$$\begin{Bmatrix} \mathrm{FR}_1 \\ \mathrm{FR}_2 \\ \mathrm{FR}_3 \end{Bmatrix} = \begin{bmatrix} X & 0 & 0 \\ 0 & X & 0 \\ 0 & 0 & X \end{bmatrix} \begin{Bmatrix} \mathrm{DP}_1 \\ \mathrm{DP}_2 \\ \mathrm{DP}_3 \end{Bmatrix} \tag{1}$$

An uncoupled design is an ideal design, as changes in one DP do not alter more than the only designated Functional Requirement. This is ideal as it satisfies the independence axiom: "to maintain the independence of the Functional Requirements (FRs)" [3].

3.3 Design Process

The first design had no passive closing components in mind. All movement of the claws was produced by the movement of the central driving shaft (Lead screw) that would drive the bottom plate which was attached to the claws. The claw was then attached to the upper plate with a linkage. This design can be seen in Figs. 4a and 4b. This design did not include a net but had holes for a net attachment.

3.4 Final Design

The final design consists of a single centerpiece on which 6 gripper claws are attached and passively held open with the use of springs. The final design also incorporates a net that is used to help secure the desired object once grabbed. The gripper claw is closed by pulling a string through the shaft in the middle of the claw. This closes the gripper as the other end of the strings is securely attached to the end of each claw. The final design can be seen in Figs. 5 and 6.

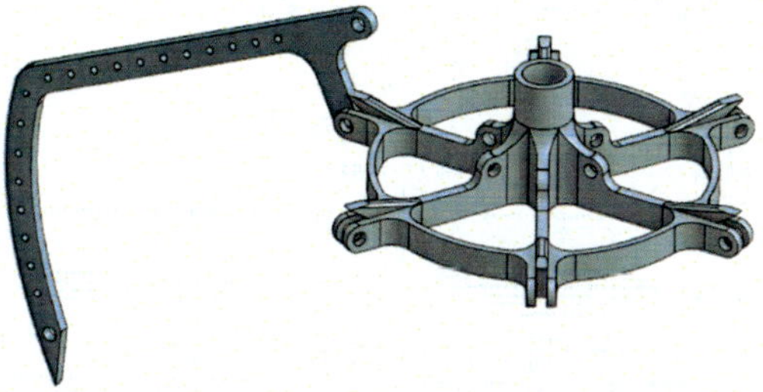

Fig. 5. CAD model of the final design, with the mounting holes for the springs and net, have been added to the claw and the number of claws is increased to 6. Additionally, a physical stopper has been added to the base plate located above the claw mounting hole.

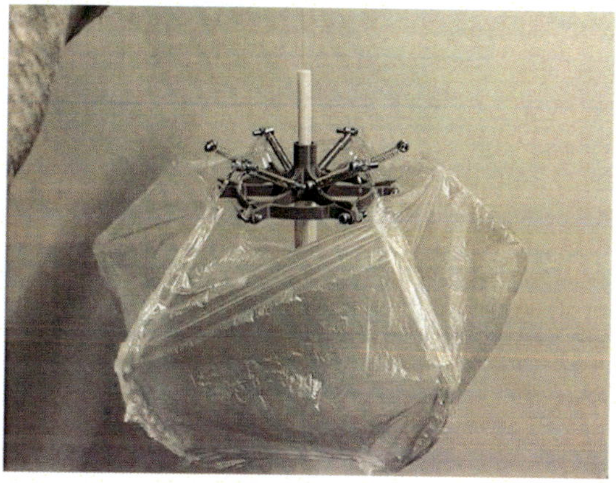

Fig. 6. The final state of the prototype with the plastic bag simulated the net functionality.

3.4.1 Geometry When deciding on the geometry of the centerpiece the main focus was on the following 5 aspects, weight, the number of arms, attachment points for the springs, a hole for the strings, and physical stoppers for the springs.

As the total weight of the gripper was not specified by the customer as a customer need no specific weight goal was held in mind when the center element was designed.

When the string was added to the 4 claw design of the gripper a weak point of the design came to light. This weak point was that an angle of 45° was between the string and each claw resulting in unsymmetrical stresses being placed on each of the claws. Additionally, too much resistance and wear would be placed on the wire itself providing a weak point in the system. Thus the decision to add more claws was made.

Using 6 claws fixed this problem as it reduced the wear and resistance in the wire and provided a more symmetrical closing pattern due to the increase of the angle between the wire and claw from 45° to 60°. The claws were redesigned to incorporate a spring attachment capability between the claw and the center plate. This spring addition provided the benefit of having the claw passively open and actively closed with the wire. Each claw had holes down the whole claw providing attachment points for the net. Finally, a physical stopper was added to the centerpiece preventing the springs from opening the gripper too much and possibly damaging itself.

The geometry of the claws had to take into account the following 3 aspects, the ability to grab/scoop the sample up, mounting holes for the net, and a reduced diameter when closed. To accomplish this, the bottom section of the claw is curved, allowing for a good scooping/grabbing ability while the rest of the claw is straight to reduce the diameter of the gripper when closed. Finally, mounting holes were added to the claw to securely attach the net to the claws.

3.4.2 Manufacturing Manufacturing the final prototype, seen in Fig. 6 requires the following 4 materials, PLA plastic, Plexiglass, Nylon, and a Plastic bag.

The centerpiece is 3D printed using a standard 3D print PLA plastic while the gripper claws are made by cutting 4 mm thick plexiglass into the desired shape with the use of a laser cutter. The strings used to close the claws were Nylon fishing strings and the "net" was made by gluing a plastic bag around the gripper arms using hot glue. Finally, the springs, bolts, and nuts were added and the prototype was assembled.

4 Analysis

When testing the MARS-DOG, the testing criteria were split into the 4 following groups. The ability to grab samples off the ground, the ability to securely hold the grabbed sample, the ability to release the sample once grabbed, and finally tests the design against a heavier load. These 4 groups were broken down into 6 tests and each test was conducted 30 times to ensure reliable and accurate test results. The sixth test was conducted 10 times for 5 s each. Below is a list of tests that were done.

1. Grabbing the sample in different angles (pitch) (FR_2), Sect. 4.1.
2. Grabbing the sample in different orientations (yaw) (FR_1), Sect. 4.2.
3. Grabbing the sample in different orientations and angles (yaw and pitch) (FR_1 and FR_2), Sect. 4.3.
4. Releasing the sample from the gripper. (FR_3), Sect. 4.4.
5. Grabbing the sample in a dusty environment(CON_1), Sect. 4.5.
6. Grabbing samples of a larger weight than 57g (CON_2), Sect. 4.6.

(a) Pitch of 0°

(b) Pitch of 15°

Fig. 7. Test setup for grabbing the sample at different angles (pitch) on a flat surface.

4.1 Grabbing the Sample in Different Angles (Pitch)

The first test completed tested grabbing the 3D printed sample tube Fig. 1 in different vertical angles (pitch). The goal of this test was to see if the angle of the sample gave different results in the grabbing functionality. The sample was placed on a flat surface that was tilted at 5° intervals from 0° to 15°. The setup of the test can be seen in Fig. 7. The test was deemed a Pass when the gripper managed to grab the sample tube and fully or partially enclose it within the net. In all 30 trials, the gripper was successful.

The test results show that the gripper is capable of picking up the sample tube regardless of the pitch of the sample. A pitch that was larger than 15° was deemed unnecessary since the sample slid/rolled off the testing surface and thus would not stay still long enough for the gripper to pick it up.

4.2 Grabbing the Sample in Different Orientations (Yaw)

The second test completed was the grabbing of the 3D printed sample tube Fig. 1 in different orientations of the sample to the gripper. This was done by placing the sample on a flat surface and rotating it in 45° increments from 0 to 135°. A red line was marked on one of the claws and for each of the tests, that red mark had to be aligned with the 45° marker. The setup of this test can be seen in Fig. 8. The test was deemed a Pass when the gripper managed to grab the sample tube fully or partially enclose it within the net. As in the previous test, the gripper was able to succeed in all 30 trials.

The test results show that the gripper is capable of picking up the sample tube regardless of the yaw of the sample tube to the gripper itself.

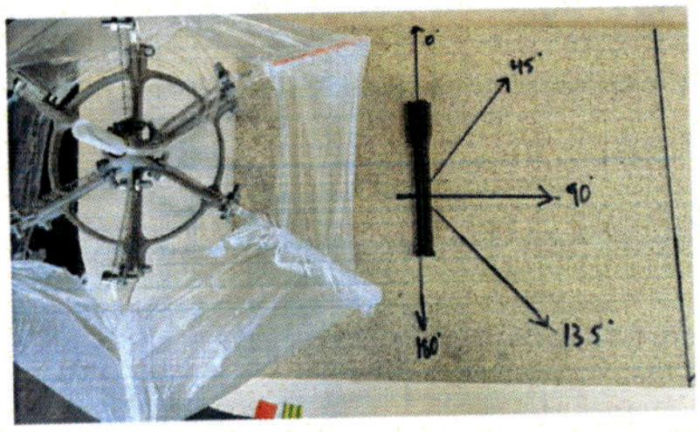

(a) 0° yaw

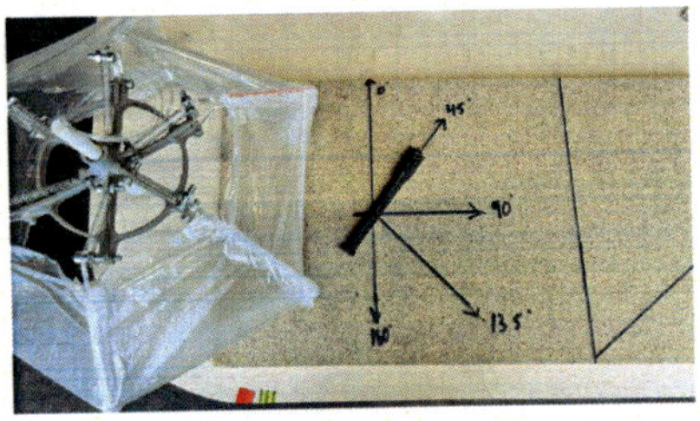

(b) 45° yaw

Fig. 8. Test setup for different (horizontal) yaw values to the gripper.

(a) Pitch of 0° and yaw 45°

(b) Pitch of 15° and yaw 45°

Fig. 9. Testing grabbing the test sample at different angles (pitch) on a flat surface with 45° yaw rotation relative to the gripper.

4.3 Grabbing the Sample in Different Orientations and Angles (Yaw and Pitch)

The third test to be completed was a combination of both tests Sects. 4.1 and 4.2. The 3D-printed sample tube Fig. 1 was placed on a flat surface. That surface was incrementally inclined by 5°. The difference between this test and test Sect. 4.1 is that the yaw rotation of 45° relative to the gripper itself as in test Sect. 4.2 is applied. The setup of this test can be seen in Fig. 9. Once again, the gripper was able to complete the test 30 times.

The test results show that the gripper is capable of picking up the sample tube regardless of the pitch or yaw of the sample tube to the gripper itself.

4.4 Releasing the Sample from the Gripper

This test checks whether or not the gripper can reliably release the sample. This test was performed when tests Sects. 4.1, 4.2, 4.3 and 4.5 were performed. After each test mentioned above was performed, the claws of the gripper were opened with the actively opening mechanics provided by the springs and noted down if

(a) Initial conditions (b) Successful grab

Fig. 10. Testing grabbing the sample tube in a dusty environment Sect. 4.5.

the sample tube fell out on its own or not. In total, the test was performed 390 times and managed to drop the sample in each test.

4.5 Grabbing the Sample in a Dusty Environment

This test checks whether or not the gripper can reliably grab the sample that is covered in dust. This was done by covering the sample in sawdust and then attempting to grab it out of the sawdust. The setup of this test can be seen in Fig. 10. The test was deemed a Pass when the gripper managed to grab the sample tube fully or partially enclose it within the net while covered in dust. The gripper was successful in all 30 trials.

This test's results show that the gripper is capable of picking up the sample tube regardless of the dust on it.

4.6 Grabbing Samples of a Larger Weight Than 57 g

The final test performed checks whether or not the gripper can reliably grab objects that are heavier than the 57 g. This test was performed by taking weights ranging from 1 to 6 kg and picking them up with the gripper. These dumb-bell style weights were borrowed from the sports science lab at Reykjavík University. The setup of this test can be seen in Fig. 11. The test was deemed a Pass when the gripper managed to grab one of the weight's ends, lift it and hold it for 5 s. Each test was performed 10 times and the gripper was passed each one.

The test results show that the gripper is capable of picking up weights ranging from 1 to 6 kg. Additionally, The gripper was able to hold each of these weights for 5 s indicating that the structure of the design is capable of holding the 57 g sample tube.

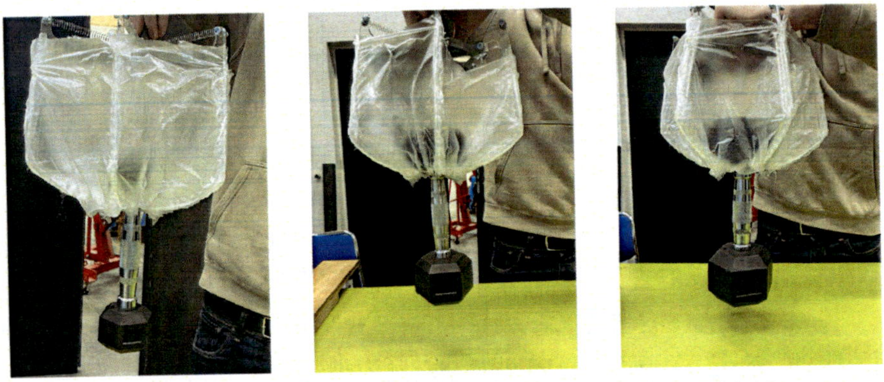

(a) 1 kg (b) 3 kg (c) 6 kg

Fig. 11. Grip strength testing with larger loads (Sect. 4.5)

5 Summary and Conclusion

When testing, the MARS-DOG was shown to be successful in lifting items weighing up to 6 kg and securing them with a pickup chance of 100% and a 0% drop rate. These tests were done with the sample in various orientations and angles. The initial goals of the product are stated in the functional requirements in Sect. 2.6. The comparison with the capabilities of the final product can be seen below:

FR_0 result: Tests for FR_0 were successful as the tests for FR_0-FR_3 where all successful and tests Sect. 4.3 (a combination of test Sects. 4.1–4.2) and 4.5 where the sample is grabbed in a dusty environment were also successful.

FR_1 result: After conducting tests where the sample was grabbed 30 times with different yaw, the gripper ended up having a 100% success rate and 0% failure rate.

FR_2 result: A test was conducted where the pitch of the sample was increased in 5° increments from $0 - 15°$. The gripper was then used to grab the sample 30 times on each incline. In each incline, the gripper had a 100% success rate and 0% failure. No tests were conducted at more than a 15° angle as the sample and gripper slid off the test setup with that incline.

FR_3 result: No special tests were conducted for FR_3 as the sample was released in all the other tests. In those tests, the gripper was able to release the sample with a 100% success rate.

It is thus shown that the MARS-DOG successfully met all its functional requirements and is thus considered a success since the main customer need Sect. 1.2 was met of the functional requirements were met. The device was deemed to work and can pick up the sample tube from a flat surface. Further testing is needed to determine whether the device is capable of picking up the sample tube from a non-flat surface.

RIOT lab, the customer from Reykjavik University was excited about the product. The collaboration was considered a success as the main customer need (Sect. 1.2) was met along with the specified requirements.

References

1. NASA/JPL-Caltech: Perseverance Sample Tube 266. https://mars.nasa.gov/resources/26218/perseverance-sample-tube-266. Accessed 10 Nov 2022
2. JPL/Caltech: Mars 2020 Sample Tube 3D print files. https://nasa3d.arc.nasa.gov/detail/Mars-2020-Sample-Tube-3D-print-files. Accessed 10 Nov 2022
3. Suh, N.P.: Axiomatic Design: Advances and Applications. The MIT-Pappalardo Series in Mechanical Engineering. Oxford University Press, New York (2001)
4. Hautaluoma, G., Johnson, A.: A Martian Roundtrip: NASA's Perseverance Rover Sample Tubes | NASA (2020). https://www.nasa.gov/feature/jpl/a-martian-roundtrip-nasas-perseverance-rover-sample-tubes. Accessed 11 Oct 2022
5. Alexnld: Small Hammer SNM2500 3D Print DIY RC Robot Arm Gripper — Alexnld.com. https://alexnld.com/product/small-hammer-snm2500-3d-print-diy-rc-robot-arm-gripper/. Accessed 30 Aug 2022
6. PRODRONE: PRODRONE Unveils the World's First Dual Robot Arm Large-Format Drone | Prodrone | Revolutionary Drones for Professionals (2016). https://www.prodrone.jp/en/archives/1420/. Accessed 30 Aug 2022
7. NOAA: Fishing Gear: Purse Seines | NOAA Fisheries (2022). https://www.fisheries.noaa.gov/national/bycatch/fishing-gear-purse-seines. Accessed 13 Oct 2022
8. Saferbrand: Ogre-Faced Spider Facts | Insects in Education (2022). https://www.saferbrand.com/advice/insect-library/insect-education/ogre-faced-spider. Accessed 30 Aug 2022
9. Robotiq: Robotiq Epick | Vacuum Gripper Collaborative Robots — Cobot Webshop. https://wiredworkers.io/product/robotiq-vacuum-e-pick/ Accessed 11 Oct 2022
10. Suh, N.P., Cavique, M., Foley, J.T. (eds.): Design Engineering and Science. Springer, Cham (2021). OCLC: 1280604069. https://doi.org/10.1007/978-3-030-49232-8
11. Slocum, A.H.: FUNdaMENTALS of Design. http://pergatory.mit.edu/resources/FUNdaMENTALS.html. Accessed 14 Sept 2022

Conceptual Design and Learning Axiomatic Design

On Dynamic Axiomatic Design or Projections of System Control Theory on Axiomatic Design

Vasilii Kaliteevskii[1](✉), Yuri Borgianni[2], and Leonid Chechurin[1]

[1] School of Engineering Science, LUT University, Yliopistonkatu 34, 53850 Lappeenranta, Finland
vkalit@gmail.com

[2] Faculty of Engineering, Free University of Bozen-Bolzano, Universitätsplatz 1 - Piazza Università, 1, 39100 Bozen-Bolzano, Italy

Abstract. The Axiomatic Design (AD) instruments provide a valuable insight for qualitative design evaluation. After having defined the inputs and outputs for the designed system, one can quickly check if it is convenient for the user to address the outputs by the inputs. Thus, AD helps to evaluate the usability of the design or the user interface, although the user interface can be given much wider sense, evaluating the interactions of the design during manufacturing, installation and maintenance. At the same time Axiomatic Design has its restrictions of applicability, which we focus on here. First, AD only shows what design is good, but it provides less guidance on how to make a good/independent design from the existing structure. Second, AD covers the only case where the number of Design parameters is equal to the number of Function Requirements. Finally, AD assumes that the mapping from Design parameters (DPs) to Function Requirements (FRs) is static, whereas very often the influence of DPs on FRs is time-dependent and it also neglects possible valuable resource – the time. In the research these drawbacks have been addressed from Control/System Theory prospective. First of all, we reveal a formal procedure how to construct the independent design for a linear system when the number of DPs is equal to FRs; the idea is based on eigen-decomposition of a matrix. Then, we generalize AD with the time domain, thus making it possible to address dynamic systems and use all the operators of Control Theory. We show that a controller can be added for the system design to make it independent. Finally, we extend the definition of independence to the case where the number of FRs is not equal to the number of DPs.

Keywords: Axiomatic Design · Dynamic Systems · Control Theory · Dynamic Axiomatic Design

1 Introduction

Axiomatic Design (AD) is a framework developed by Suh [1], which is based on two axioms, namely Independence and Information Axioms, which portray a relevant body of knowledge to support the definition of good designs. Recognition and application of AD principles in practice are known despite AD has been mostly developed at a

E. Puik et al. (Eds.): ICAD 2023, LNNS 849, pp. 167–178, 2024.
https://doi.org/10.1007/978-3-031-49920-3_11

theoretical level with limited quantitative evaluations of design results. Nevertheless, AD has gained a certain share of interest in the scientific community as a research topic. According to the SCOPUS database, the total number of papers with "Axiomatic design" in Title, Abstract of Keywords fields is about 1500 with the steady dynamics of around 100 publications per year. It is worth comparing this number with the most known alternative approach in the field of systematic inventive design: The Theory of Inventive Problem Solving, TRIZ, which has a similar number of publications but has more than doubled in volume dynamics of publication.

In the authors' best knowledge, Kulak et al. represents the most recent and cited attempt to survey and analyze the use of AD systematically [2]. It explains which AD-oriented practices are most successful and diffused, shedding light on the overwhelming majority of case studies in which just the Independence Axiom is used. More seldom, the classical version of the Information Axiom is exploited, but no article actually implements the full procedure prescribed by AD theory based on our literature exploration. By analyzing the illustrated product design examples insightfully, Borgianni and Matt summarize the emerging drivers and targets of AD into [3]:

- simplification and decomposition of complex systems;
- "optimization" tasks carried out in order to maximize/minimize certain effects with the recurrent goal of enhancing operability and safety.

In addition, the increasing integration of manufacturing and design areas might favor the attractiveness of AD, whose industrial focus might be seen at the border between the two disciplines [4].

Huge efforts are paid to merge the method and/or philosophy of AD with other approaches or principles. Robustness and uncertainty of AD is discussed in the light of possibly uncertain relationships between Design Parameters (DPs) and Functional Requirements (FRs) in the branch of literature concerning Fuzzy and Crisp AD [5, 6], which supports decision-making by managing imprecise and hardly predictable effects relevant for the application of the Information Axiom [7]. More articulated schemes for decision-making include considerations that pertain customer requirements to be fulfilled [8]. Complexity definition and AD is the focus of some literature contributions, e.g., [9]. Other instruments of quantitative analysis are used, from fuzzy logic to artificial intelligence [10].

The relation of AD with design theory represent another AD strength: for example, the link between modular design and AD is acknowledged [11], as independence is fulfilled through the introduction of different specialized sub-systems. This aspect represents a further element of compliance with the acknowledged design principles for Industry 4.0 [12].

The present paper and the literature overview that follows are by no means intended to deliver a comprehensive investigation of current efforts in AD. They rather highlight the research background and interesting trends to make AD (alone or along with other methods) more applicable to design problems.

2 Application Issues in Axiomatic Design

The reduction of complexity appears to be supported by the axioms by both removing intertwined relationships within systems and attenuating the impact of uncertain events on design. If a recent review of complexity in manufacturing and design is considered, these two objectives can be seen as the attempt to limit complexity at the functional level and in information-related terms [13]. At the same time, still with reference with the design of systems that characterize the new industrial revolution wave, the increasing number of sensors, controls and interfaces embodied by products does not comply with the attempt to reduce complexity [14], as the quantity of components is an acknowledged dimension of complexity. Still with a focus on such a dimension, the reduction of complexity results conflicting also with the common strategy to ensure independence through AD, i.e. the introduction of new and specialized modules.

A similar paradigm emerges also when considering other fundamental dimensions of product and system design. Limited complexity is seen as a catalyst for changeability [15], which, in turn, represents another hallmark of Industry 4.0-oriented designs. In this respect, the principles of Design for Changeability have been introduced explicitly in [16], which indeed include simplicity and low complexity. Interestingly, this contribution identifies a conflict between the conditions achieved through the Information Axiom (little information content fostering change), and the independence concept defined within AD. Indeed, according to [16], the first axiom addresses modular designs, supposedly featured by a larger number of parts and greater complexity, while the second one favors the development of integral designs.

The same dichotomy complies with the discussion about the compatibility between the theoretical background of AD and TRIZ. It is worth emphasizing how often the latter is integrated with AD [17–19], in order to overcome the former's limitations in terms of identifying practical solutions to emerging design problems; a relatively recent contribution that combines the two design methodologies can be found in [20]. It is claimed that integral design is the "natural" outcome of the TRIZ ideality concept, in contrast with the first axiom. A possibility to overcome this contradiction is represented by leveraging independence without increasing the number of components. A case in point is the faucet example, which presents aspects of both AD independence and TRIZ ideality, as the new system architecture is more compact than its predecessor. Rather than introducing new parts that fulfill a specific function, different DPs of a single component (two distinct rotation angles of the mixer faucet) make it possible to pursue independence.

While the faucet case can be seen as a win-win situation, similar results could not be achieved in different circumstances, as the classical AD example includes just two FRs. A possible alternative is the partial infringement of some design fundamentals. As Frey et al. [21] suggest with regard to the evolution of jet engines, a certain degree of not-ideality should be accepted in the design practice, as multiple cases are shown in which either TRIZ ideality or the first axiom fail to explain changes in successful designs. Besides, Ibragimova et al. [22] admit and tolerate the existence of certain degrees of non-ideality in terms of lacking adherence to AD principles. More explicitly, Cebi and Kahraman [6] point out that some designs are to be considered satisfying, even if they do not comply with the Independence Axiom. Moreover, according to Tang et al. [23], the interactions (and independence) among DPs cannot be ensured in the initial design phases in which AD is commonly applied.

In addition to the above shortcomings, other inherent AD limitations concern dynamic systems and transitory effects. Time is never leveraged in AD, existing AD-based DPs vs. FRs matrices never include time-dependent factors [24], as well as Suh's design theory admittedly shows limitations with regard to time-dependent problems [25]. This is somehow reflected by the attempt of transforming time-dependent systems into designs with periodical functionality, whose predictable effects allow for comparing them with static systems in terms of stability [26]. However, different forms of time dependency associated to DPs are likely to emerge in engineering systems [27]. Moreover, the attempt of steering towards periodical functionality cannot be considered compliant with the objectives of Industry 4.0, where consumer-related information is massively introduced through big data [28]. It can be noted that customer preferences are hardly predictable and show markedly irregular patterns [29]. As a result, the need to adapt AD to dynamic models that include the social dimension is remarked in [30].

Overall, the applicability of AD seems to be limited to certain circumstances, which cannot be always met when designing, especially in the incumbent highly automated and digital technology-driven industrial framework, despite the supposed suitability of AD to face certain inherent design challenges. The ideal conditions for employing AD include absence of any time-dependent phenomena, full controllability of the DPs in play, existence of the possibility to decouple systems (despite good designs exist that do not fulfill independence) without increasing the number of components and modules. In this sense, the present research illustrates a way to include non-predictable time behaviors within the parameters that feature AD-based design processes and a procedure that favors the relaxation of the limiting one-to-one mapping between DPs and FRs.

3 Axiomatic Design Extension for Dynamic Systems

We address here the limitation of required equality of functional requirements and design parameters.

We need to revisit the Control theory as it has already developed very useful analytic tools for system analysis. For those looking for foundations of Control Theory, which are taken here for granted, we recommend any classic control textbook, e.g. [31]. Thus, we first consider a general linear system that is seen as a transformation (called Transfer matrix) of the vector of inputs to the vector of outputs. The transformation is also called "multi input multi output (MIMO) transfer function". Let the MIMO system be described by a set of linear differential equations, so we can give it the operator form $y = G(s)u$. The variable s is Laplace operator and its simplified meaning is that the relationship between y and u is dynamic or the system has a memory or the inputs are transformed into outputs with a time delay.

For the MIMO system $G(s)$ is $m \times r$ matrix of transfer functions from the i-th element of the input vector u to the j-th element of the output vector y. The definition of output controllability from Control Theory is the following. Output controllability (OC) is the related notion for the output of the system (denoted y in the previous equations). The OC describes the ability of an external input u to move the output from any initial condition to any final condition in a finite time interval.

In other words, if the system has the property of OC, for any given output position $y(T)$ it can be found an input signal $u^*(t)$ that takes the output $y(t)$ to the desired position in the time interval T (from any initial state $y(0)$). In the case of static systems, the $G(s)$ degrades into a static gain matrix $G_{m \times r}$ and output controllability simply requires that G is to be square $m = r$ and invertible $det(G) \neq 0$. If we approach this "design" in an AD perspective, the system design would have been called "independent", because the number of inputs is equal to the number of outputs and each output is affected by its corresponding input.

But if we consider the general case where $G(s)$ represents a linear dynamic system, the latter requirements can be relaxed. First, the number of inputs can be less than number of outputs while the system is still under the conditions of OC. It is formulated in the following theorem [31]:

For a linear continuous-time system, described in time domain by a quadruple of state space matrices A,B,C and D, the matrix of the system operator $G(s)$ can be directly found

$$G(s) = \mathrm{C}(sI - A)^{-1}B + D. \tag{1}$$

We skip the derivation of the output controllability condition here that can be found again in [31]. Its simplified interpretation is the following: there is a specific algebraic matrix, the OC matrix, that can be found if we know system parameters (or, what is the same, A,B,C, D quadruple). This matrix has a full row rank if and only if the system follows OC.

Qualitatively projecting the statement above into AD evaluation principles, one can conclude that if the system is dynamic and "well designed", then any desired output position can be reached by manipulation with inputs, even in the case the input is scalar. In the design practice, this means that making the system dynamic may reduce the number of control inputs (i.e., making the control interface easier) while maintaining the full control of output.

The classical faucet example from AD is based on the statement that there are two DPs, (either hot and cold water valves or flow and temperature valves if the design is independent). If the faucet design is dynamic, we can reduce the number of valves (controls) to one. Indeed, the practical realization could for example mean the single valve that is first used to assign the flow and, after a period of time, is used to assign the temperature. Therefore, it may be seen that a DP controls both FRs when time is considered in such a model.

Then the interface concept of such a valve could have a look as in the Fig. 1. The first raw of numbers (black) can represent the required temperature while the second (blue) represents the flow intensity. It is assumed that the first move of the valve is used for temperature assignment and the second move (either immediately or after a short period of time) is used to assign the flow. The realization of such a mechanism does not seem to be a complicated problem, the same idea has been used for example for mechanical code locker in safes: the rotation of the same code wheel is used to input various number of the opening code.

It is worth mentioning that the variability of the transformation of DPs into FRs with time is already a point of discussion in the developments of AD [36]. Suh introduces the

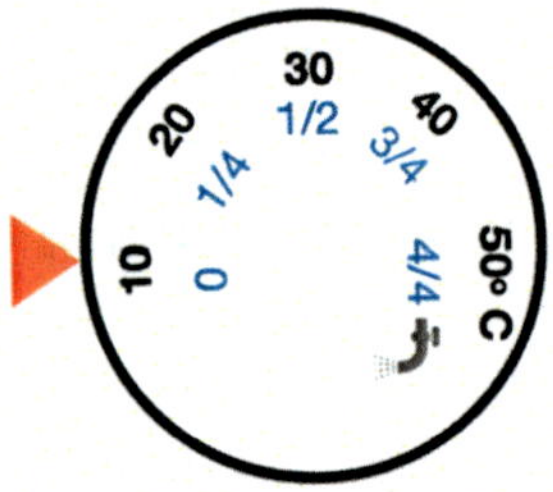

Fig. 1. A dynamic valve with time-sensitive scale.

concepts of time-dependent combinatorial and periodic complexity. At the same time, it can be argued that these concepts and their definitions are used to illustrate complexity instead of seeing them as the source for design solutions.

4 Axiomatic Design Extension for Dynamic Systems

The classical multivariable control theory affirms that a dynamic linear MIMO system with equal number of inputs and outputs, in which changes in each single input result in changes in all outputs can be decoupled and given an independent (according to AD) form by assigning proper feedback controllers. The idea is illustrated in the Fig. 2. A MIMO plant with the transfer matrix $G(s)$ is uncoupled. Under certain conditions, it can be given a feedback controller described by the transfer matrix $C(s)$, that the relationship between new inputs to the system w_i and outputs y_j is decoupled. So, the closed loop system, or the new system block (that is given the gray background in Fig. 3) has diagonal structure.

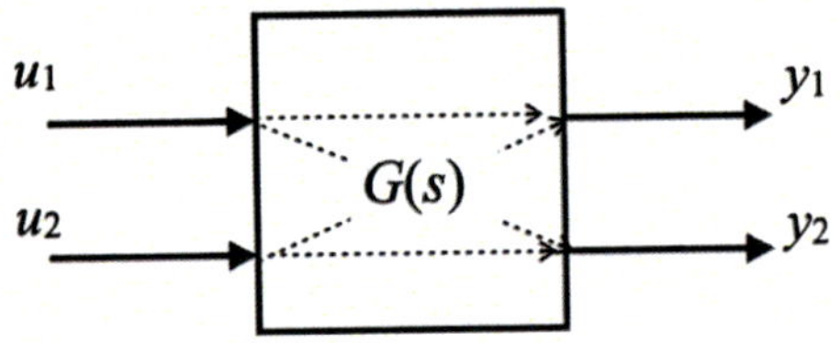

Fig. 2. Decoupling by feedback. Original plant.

This theoretical result, most probably first delivered by Gilbert, can be seen as a hidden inversion of the plan $G(s)$ with the help of feedbacking [32].

A note about linearity is needed here: an obstacle for the applicability of the formal derivations above is the fact that input-output relationships are nonlinear in general. This means that the relationship between inputs and outputs has a more complicated form than $y = G(s)u$, so the matrix $G(s)$ cannot be extracted. Nevertheless, it should not be a conceptual deal breaker: first, linearization is always possible, so in the vicinity of a certain point (for example, near the basic operation point) we can approximate the reality by linear input-output relationships. Second, there are classes of nonlinear systems that can be decoupled [33].

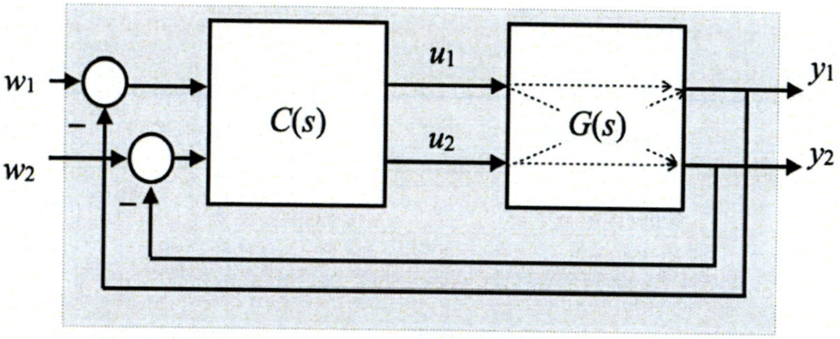

Fig. 3. Decoupling by feedback. Plant with feedback controller.

To give this general fact an application example we choose the classical distillation column decoupling design [34] (see Fig. 4 and 5). There are three inputs (valves' position u_i) and three outputs (distillate fraction concentration y_j) shown in Fig. 4. The relationship between the input and the output is dynamic and described by the linear transfer matrix $G(s)$. We take the valve position as DP and the concentration of the distillate as the FR for this system. The nature of the distillation process is such that changes in the position

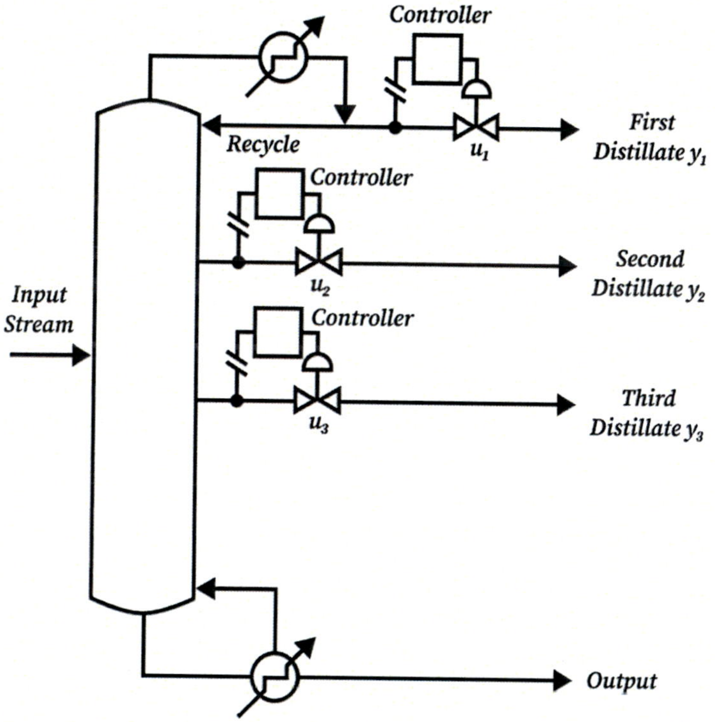

Fig. 4. Example of parameter-dependent design of a distillation column.—u_i input parameters (valves) and—y_j output parameters (distillate fractions).

of any control valve affect all three concentrations of the output, which means that the matrix $G(s)$ is non-diagonal. The system design is coupled while we would like to be able to assign the concentration of any output independently. In other words, we would like to make the design independent. It is hardly possible to change the design of the operating device in most real industrial cases. But if the feedback control paradigm is admissible, the problem can still be solved. A new system architecture is illustrated in Fig. 5. It can be shown that, given the model of the process $G(s)$, the controllers $g_{ij}(s)$ can be tuned in such a way that the closed loop input-output behavior is autonomous. So, new input valve variations Y_{di} would affect the corresponding output concentration only.

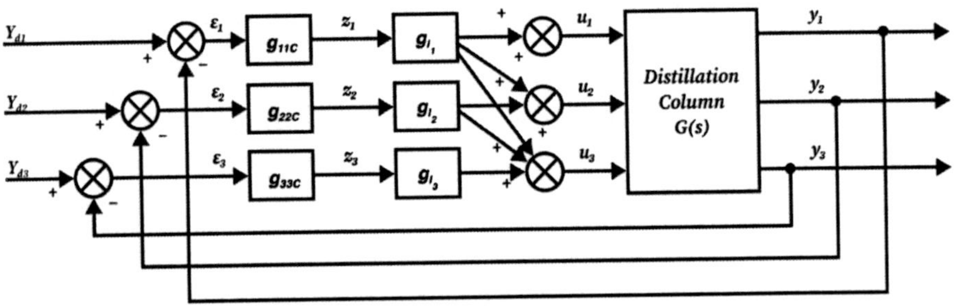

Fig. 5. Example of the structure of a decoupling feedback controller.—u_i input parameters (valves) and—y_j output parameters (distillate fractions),—$G(s)$ feedback control parameters.

5 Discussion, Crosstalk of AD, TRIZ and Control Theory

It has been already mentioned in the introduction that the authors see TRIZ as one of the instruments closest to AD in inventive design. This section is not intended to extend the AD theory with TRIZ tools, but to point out how TRIZ is relevant to AD and also how TRIZ might be linked to the AD in the context of the Control Theory. More specifically, out of the basic concepts or formal tools of TRIZ, Contradiction analysis is directly addressing the uncoupled design situation with two important differences. First, TRIZ analysis is limited by the case of 2x2 design matrix. This is a serious drawback for the analysis of complex systems, as we need to reduce it to pair subsystems, which is not always possible. The second limitation is that, while in AD the FRs are mostly parameters or variables, the function requirements are always design criteria in TRIZ, so they are always fitness indexes that can be used to compare designs. In other words, TRIZ contradictions are correctly formulated when changes in one DP is good for one FR but bad for another. However, the advice from both AD and TRIZ is identical: change the design in such a way that the system is decoupled, that means the contradiction is resolved in such a way that each FR can be addressed (obviously, improved) independently by separate DPs.

However, the formal application of TRIZ can be extended even after the achievement of a decoupled design. The necessity to address two (or even several) FRs with only one DP can also be seen as a contradiction in TRIZ school of thought. Indeed, the situation when conflicting requirements are exposed to the same element of the system is called physical contradiction [35]. Coming back to the faucet example and the idea of Fig. 1 solution: we have here a DP,"a valve" and two requirements, namely, to use it for temperature assignment and for water flow regulation. TRIZ suggests separation principles as the systematic way to ideate a possible design improvement. Interestingly, separation means exactly finding a new design which is decoupled. In contrast to AD, TRIZ goes a bit beyond the recommendation and also suggests considering specific ideas for separation: separation in time and separation in space, among others. Using the time domain for decoupling, the design of faucet in Fig. 1 can be seen as a possible implementation of separation in time. One can also try to apply "separation in space": a possible result could be embodied by a sector of the valve used for flow control and another sector for temperature assignment. Actually, this design is almost a standard for shower faucets in North America, where there is a single valve and a user is unable to get his/her warm shower without being exposed to a portion of cold water.

Finally, the ideality concept of TRIZ should force a designer to think of even simpler system interfaces. With this objective in mind, having ensured the possibility to address the FRs, the fewer DPs we have to deal with the better. Simply speaking, the fewer "valves" we have to control the better and the design with one valve is "more ideal" according to TRIZ. Moreover, the aggressive application of ideality concepts would model a controllable faucet without any valves at all (think of an AI-supported faucet that can predict what kind of flow/temperature combination will be needed). Again, these design ideas would definitely show how TRIZ-driven design differs from AD-driven design (and its possible combination with Control Theory), but for the practicing engineer the instrument is not as much important as the result.

A final comment needs to be added in regards to the quantitative system Control Theory. One can see a problem in the decoupled design that relates to system robustness. Having decomposed the system, we lose any possibility to influence a FR if the corresponding DP is not working well, the subsystem is broken. In other words, if the decoupled faucet has a problem with the water flow channel (e.g., it is clogged), changing the temperature valve does not yield anything, we are not able to get any "water". While in the "old fashion" faucet, even having one of the valves broken, both FRs would still be partially delivered. This problem can also be tackled with the MIMO dynamic paradigm from Control Theory. In simple words, the theory says that input-output controllability can be achieved for the systems by introducing feedback controllers in which the number of inputs is less than the number of outputs.

6 Conclusion

AD instruments provide a valuable insight for qualitative design evaluation. Indeed, having defined the inputs and outputs for the designed system, one can quickly check whether it is convenient for the user to address the outputs by the inputs. Thus, AD helps to evaluate the usability of the design or the user interface, although the user interface can be given much wider sense, evaluating the interactions of the design during manufacturing, installation and maintenance.

At the same time, AD has its restrictions in terms of applicability. First of all, AD well identifies a good design, but it is more limited in terms of guiding the design process towards a good/independent design from the existing structure. Second, AD is particularly suitable just in those cases where the number of DPs is equal to the number of FRs. Another limitation is the fact that the mapping from DPs to FRs is static, whereas very often the influence of DPs on FRs is time-dependent; here, we have argued that AD also neglects possible valuable resource – time.

In current research these drawbacks have been addressed from Control/System Theory prospective, which is the original contribution of the present paper. We have speculated through some examples how Control Theory could help overcome some of the above limitations. More research is clearly needed to validate the benefit of juxtaposing AD and Control Theory, also in light of the possible difficulties in using the latter proficiently for designers. We have shown, yet through some examples, that some results could be similar if AD is integrated with TRIZ, where this combination is abundantly described in the literature. However, results can vary and the choice of the methodology might depend on design objectives and priorities.

It is in the authors' intention to develop a full integration of AD, TRIZ and Control Theory, which can possibly face the deficiencies of the used design methodologies.

However, already in the current integration of AD with Control Theory, it is possible to generalize AD with the time-domain, thus enabling tackling dynamic systems through the use all the operators of Control Theory. And even though the use of Control Theory causes the introduction of the time dimension with a consequent increasing complexity of AD, the new opportunities of such an approach make the applicability of the framework wider and more generalizable. Thus, we showed that a controller can be added for the system design to make it independent, which is one of the main challenges stated in AD. As an additional contribution, we extended the definition of independence to the case where the number of FRs does not correspond to the number of DPs.

References

1. Suh, N.P.: The Principles of Design. Oxford University Press, Oxford, No. 6 (1990)
2. Kulak, O., Cebi, S., Kahraman, C.: Applications of axiomatic design principles: a literature review. Expert Syst. Appl. **37**(9), 6705–6717 (2010)
3. Borgianni, Y., Matt, D.T.: Applications of TRIZ and axiomatic design: a comparison to deduce best practices in industry. Procedia CIRP **39**, 91–96 (2016)
4. Borgianni, Y., Matt, D.T.: Ideality in Axiomatic Design and beyond. Procedia CIRP **53**, 95–100 (2016). https://doi.org/10.1016/j.procir.2016.07.029
5. Kulak, O., Kahraman, C.: Fuzzy: multi-attribute selection among transportation companies using axiomatic design and analytic hierarchy process. Inf. Sci. **170**(2–4), 191–210 (2005)
6. Cebi, S., Kahraman, C.: Extension of axiomatic design principles under fuzzy environment. Expert Syst. Appl. **37**(3), 2682–2689 (2010)
7. Cheng, J., Feng, Y., Lin, Z., Liu, Z., Tan, J.: Anti-vibration optimization of the key components in a turbo-generator based on heterogeneous axiomatic design. J. Clean. Prod. **141**, 1467–1477 (2017)
8. Kumar, P., Tandon, P.: A paradigm for customer-driven product design approach using extended axiomatic design. J. Intell. Manuf. **30**(2), 589–603 (2019)

9. Modrak, V., Bednar, S.: Using axiomatic design and entropy to measure complexity in mass customization. Procedia CIRP **34**, 87–92 (2015)
10. Li, W., Suh, C.S., Xu, X., Song, Z.: Extenics enhanced axiomatic design procedure for AI applications. Artif. Intell. Eng. Design Anal. Manuf. AI EDAM **36**, e23 (2022)
11. Gershenson, J.K., Prasad, G.J., Zhang, Y.: Product modularity: definitions and benefits. J. Eng. Design **14**(3), 295–313 (2003). https://doi.org/10.1080/0954482031000091068
12. Hermann, M.P., Otto, T.B.: Design principles for Industrie 4.0 Scenarios. In: 49th Hawaii International Conference on System Sciences (2016)
13. ElMaraghy, W., ElMaraghy, H., Tomiyama, T., Monostori, L.: Complexity in engineering design and manufacturing. CIRP Ann. **61**(2), 793–814 (2012)
14. Vellmar, J., Gepp, M., Schertl, A.: The future of engineering—scenarios of the future way of working in the engineer-to-order business. In: 2017 Annual IEEE International Systems Conference (SysCon), pp. 1–5. IEEE (2017)
15. Francalanza, E., Borg, J., Constantinescu, C.: Deriving a systematic approach to changeable manufacturing system design. Procedia CIRP **17**, 166–171 (2014)
16. Fricke, E., Schulz, A.P.: Design for Changeability (DfC): principles to enable changes in systems throughout their entire lifecycle. Syst. Eng.**8**(4) (2005). https://doi.org/10.1002/sys.20039
17. Shirwaiker, R.A., Okudan, G.E.: Triz and axiomatic design: a review of case-studies and a proposed synergistic use. J. Intell. Manuf. **19**(1), 33–47 (2008)
18. Borgianni, Y., Matt, D.T.: Axiomatic design and TRIZ: deficiencies of their integrated use and future opportunities. Procedia CIRP **34**, 1–6 (2015)
19. Chechurin, L., Borgianni, Y.: Understanding TRIZ through the review of top cited publications. Comput. Ind. **82**, 119–134 (2016)
20. Ko, Y.T.: Modeling a hybrid-compact design matrix for new product innovation. Comput. Industr. Eng. **107**, 345–359 (2017)
21. Frey, D., Palladino, J., Sullivan, J., Atherton, M.: Part count and design of robust systems. Syst. Eng. **10**(3), 203–221 (2007)
22. Ibragimova, E., Pena, M., Thompson, M.K.: The evolution of Sihwa dam: A formal design perspective. In: The 8th International Conference on Civil and Environmental Engineering. ICCEE (2009)
23. Tang, D., Zhang, G., Dai, S.: Design as integration of axiomatic design and design structure matrix. Robot. Comput.-Integr. Manuf. **25**(3), 610–619 (2009). https://doi.org/10.1016/j.rcim.2008.04.005
24. Thompson, M.K., et al.: In Global Product Development, Springer, Berlin, Heidelberg, pp. 61–72 (2016)
25. Lu, S.C.Y., Suh, N.P.: Complexity in design of technical systems. CIRP Ann. **58**(1), 157–160 (2009)
26. Suh, N.P.: On functional periodicity as the basis for long-term stability of engineered and natural systems and its relationship to physical laws. Res. Eng. Design **15**(1), 72–75 (2004)
27. Weber, C.: Modelling products and product development based on characteristics and properties. In: Chakrabarti, A., Blessing, L.T.M. (eds.) An Anthology of Theories and Models of Design, pp. 327–352. Springer, London (2014). https://doi.org/10.1007/978-1-4471-6338-1_16
28. Saldivar, A.A.F., Goh, C., Chen, W.N., Li, Y.: Self-organizing tool for smart design with predictive customer needs and wants to realize Industry 4.0. In: 2016 IEEE Congress on Evolutionary Computation (CEC) , pp. 5317–5324. IEEE (2016)
29. Löfgren, M., Witell, L., Gustafsson, A.: Theory of attractive quality and life cycles of qualit attributes. TQM J. **23**(2), 235–246 (2011)

30. Sheikh, R., Abbasi, M., Talaei, A.A., Tahmasbi, M.: Dynamic Axiomatic Design (DAD): applying the independence axiom in the design of social systems. Procedia CIRP **34**, 125–130 (2015)
31. Ogata, K.: Modern Control Engineering, NJ: Prentice Hall, Upper Saddle River, pp. 6142–6143 (2009)
32. Gilbert, E.G.: The decoupling of multivariable systems by state feedback. SIAM J. Control **7**(1), 50–63 (1969)
33. Claude, D.: Decoupling of nonlinear systems. Syst. Control Lett. **1**(4), 242–248 (1982)
34. Ray, W.H.: Advanced Process Control. McGraw-Hil, New York (1981)
35. Altshuller, G.S.: Algorithm of inventive problem solving (1956)
36. Suh, N.P.: Complexity: Theory and Applications. Oxford University Press, New York (2005)

Axiomatic Design as a Tool to Develop a Global Production Strategy in Transportation Industry

Johanna Lechner[1](✉), Erwin Rauch[2], and Tobias Reggelin[3]

[1] Hübner GmbH, Heinrich-Hertz-Straße 2, 34123 KasselKassel, Germany
johanna.lechner@hubner-group.com
[2] Free University of Bozen-Bolzano, Piazetta del Università 1, 39100 Bolzano, Italy
[3] Otto-Von-Guericke-University Magdeburg, Universitätsplatz 2, 39106 Magdeburg, Germany

Abstract. Manufacturers of almost every industry sector face the challenge to develop and constantly adapt their production strategy according to a changing political and economic environment. A well aligned production strategy is the basis to satisfy customer demands, create competitive advantage and therefore expand the market share. In an increasingly globalized economy, global production strategies and the design of global value chains become more and more important to gain sustainable competitive advantage in global markets. This is, to a large extend, a management challenge that inheres a change in strategy in terms of network design, process technology and manufacturing, leadership, and training, and forecasting and planning. To master these challenges Axiomatic Design (AD) was used to derive concept design parameters based on the analysis of customer attributes (CAs). Coming from the definition of internal stakeholders of the production strategy, CAs were defined and matched with system and company internal constraints. The remaining CAs were grouped into a set of functional requirements and broken down to an implementable set of design guidelines for a global production strategy. The results show a global production strategy approach based on five independent pillars while the decomposition allows to point out important interdependencies and helps to schedule an implementation process of the strategy.

Keywords: Axiomatic Design · Global Production Strategy · Sustainable Manufacturing · Industry 4.0

1 Introduction

Complex systems require special treatment and attention in every phase of their life cycle. Especially in the design phase crucial elements, stakeholders and limitations need to be identified to create a holistic view on the target. The tradeoff between market, product and process requirements is the starting point of the analysis of production systems. Based on these dimensions and the individual company profile a first impression on the complexity and diversity of a future global production strategy is created.

The motivation to create a suitable production strategy is five-dimensional. The first and leading element of the motivation is **profit maximization**. This can be seen as the

E. Puik et al. (Eds.): ICAD 2023, LNNS 849, pp. 179–189, 2024.
https://doi.org/10.1007/978-3-031-49920-3_12

overall company goal independent from the industry sector. Going into more detail, increasing, or at least retaining, **market share** is crucial in public transportation due to the constant pressure from competitors. Flexibility and adaptability in critical to fulfill the latest market requirements. For this reason, it is highly aimed to provide **individual customer solutions**. This implies the product design as well as the lead time from engineering to delivery. Through these individual solutions, **competitive advantage** can be created. Being close to the customer, physically in terms of production and virtually in terms of administration, is key. The fifth pillar therefore is the increase of **internal flexibility** by well trained staff that can be employed more flexible when demand changes. To master this multi-dimensional task, a proper production strategy needs to be worked out and aligned to internal and external requirements.

2 Theoretical Background

The design of an adequate production strategy including the applied tools, mechanisms, and strategic measures as well as an aligned global production network play a significant role for a successful production.

2.1 Production Strategy and Systems

Since the 1960's strategic management is treated as a separate discipline in business administration which has developed and continuously evolved over the past decades from being based on empiric values to scientific modelling and market-specific approaches [1]. Production strategy often is defined as the exploration and exploitation of production capabilities by structural and infrastructural decisions to achieve a unique strategic position in the market, which matches overall business objectives [2–4].

To define a production strategy, internal and external analysis need to be executed. The external analysis helps to determine the position of the company in the market. The internal analysis has the purpose to set strategic goals for the production. Additionally, companies face the challenge to act agile and flexible concerning a changing business environment and customer demands [5]. A traditional SWOT (Strengths, Weaknesses, Opportunities and Threats) Analysis is a common tool in this first phase of the strategy design process [6]. Opportunities and Threats represent the two dimensions for the external, Strengths and Weaknesses the perspectives of the internal analysis.

2.2 Manufacturing Industry for Public Transportation

The following information has been collected through interviews with managers and experts in the rail sector.

Movement on tracks is a widely accepted mode of transport which is globally established since decades. The energy consumption with respect to the transported volume is significantly lower compared to other modes of transport. The rail industry is dominated by several big players in the market which either produce most of the components in house or rely on suppliers to produce the rail setup.

The rail industry is majorly characterized by project business. Public tenders determine the award process and are mostly driven by local governmental interests. Local content requirements, customer specific solutions, and governmental budgets request highly flexible production structures of OEMs and their suppliers and therefore increase product complexity and variances. The production of this product portfolio is mostly synchronized in one production line for optimized utility which further increases complexity and the need for agile production structures.

Despite the ongoing trend of urbanization in cities and the continuous interconnection on a national and international level, forecasts in the rail industry are hardly possible. Investments in infrastructure can affect several types of local or long-distance transport. These factors impede the overall forecast. As the railroad sector usually is state-owned, advancement and development are depending on political developments.

Additionally, requirements concerning safety are highly important in the industry as the transportation of passengers is risk critical. Further criteria are stability, resilience, robustness, and longevity as rail vehicles are commonly used up to 40 years.

2.3 AD for Global Production Strategies

The Axiomatic Design (AD) method is a top-down approach that belongs to systems theory and was invented by the scientist Nam P. Suh in the mid-seventies [7]. The purpose of AD was to find a method to control complex and interconnected systems [8].

The basis of AD are four domains in which aims, and requirements are specified. Within the customer domain (for production strategy: internal and external stakeholders) the aims of the customers (customer attributes - CAs) are defined [9]. The functional domain shows a deduction of the CAs and translates those attributes into functional requirements (FRs). FRs indicate the required function of the production strategy to satisfy a certain CA [10].

The physical domain consists of design parameters (DPs) and guidelines that provide a solution for the defined requirements. The process domain is the final domain within AD that converts design guidelines into measurable process variables (see Fig. 1).

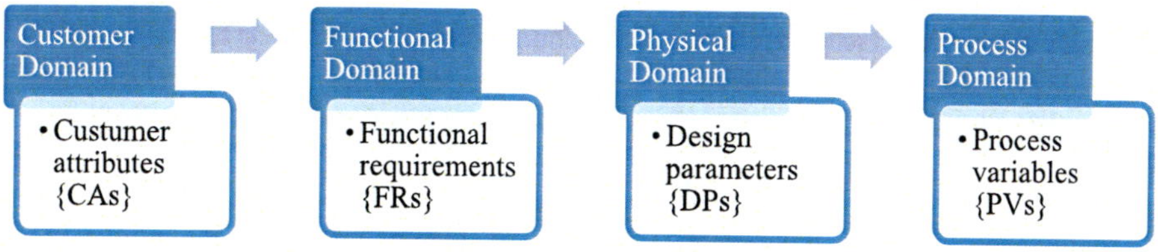

Fig. 1. The four domains in Axiomatic Design [11]

FRs and DPs are intended to build pairs. While FRs describe the requirement in an active way, the DPs represent the implementable result to fulfill the FRs. Within the matching process of FRs and DPs two axioms need to be considered [12].

Axiom 1 - Independence axiom: FRs are independent from each other, meaning that requirements of the functional domain should be satisfied by one design element of the physical domain [8]. The system is called ideal when there is no correlation between the elements [13].

Axiom 2 - Information axiom: The second axiom conveys that if there is more than one solution (DP) to a requirement, the one offering the least informative content should be preferred [8].

The requirements and design elements within the AD approach are becoming more detailed as the so-called decomposition of the top FRs is executed. A hierarchical tree structure evolves until an implementable level of design elements, the design guidelines, is reached [7]. DPs are mapped to FRs and with the "zigg-zagging" unspecified DPs are developed further to more detailed FRs [13]. The result of the decomposition process is the design matrix (DM) set up by the following Eq. 1 [12].

$$\{FR\} = [DM] * \{DP\} \tag{1}$$

3 Concept Design for a Global Production Strategy

With the aid of the AD approach the concept design for a global production strategy is worked out. After the deduction of CAs and system constraints the leading FR's are derived. The presented concept design can be applied to other production strategy development initiatives with similar characteristics.

3.1 Deduction of Customer Attributes

Within a conducted workshop the stakeholders of a global production strategy were analyzed. The focus was on requirements towards strategy including the company vision and the business unit goals. These requirements represent the CAs of the AD approach. Additionally, system constraints were formulated that may interfere with strategy objectives and therefore might influence the design guidelines of the strategy.

It was agreed that, besides local governments, only internal stakeholders will be considered and can be subdivided into managing instances and affected departments. The requirements of OEM customers are represented by the sales department. The following list shows the stakeholders that are considered for further analysis (see Table 1).

From these stakeholders CAs are be derived. An abstract of the most relevant attributes is displayed in Table 2.

Coming from the overall objective the following superior FR for the development of a global production strategy is defined:

FR0Create organizational and functional framework to maximize profit.

To meet this requirement the corresponding solution approach on the physical design domain DP0 is dedicated:

DP0Global Production Strategy.

Table 1. Stakeholders of a global production strategy

Managing instances	Affected departments
• Trustees / Owner	• Global and Local Sales
• Managing directors	• Procurement
• Business unit managers	• Operations
• Site managers	• Quality Management
Others	
• Local governments	

Table 2. Abstract from CAs derived from stakeholder analysis

No	Stakeholder	Customer Attribute
CA_1	Trustees // Owner	Long-term competitive advantage
CA_2	Business unit managers	Maximize profit over all sites
CA_3	Business unit managers	Use capacities of all sites
CA_5	Site managers	Maximize profit at own site
CA_6	Site managers	Use capacities of own site
CA_8	Trustees / Owner Managing directors	Increase or retain market share
CA_{10}	Procurement	Transparency of long-term demands
CA_{13}	Global and Local Sales	Fulfill local content requirements
CA_{16}	Quality Management	Produce at local quality requirements
CA_{17}	Workers Council	Safe jobs and employment
CA_{18}	Operations	Generate high productivity
CA_{20}	Operations	Be able to react to demand volatilities

3.2 Determination of Top-Level Functional Requirements

The overall goal to maximize profit (first level of FR and DP pairs) can be deducted within a systematic and structured decomposition process (see Table 3). In addition to the previously defined CAs also the first level of FRs and DPs was aligned within the interviews with strategically important stakeholders of the production strategy (see Tables 1 and 2).

The result is the categorization into the five pillars of the production strategy. Each pillar can be seen as an individual strategy to serve the previously defined FR-DP_0 requirement.

These five categories build the foundation of the production strategy. The pillars Production Network, Operations and Manufacturing, Know How and Training, Forecasting and Planning, and Process Technology are independent, and their design-matrix

Table 3: Decomposition FR_0 - Level 1

FR1	*Create an efficient production network*	*DP1*	*Production Network Strategy*
FR2	*Produce requested quality at minimum cost*	*DP2*	*Operations and Manufacturing Strategy*
FR3	*Ensure high education of employees on every level*	*DP3*	*Know-How and Training Strategy*
FR_4	*Ensure accurate forecasting and planning on different hierarchy levels for production*	DP_4	*Forecasting and Planning Strategy*
FR_5	*Push innovation, digitalization, and automation in production*	DP_5	*Process Technology Strategy*

shows an uncoupled structure (see Eq. 2) which was confirmed during the interview and workshop sessions held with the stakeholders.

$$\begin{pmatrix} \mathrm{FR1} \\ FR2 \\ FR3 \\ FR4 \\ FR5 \end{pmatrix} = \begin{bmatrix} X\ O\ O\ O\ O \\ O\ X\ O\ O\ O \\ O\ O\ X\ O\ O \\ O\ O\ O\ X\ O \\ O\ O\ O\ O\ X \end{bmatrix} * \begin{pmatrix} DP1 \\ DP2 \\ DP3 \\ DP4 \\ DP5 \end{pmatrix} \text{"}uncoupled\text{"} \tag{2}$$

3.3 Decomposition of Functional Requirements

For demonstration reasons the first pillar is decomposed in the following section. The analysis of the first pillar (FR1) shows that three partial requirements can be derived (see Table 4).

Table 4. Decomposition FR_1 - Level 2

FR11	*Generate a clear understanding of the production landscape*	*DP11*	*Criteria and categorization of production sites*
FR12	*Decrease production cost*	*DP12*	*Grouped and connected production sites for a network*
FR13	*Decrease logistics cost*	*DP13*	*Standard allocation process for customer orders*

The DM for this section of the decomposition shows a decoupled design as for increased efficiency (DP_{12}) and the optimal choice of a production site (DP_{13}) the production site set-up needs to be analyzed and assessed (DP_{11}). Additionally, the efficiency effects within a designed production network have a heavy impact on logistics cost, which is explained in lower levels of detail in the following paragraphs (see Eq. 3).

$$\begin{pmatrix} FR11 \\ FR12 \\ FR13 \end{pmatrix} = \begin{bmatrix} X & O & O \\ X & X & O \\ X & X & X \end{bmatrix} * \begin{pmatrix} DP11 \\ DP12 \\ DP13 \end{pmatrix} \text{"decoupled"} \tag{3}$$

Both pairs (FR-DP_{12} and FR-DP_{13}) require a further decomposition which is not shown in the following section of this paper. The procedure of the decomposition is exemplarily shown for the pair FR-DP_{11}). The following figure (Fig. 2.) shows an abstract of the tree chart to visualize dependencies and levels of detail within the decomposition starting from FR/DP_{11}.

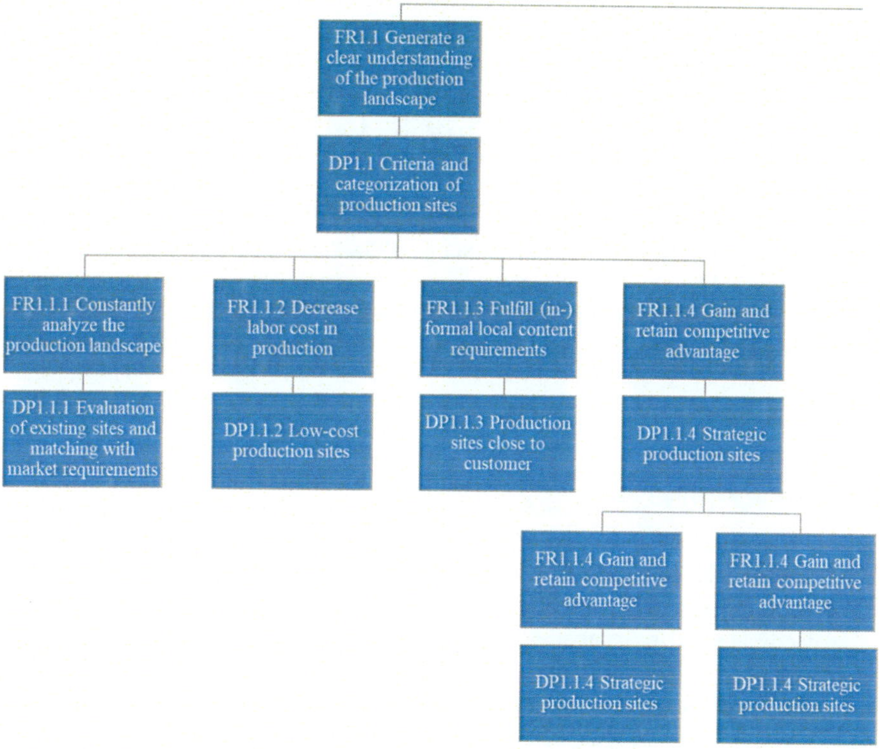

Fig. 2. Exemplary Decomposition of FR_{11}

All three second level subcategories can be split up further into the next level of partial requirements. The constant screening and evaluation of existing sites is essential to react to a changing market environment (FR/DP_{111}). In order to remain competitive and decrease labor cost, it is recommendable to install low-cost production sites (FR/DP$_{112}$). Local content requirements and short delivery lead times can be generated by production sites close to important customers (FR/DP$_{113}$). An additional leverage is possibility of geographical proximity is to create stronger business relationships. To gain und to retain sustainable competitive advantage the establishment of strategic production sites is advisable (FR/DP$_{114}$).

Again, the DM shows a decoupled design as for the configuration of low-cost, geographically close, and strategic production sites the evaluation of all sites in necessary (see Eq. 4).

$$\begin{pmatrix} FR111 \\ FR112 \\ FR113 \\ FR114 \end{pmatrix} = \begin{bmatrix} X & O & O & O \\ X & X & O & O \\ X & O & X & O \\ X & O & O & X \end{bmatrix} * \begin{pmatrix} DP111 \\ DP112 \\ DP113 \\ DP114 \end{pmatrix} \text{"decoupled"} \tag{4}$$

Strategic production sites have two different purposes in the described framework. This is detailed further in the next decomposition step. Strategic production sites come in two different shapes. One aim can be to dispel competitors from strategically important markets (FR/DP$_{1141}$). The other is to shorten development and engineering times ("time-to-market") by the establishment of technology factories with focus on R&D activities and product development (FR/DP$_{1142}$). The DM for this section is uncoupled.

4 Application Use Case: Global Production Strategy Design in Transportation Industry Sector

The final level of DPs in each pillar of the production strategy is considered the set of design guidelines that need to attain the desired effects and finally satisfy the previously defined CAs. In the following section all design guidelines are summarized and grouped in work packages.

4.1 Design Guidelines

Production Network Strategy

The Production Network Strategy consists of three different work packages. At first the **production site classification** is executed by the evaluation of the existing production landscape and the categorization of existing production sites towards their purpose (low-cost, closeness to customer, strategic, lead factory).

The next step is the **set-up of a production network**. It needs to be decided whether the entire production landscape or only several sites are linked within in a network. Regular and automated data exchange is installed between the sites of the production network and a superior instance to manage and control the network is established. Additionally, continuous reviews of the network and the remaining sites are performed to adapt the production according to changing internal and external influences.

The third package is about a **standard allocation process**. Once defined, it should ease the allocation of production orders (especially within the network) in terms of time and cost and should consider all relevant stakeholders.

Operations and Manufacturing Strategy
The Operations and Manufacturing Strategy is subdivided into three work packages that basically cover the fields of **lean management process tools** with the implementation of global and local process experts as well as the integration and documentation of standard production processes. Additionally, **best practice approaches** need to be pushed by global and local experts and incentives need to be set at sites to pursue group goals not site goals.

Know-How and Training Strategy
For the Know-How and Training Strategy two different dimensions need to be covered. The first one is to **enable management** employees to communicate globally and create a mutual understanding of the production processes and objectives. Additionally, know-how losses due to employee fluctuation need to be prevented. The second dimension is about the **production know-how** of manufacturing employees and the development of a training program. Furthermore, the employee training needs a proper location and set-up to be efficient.

Forecasting and Planning Strategy
The Forecasting and Planning Strategy is a two-dimensional workspace covering **regular exchanges** with sales and the set-up of a comprehensive production execution and planning. The continuous **tracking of capacities and capabilities** over all sites as well as the planning of future production activities starting from this general overview is core of this work package.

Process Technology Strategy
The Process Technology Strategy again is a two-dimensional workspace. At first the **basis for automation and innovation** needs to be created by comprehensive and structured master data care and the capturing of all essential production data. These are translated into a reporting system where first control relevant indicators are defined and then transferred into a KPI reporting. With an established KPI reporting, the basis for site comparisons is given and a benchmarking system can be implemented.

Production optimization takes place in three separate dimensions. The first one is the immediate instance responsible for trouble shooting of urgent requests. The second dimension is the standardization of processes and components, and the third dimension covers the product and process optimization.

4.2 Timeline

A big advantage of the AD method is the possibility to derive the timeline directly from the strategic content. The previously defined independence of the five pillars technically indicates, that the implementation of all five pillars can start simultaneously. By applying this approach to the use case, it becomes clear that (in this specific case) certain design elements need to be established prior to others. Within a strategic pillar the DMs clearly determine the order of work packages for a successful implementation.

Therefore, it is recommended to start with the first and second pillar of the production strategy. It is important to note that due to the independence axiom the work packages must be processed in the given order to avoid negative correlations.

5 Discussion and Conclusion

The developed model of a five-pillar global production strategy is a comprehensive and holistic approach that supports medium-sized companies with global production and a high ratio of manual production processes to formulate their individual production strategy. With little assumed fundamentals the developed work packages allow the user to build a company strategy from top management goals. Besides the consideration of the global production network and the integration of production sites to increase efficiency, also site internal topics are touched. The emphasis on standardized processes and clearly defined responsibilities is the heart of the operations and manufacturing. New to this approach is the emphasis on the people's education and development which is essential for highly manual manufacturing. The enabling of workforce on every company level to act in an international context as well as executing defined production processes with highest precision is key to successful operations. Forecasting and planning is essential to provide a comprehensive view in the global production strategy. Process technology in the sense of digitalization, automation, and innovation technically is not new to the theory of production strategies. However, multiple (especially medium-sized) companies struggle with the integration of innovative activities into the live production. This gap is closed by the two-dimensional design of this production strategy pillar, data base and competitive advantage. Bringing those together is a challenge for every company that tries to implement international structures and additionally requires a change in the mindset of people.

To use the methodology of axiomatic design within the production strategy development has several advantages. Besides the delivered top-down approach the approach forces the relevant stakeholders to consider all production related topics and interfaces and additionally implement the company strategy and mission. Furthermore, axiomatic design supports the strategy design workshops by a systematic guidance for workshops and interviews and provides a structure during the process. This additionally applies to the implementation phase after the strategy development as the sequencing of tasks or work packages is provided through the correct application of axiom 1 and the DM.

Even though the developed concept for a global production strategy considers different aspects and dimensions, the model still lacks the coverage of a few relevant topics. This is mainly driven by the fact that the model is developed for small and medium sized companies with no or only rough strategic elements in the production environment. It therefore focuses on laying a foundation for other strategic add-ons. It is considered most important to first create a proper basis and ensure the strategic readiness of the company before focusing on higher level design elements of the strategy. For this reason, concepts for agile manufacturing and implementation approaches for mobile production solutions are not considered in the developed model. Additionally, there is little focus on automation in production. Automation in production in only touched in the last pillar bringing in a future oriented perspective but this is not worked out in detail. Furthermore, the described and required organizational and processual changes require a high level of sensitivity with respect to a proper change management. Comprehensive communication about the desired changes and the managerial idea behind those changes is highly important for a successful implementation, as people are the main driver and therefore the most critical blockers of change.

References

1. Schuh, G., Kampker, A., Boos: Instrumente des strategischen Managements: Handbuch Produktion und Management 1. Springer-Verlag, Berlin, Heidelberg, 64 (2011)
2. Hayes, R. H., Gary, P.: Beyond_World_Class.pdf., pp. 77–78 (1994)
3. Miltenburg, J.: Setting manufacturing strategy for a factory-within-a-factory. Int. J. Prod. **113**, 307–323 (2008). https://doi.org/10.1016/j.ijpe.2007.09.001
4. Platts, K.W., et al.: Testing manufacturing strategy formulation processes. Int. J. Prod. Econ. **56**, 517–523 (1998). https://doi.org/10.1016/S0925-5273(97)00134-5
5. Spath, D., Ganschar, O., Gerlach, S., Hämmerle, M., Krause, T., Schlund, S.: Produktionsarbeit der Zukunft – Industrie 4.0. Fraunhofer Verlag, Stuttgart, 18 (2013)
6. Dombrowski, U., Krenkel, P. (eds.): Ganzheitliches produktions management: strategischer rahmen und operative umsetzung. Springer Berlin Heidelberg, Berlin, Heidelberg (2021). https://doi.org/10.1007/978-3-662-62452-4
7. Suh, N.P.: Axiomatic Design: Advances and Applications. Oxford University Press, New York (2009)
8. Park, G.-J.: Analytic Methods for Design Practice. Springer, London (2007)
9. Suh, N.P., Gebala, D.A.: Research in Engineering Design: An Application of Axiomatic Design. Springer-Verlag, New York, 150 (1992)
10. Cochran, D.S., Eversheim, W., Kubin, G., Sesterhenn, M.L.: The application of axiomatic design and lean management principles in the scope of production system segmentation. Int. J. Product. Res. **38**(6), 1377–1396 (2000). https://doi.org/10.1080/002075400188906
11. Suh, N.P.: Complexity Theory and Applications. Oxford University Press, New York (2005). https://doi.org/10.1093/oso/9780195178760.001.0001
12. Fiege, R.: Axiomatic Design. Eine Methode zur serviceorientierten Modellierung. Springer Gabler, Wiesbaden (2009). https://doi.org/10.1007/978-3-8349-8468-5
13. Rauch, E.: Konzept eines wandlungsfähigen und modularen Produktionssystems für Franchising-Modelle. Stuttgart (2013)

Enabling Vocational Students to Develop Axiomatic Design Application Knowledge

Marco Foggi[1,2](✉) and Erik Puik[1,2]

[1] HU University of Applied Sciences, Padualaan 99, 3584 CH Utrecht, The Netherlands
m.m.c.foggi@gmail.com

[2] Fontys University of Applied Sciences, de Rondom 1, 5612 AP Eindhoven, The Netherlands

Abstract. This paper discusses the use of Axiomatic Design (AD) as an instructional strategy for third-year engineering students at the Utrecht University of Applied Sciences. AD is a systematic approach to solving design challenges that is not commonly taught in Higher Vocational Education institutions, particularly in the Netherlands. The paper outlines how AD can be adapted to suit the needs of students with limited academic preparation and conceptualization ability using visual metaphors such as bookcases, shelves, and books to explain abstract concepts. The paper highlights the effectiveness of AD when tailored to the needs of vocational students in developing design knowledge and communication skills in engineering education. By incorporating practical examples and exercises, AD theory can be linked to disciplinary knowledge and motivational interests, mitigating preconceptions about the utility of learning the AD approach. Overall, this paper demonstrates the potential of AD as a valuable instructional strategy for engineering education, particularly for students who may struggle with traditional approaches to design problem-solving.

Keywords: Axiomatic Design · Instructional Strategy · Visualization

1 Introduction

Higher complexity, increasingly demanding quality requirements, and shorter time to market characterize the current technology market, putting pressure on determining error-free product and process designs as much as possible. Costly design errors hinder professional organizations from innovating and remaining competitive, or even surviving. Roughly 85% of problems with new products are due to poor design [1], resulting from insufficient or incorrect application of knowledge [2] to the Engineering Design Process (EDP). For this reason, technology companies expect highly skilled professionals capable of defining, developing, and managing the entire EDP from the early stage of conceptual design, as nearly 80% of a product's lifecycle costs are determined during this stage of the EDP [3]. Consequently, product development teams are challenged to deal with uncertainties and a scarcity of design information to achieve project yield at the lowest possible cost.

E. Puik et al. (Eds.): ICAD 2023, LNNS 849, pp. 190–206, 2024.
https://doi.org/10.1007/978-3-031-49920-3_13

The Utrecht University of Applied Sciences (HU) aimed to improve the preparation of its engineering students to address these challenges. Teaching experiences revealed that a lack of interdisciplinary teamwork and a critical attitude were the major causes of project flaws during the realization of product concepts within an educational project context. HU engineering students tended to rely on a 'trial-and-error' design approach. They were unable to clearly define and communicate the design intent due to a lack of design experience and inadequacy in acquiring and processing design information. Relevant design issues were overlooked, and flexibility to design changes and improvements was lost due to their tendency to 'immediately seek solutions' [4], rather than starting with a functional problem analysis and determining the most ap-propriate strategy to solve the problems [5]. The Axiomatic Design (AD) approach was identified as a possible solution to teach third-year HU engineering students the right skills to conceptualize around the Independence Axiom, structure and control the EDP systematically towards reliable design criteria. However, the students' insufficient abstract reasoning ability and skepticism about the usefulness of a "too theoretical" approach to design prevented the acceptance of the new instructional approach and its further implementation. As a solution, Puik designed and proposed the 'Billy Kasten' (BK) instructional method to make the AD approach accessible to his third-year Industrial, Mechanical, and Electrical & Industrial Automation HU engineering students. Visualization of complexities, providing clear instructions, and linking the AD to disciplinary knowledge and motivational interests are key to making the AD approach accessible to HU engineering students. Did it enable assimilation and application of the AD principles by third-year HU engineering students with different study backgrounds?

This paper assesses the effect of the BK instructional method on the assimilation of AD principles in third-year engineering students. Strengths, weaknesses, limitations, and proposed enhancements are discussed.

This paper is organized as follows: Sect. 2 explains the background of the BK Method. Section 3 explains the methodology of investigation and the case studies in which the BK Method is applied. Section 4 reports the results. Section 5 discusses the findings, while Sect. 6 draws conclusions. Finally, Sect. 7 suggests possible further improvements.

2 Background

Early recognition and prioritization of design, development, and manufacturing risks are key to making uncertainties manageable. Risks can be comprehensively characterized by gaining an understanding of what is known and not yet known about the design and by focusing on issues involving multiple domains and contexts. The BK method builds on the AD methodology and takes inspiration from the 'Constituent Roadmap of Product Design' (CRPS) framework of Puik and Ceglarek [6], particularly the 'Check Matrix' used to monitor knowledge progression without using scores and the V-Model managerial model included in the framework. Figure 1 shows how Development and Design Engineer's Knowledge progression is harmonized with the V-Model.

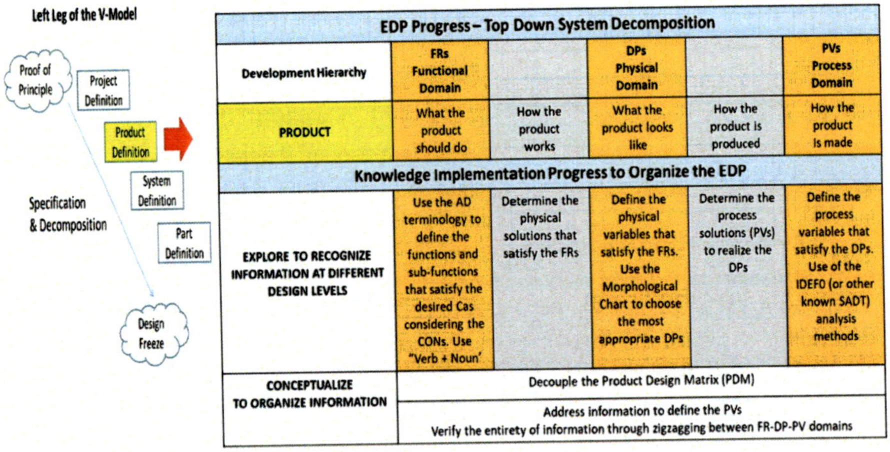

Fig. 1. Development and Design Engineer's Knowledge progression is harmonized with the V-Model.

The 'Billy Kasten' (BKs) are inspired by the 'Billy Bookcases™', a trademark product of IKEA. Four Billy Kasten respectively represent the AD Customer, Functional-, Physical-, and Process-domains containing, respectively, the Customer Attributes (CAs), Functional Requirements (FRs), Design Parameters (DPs), and Process Variables (PVs) information that determine the state of design. Application knowledge (information and decisions about design) is stored in 'books', placed in shelves at different hierarchical levels of the bookcases. In essence, the BKs contain information that reflects the design situation. Figure 2 shows how domains and hierarchies are organized according to the AD theory and to the BK method of Puik (2015).

The HU engineering students learn conceptualization, which involves understanding how to define the main FRs and DPs, satisfying Axiom 1 by decoupling the product design matrix. By doing so, they can achieve better product performance and flexibility to changes with the least investment. Essentially, the goal for HU engineering students is to fill the FRs and DPs bookcases completely with 'books' containing unequivocal FR-DP solutions to pass the conceptual phase of the EDP. Figure 3 shows the position of the engineer's Conceptual Application Knowledge.

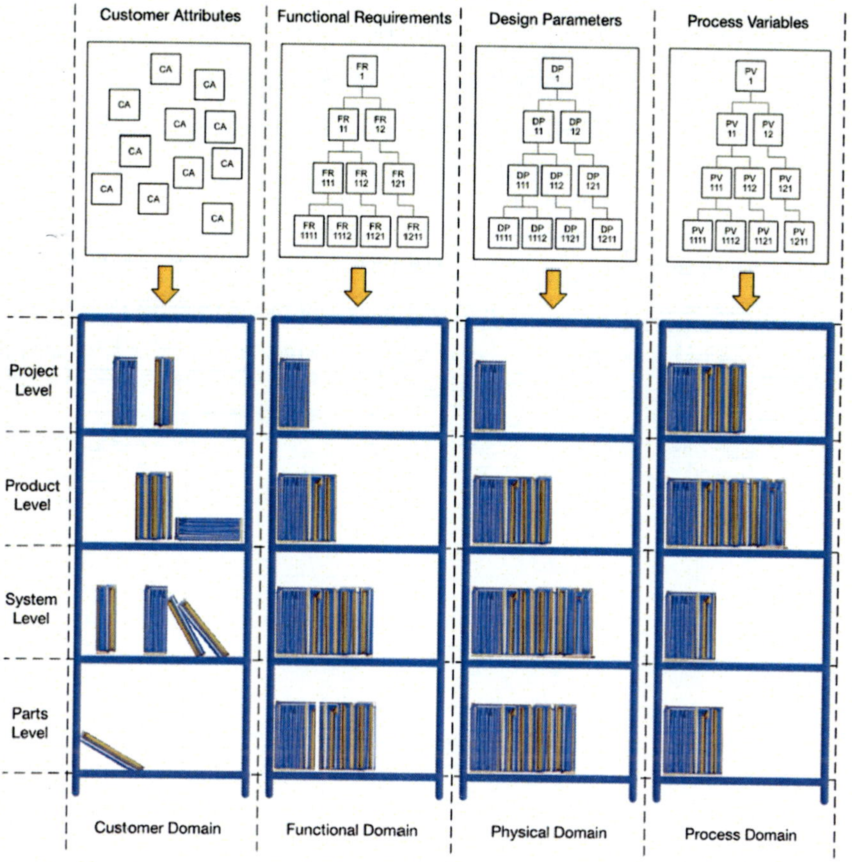

Fig. 2. Top: Domains and hierarchies according to AD theory. Bottom: Domains represented according to the BK Method. (Inspired and adapted from instructional material of Puik, 2015).

The logical relation between the FRs and the DPs is related to the Independence Axiom as defined by AD or 'Doing the right things'. Contextually, Business Engineering (TBK) students play the role of marketers, while Mechanical Engineering (WTB) and Electrical & Industrial Automation Engineering (ELT&IA) students are product designers or process engineers, the domain knowledge owners who respectively must answer the questions 'What is the customer looking for?', 'What does the system do?', and 'How is the system made?'. Systems engineering tools such as Quality Function Deployment (QFD), the Morphological Chart, and the Integrated Definition method (IDEF-0) help establish the design knowledge that assists engineering students in synthesizing solutions in the adjacent Billy Kast. In particular, the Morphological Matrix/Matrix is used to determine the DPs that satisfy the FRs. Finally, the mapping of the product design matrix is performed without the use of linear algebra. Zigzagging between the FRs and DPs, according to good practice in AD, ensures that the engineering students carry out functional decomposition in a correct manner. Figure 4 shows the framework of the engineer's application knowledge. Use of systems engineering methods during Concept

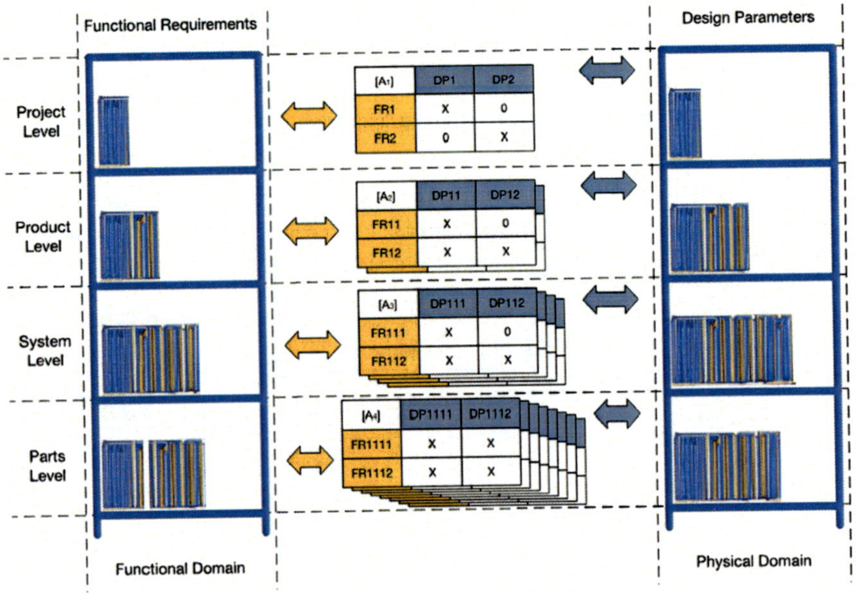

Fig. 3. Top: Position of the engineer's conceptual application knowledge.

Validation, Verification (while zigzagging and reverse zigzagging), and reference to the stage-gate process for project management.

In summary, the BK method educates engineering students to become selective with design information, from unorganized data until proof of concept. Forcing analysis compensates their tendency to start immediately working on solutions and enlarges room for creativity at the start of the project. 'Doing the right things' enables exploration noting what is not yet understood from the available data (unrecognized information as defined by Puik [6]) and cannot be synthesized yet to physical solutions. Conceptualizing consists of analysis and sorting design data from the pile of books to define FRs and DPs, and then combining and decoupling thus synthesizing to physical solutions according to functional precedence. Design Information are stored in 'books' that fill the shelves of the FR-DP bookcases. Figure 5 depicts the process of managing (design) information to organize the EDP.

3 Key Limitations

The BK instructional method does not consider what may limit or enhance the assimilation of new concepts in engineering students with different study backgrounds. For example, preconceptions [7] on the subject to be learnt harm the assimilation of new knowledge, whereas linking new theory to domain knowledge and motivational interests drive even skeptical and practical students to acceptation and learn deeply. Figure 6 shows that assimilation of new knowledge is an indirect process, often running through 'Preconceptions' that can be corrected by a tailored instructional method.

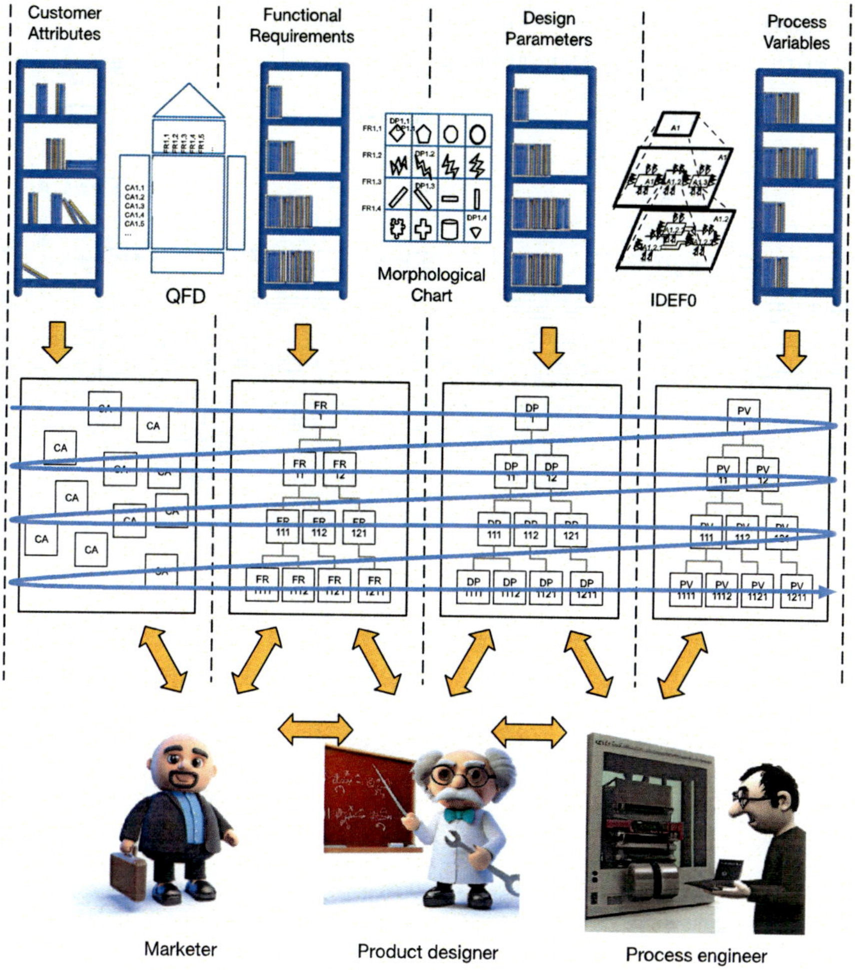

Fig. 4. Framework of the engineer's application knowledge (Inspired and adapted from Fig. 8.7 from Puik, 2016, p.203).

4 Methodology

There is no evidence covering the use of teaching and learning AD at higher vocational education institutions (in the Netherlands); hence, qualitative exploratory research is the most chosen research method at these institutions. Engineering design methods and educational models will be combined to measure the impact of the initial BK instructional strategy on the learning of the HU engineering students. This will also help determine successive verification and improvement strategies. Figure 7 shows the knowledge necessary to conduct the study. Engineering, instructional, cognitive, and pedagogical theories were consulted to design and implement the study.

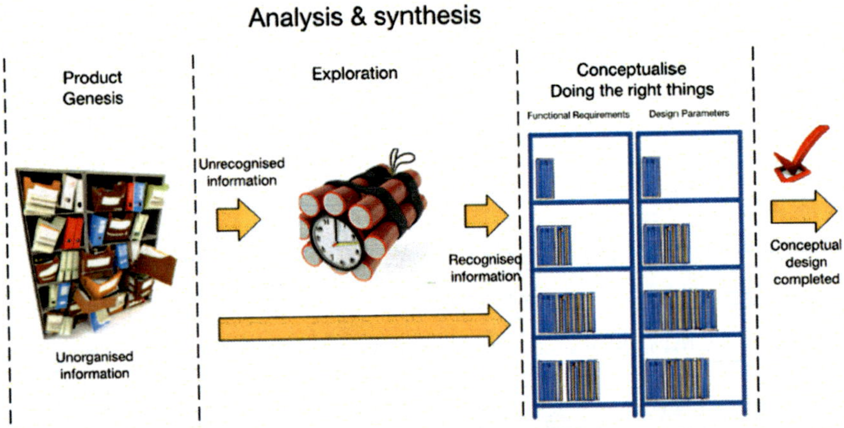

Fig. 5. Managing (design) information is organizing the EDP. The image on the left, titled 'Unorganized Information,' is an adaptation that includes the addition of the title 'Unorganized Information.

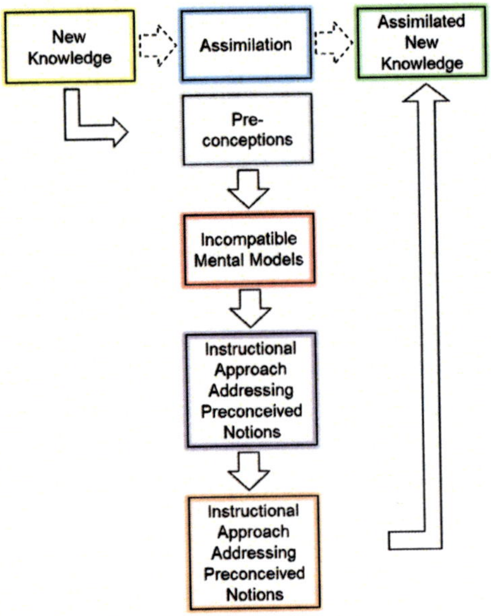

Fig. 6. Assimilation of new knowledge as an indirect process.

Accretion, restructuring, and tuning from the 'Three Modes of Learning' model [8, 9] are used to verify and implement knowledge assimilation incrementally. Three case studies (A, B, and C) are defined and used to identify the engineering students' motivations for and preconceptions about accepting the AD approach, and to deepen their knowledge. Case study A is about the integration of AD knowledge, case study B is about the differentiation of the engineering students' learning capabilities, and case

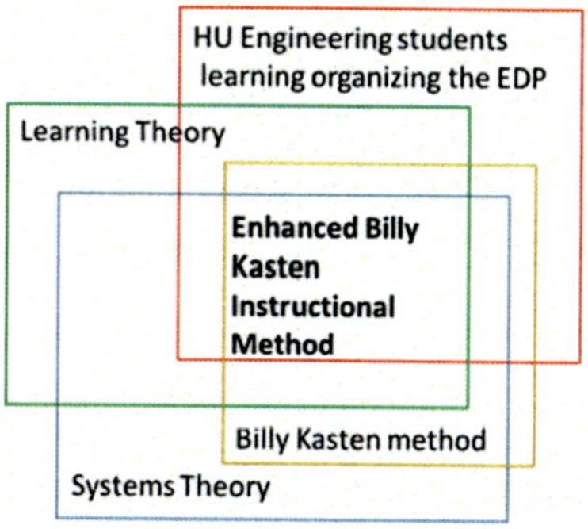

Fig. 7. Knowledge domains covered in the study and their intersections.

study C is about adapting the BK instructions to facilitate the acceptance and assimilation of the AD approach. Table 1 shows the Deductive and Inductive approaches used to define the research steps.

Table 1. Deductive and Inductive approaches used to define the research steps.

Reasoning Approaches for the Case Studies	Deductive	Inductive
A Integrate New Knowledge (Accretion)	The Billy Kasten method deepens understanding through practical examples and assignments."	Create instructional material to explain the "math behind".
B Differentiate Learning (Restructuring)	Explain matrix mathematics to accommodate diverse student learning capabilities.	Observe and evaluate barriers and enablers to assimilation. Generate instructional material to address learning differences and facilitate effective learning.
C Tailor Instructions to Study Area Commonalities (Tuning)	Explain the Axiomatic Design approach to decouple the Product Design Matrix (PDM) using mechanical and electrical components analogies.	Observe and evaluate the results, identifying patterns to validate the propositions. Formulate tentative hypotheses for recommendations and future research based on the findings.

5 Case Studies

Case Study A determines the initial educational situation, addressing the integration of new AD knowledge. It enables the detection of what unites and distinguishes the engineering students in using existing knowledge to learn and use the BK meth-od.

Explaining the AD theory behind the BK method and linking it to the practical examples used in the BK method enables knowledge integration. Students' behaviors and interactions between existing and new knowledge are observed without corrective interventions. This approach aims to bring the HU engineering students to the same level of preparation, fostering a sense of belonging to the 'same study group' and improving communication and cooperation.

Case Study B adds mathematical argumentation to the practical examples of Case Study A (i.e., linear algebra behind solving the design matrices). This evaluates how, and to what extent, existing knowledge could be influenced to accommodate complex information into the mental schemes of the HU engineering students. This also pro-vides insights into their ability to deal with complexities and different study preparations. Natural curiosity and motivational interests are further stimulated to learn and perform at a higher level. Stretching the comfort zone of the HU engineering students could provoke the externalization of skills and talents, uncover learning needs, and detect preconceptions.

Finally, Case Study C defines, implements, and tests a method that should facilitate the acceptance and assimilation of the AD approach. Domain knowledge is crucial to understand and appreciate the essence of the AD approach. For example, electro-mechanical component analogies link the existing knowledge of more practical and fact-oriented engineering students (WTB and ELT&IA) to the AD theory. Recalling the electro-mechanical component analogies and showing step-by-step how to solve the product design matrix of product concepts containing these components to the WTB and ELT&IA students could facilitate linking their existing knowledge and motivational interests to the AD theory. A better match with mental schemas would help the HU engineering students put forth effort towards the assimilation of new knowledge. In conclusion, the expectation is that at the end of Case Study C, the TBK, WTB, and ELT&IA engineering students will be brought to the same level of preparation, better supporting their effort towards the assimilation of new knowledge. Figure 8 shows the Research Framework designed and used to conduct the case studies.

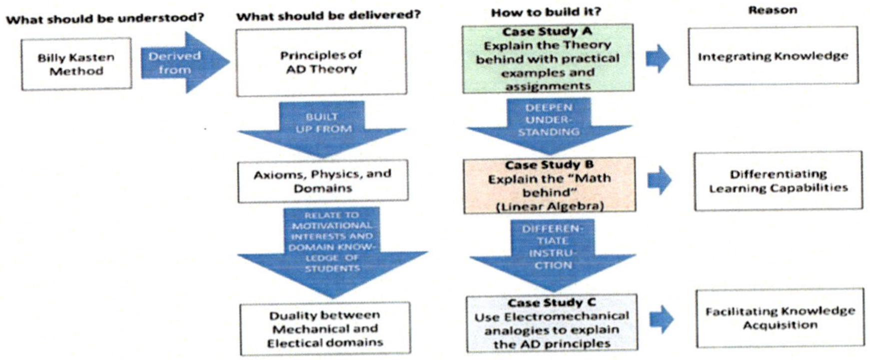

Fig. 8. Research Framework schematics as a reference to conduct the case studies.

Goals, research actions, and evaluations of results aiming at collecting data were structured as in Fig. 9 according to the Shewhart Cycle Plan, Do, Check, and Act (PDCA).

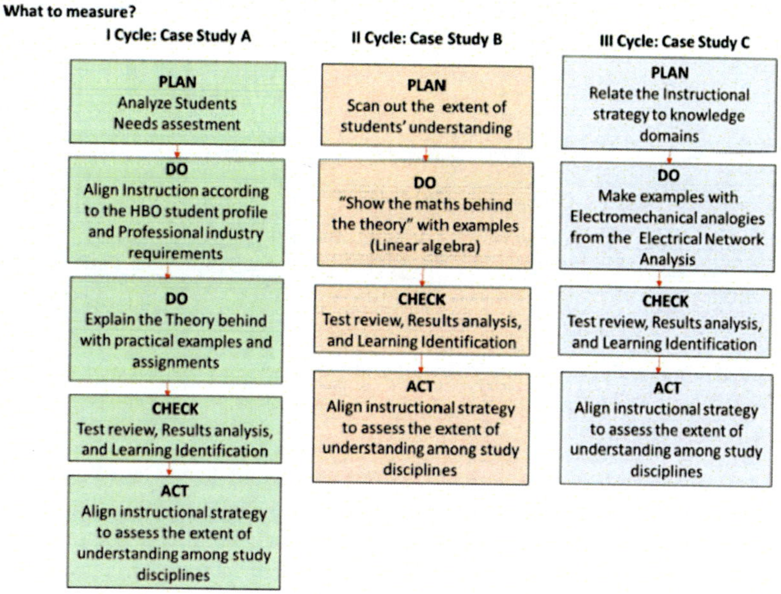

Fig. 9. PDCA applied to the Case Studies A, B, and C.

Intermediate learning achievements are assessed through open discussions with the students. Final learning achievements are assessed through structured interviews. The quality of the answers and the level of acceptation of the BK method are classified as shown in Tables 2 and 3.

Table 2. Classification concerning 'Learning'.

Classification	*Corresponding to*	*Qualification*	*Description*	*Evaluation*
--	No or Wrong Answer	Far from Expectations	The student gave no answer	The student does not meet the desired level of assimilation yet
-	Insufficient of Moderate Answer	Below Expectations	The student tried to give an answer. Large support provided by the interviewer (i.e researcher) to formulate the answer	
□	Sufficient Answer	Within Expectations	The student gave a substantially correct answer with a little support of the interviewer (i.e researcher)	The student meets the desired level of assimilation of the BK Method and the AD principles behind it
+	Good Answer	Above Expectations	The student gave an correct answer without support of the interviewer (i.e. researcher)	
++	Better than Expected	Strong Above the Expectations	The student gave an correct answer, adding interesting information to it.	The student exceeds the desired level of assimilation of the BK Method and the AD principles behind it

Table 3. Classification concerning 'Acceptation'.

Classification	*Statement Measure*
--	Strongly Disagree
-	Disagree
□	Neither Agree Nor Disagree
+	Agree
++	Strongly Agree

6 Results

6.1 Results from Case Studies

An overview of the instructional actions and results from Case Studies A, B, and C is summarized in Fig. 10. The + and –, and ± symbols in the 'Learning Identification' and 'Evaluation Learnings AxB and BxC' boxes indicate whether a particular group of students was successful, unsuccessful, or partly successful in achieving the learning goals. Figure 10 shows the approach used to conduct research.

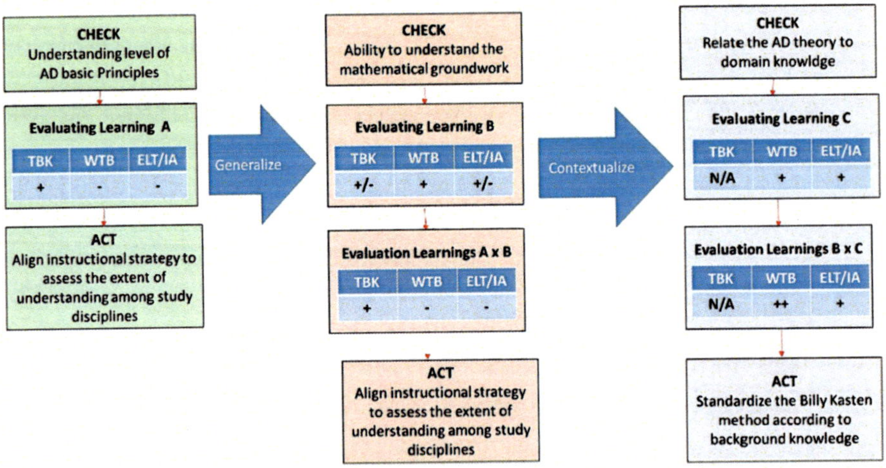

Fig. 10. Schematic of the research approach.

Case study A revealed that the TBK engineering students are the most receptive to the BK method (rated with +), whereas this is not the case for the WTB and ELT&IA engineering students (-). Actions (ACT) were designed and implemented for improvements. With case study B, it was detected that the WTB engineering students are the most eager to absorb mathematical explanations, whereas both the TBK and ELT&IA engineering students' abilities, in this case, are moderate (±). 'Evaluation Learnings AxB' summarizes the learning achievements of the HU engineering students after the case studies A and B. While the TBK engineering students have learned the most (+), both the WTB and ELT&IA students still have too little understanding about the added value of the BK approach. However, the WTB engineering students would have the potential to excel with the AD approach due to their ability to better deal with mathematics and physics. Finally, 'Evaluation Learnings BxC' summarizes the learning achievements of the HU engineering students after Case Study C. The TBK engineering students were practically excluded from this case study (N/A) because of their satisfactory AD knowledge achievements. With case study C, the use of electro-mechanical analogies had a positive impact on the assimilation of the AD approach for both the WTB and the ELT&IA engineering students (+). As a conclusive evaluation from this case study, the WTB engineering students benefited the most from the BK + method (+ +). Additionally, the ELT&IA students improved their assimilation skills (+).

6.2 Results from Final Interviews

The occurrence of answers linked to the perceived qualifications of the TBK, WTB, and ELT&IA engineering students is summarized in Fig. 11.

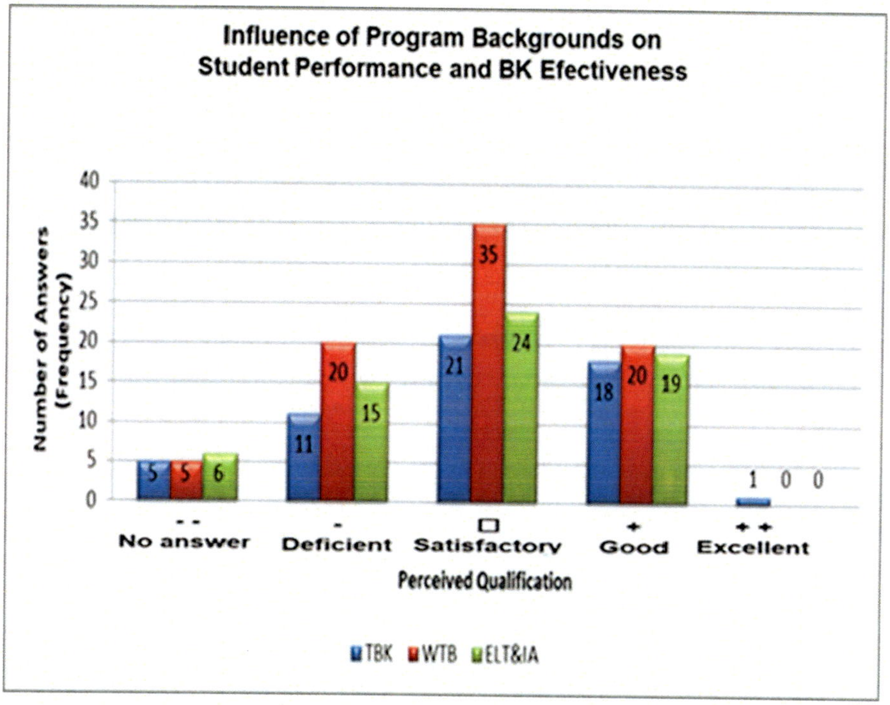

Fig. 11. Distribution of Overall Perceived Qualifications among TBK, WTB, and ELT&IA Engineering Students.

The histogram in Fig. 11 provides an overview of the answers to questions of levels 1, 2, and 3 for each group of the engineering students. The frequency distribution of the answers shows a peak centered on the Satisfactory perceived score (Sufficient, corresponding to Within Expectation). Considering the number of respondents; there were seven TBK students (28% of total participants, N = 25), ten WTB students (40% of total participants, N = 25), and eight ELT&IA engineering students (32% of total participants, N = 25). It is important to note that WTB engineering students had a dominant influence on the results due to their larger representation in the sample. However, when we focus solely on the results from the Satisfactory category and above, which includes Within Expectation, Good, and Excellent responses corresponding to Beyond Expectation, the performance of WTB engineering students stands out. They achieved 35 Within Expectation and 20 Above Expectation answers. In comparison, the TBK students had 21 Within Expectation and 18 Above Expectation answers, while ELT&IA students had 24 Within Expectation and 19 Above Expectation answers. The ELT&IA engineering students rank second, followed by TBK students. It is worth mentioning that only one TBK engineering student provided an answer of Above Expectation.

These findings highlight that when considering the Satisfactory category and above, WTB engineering students exhibited the highest performance levels, followed by ELT&IA students and then TBK students. It suggests that particularly WTB students have achieved a satisfactory level of qualification and have surpassed the expectations to a greater extent. It is important to interpret these results with caution, as the analysis is based on a limited sample size and specific context.

The engineering students' general opinion regarding whether the BK (BK +) method has contributed to better prepare them for managing development projects ranges from Neither Agree nor Disagree to Agree. On average, TBK engineering students were the most satisfied. Although WTB engineering students were reasonably satisfied as well (six 'Neither Agree nor Disagree' and three 'Agree'), one student expressed disagreement and could not fully appreciate the added value of the BK + approach for managing the EDP ('Disagree'). This student questioned the necessity of learning and using a new method when the existing tools were deemed sufficient.

In contrast, the ELT&IA engineering students were adequately satisfied with the BK + method for project management (i.e., seven Neither Agree nor Disagree and a single Agree). However, they generally do not perceive its utility for problem-solving within their domain. Overall, the BK instructional experience was enjoyed by all the engineering students, except for one WTB and one ELT engineering student (i.e., Disagree).

7 Discussion

Learning 'Why,' 'What,' and 'How' to secure one's own knowledge in the EDP and using it properly enables engineering students to develop an effective thinking process to efficiently arrive at results and communicate within and outside development teams with the same language. To achieve this, Puik's instructional strategy aims to accommodate the existing (domain) knowledge and average preparation level of HU engineering students and is creatively aided by the visualization of complexities. Firstly, it uses a 'Constructivist approach' [10]. Puik has designed and applied his instructional strategy while considering HU engineering students' skills, prior knowledge, and eagerness to learn. Secondly, it introduces the "Billy Kasten" metaphor as a playful novelty to trigger his students' attention towards dealing with de-sign information, relationships, and decoupling the design matrix. Doing so, Puik aims to enable retention, quick access, and recall of information in the long-term 'Working memory' [11]. Thirdly, it presents new knowledge (i.e., data) recalling known facts through pictures and anecdotes. New notions are immediately applied to the educational project context following the Problem-Based Learning methodology, often used in engineering. Finally, it aims to relieve the 'Cognitive Load' [12] facilitating quick association between new and existing knowledge [13]. As a result, deep learning is enabled, independently of scarce design experience and scientific preparation. The BK aims for a prompt assimilation of the (AD) knowledge into the 'working memory space' [14]. On the one hand, HU engineering students are guided to focus immediately on 'What' is essential to learn for achievements. On the other hand, the scientific nature of the method invites them to contextual and conceptual reasoning. As a result, HU engineering students become conscious that 'thinking' precedes 'doing'; therefore, they should not rush working on hardly existing "best solutions."

Initially, the BK Method did not completely fulfill its goal of bringing TBK, WTB, and ELT&IA HU engineering students to the same level of confidence with the AD approach. WTB and ELT&IA engineering students lag their TBK colleagues in the assimilation and use of the AD approach. A distinction should be made among students of different study specializations. A better match with mental schemas reduces the cognitive load and helps students put more effort towards the assimilation of new AD knowledge. Motivational interests, natural curiosity, perceiving strategies, and incentives to succeed should be included in the AD instructional strategy for vocational engineering students. While TBK engineering students have a natural aptitude for 'system approaches,' mostly due to their 'helicopter view' and the broad scope of their study specialism, 'Systems Engineering' and 'Project Management' preparatory courses would seem to suffice for assimilating the BK method in a reasonably short time. More practically oriented WTB and ELT&IA engineering students have better potentialities to learn in-depth and use the AD approach, mainly due to their predisposition to mathematics and physics. However, a difference was observed be-tween the students of the two specializations. The systems oriented WTB engineering students seemed to be advantaged over the ELT&IA engineering students. While the former were the most predisposed to thinking functionally, the latter had the strongest tendency to immediately look for solutions. This was probably due to the peculiarity of their study discipline, often focused on the detailed development of sub-parts of products, rather than whole systems. Not only did this turn out to be a disadvantage for the smooth assimilation of the AD concepts, but also for learning preconceptions on the BK method.

7.1 Strengths of the Enhanced BK Approach (BK+)

Considering and addressing students' preconceptions on the subject to be learnt is the key argument that enhances the BK method. Prejudices can be counteracted by exploiting the student's backgrounds, curiosity, motivational interests that can encourage the HU engineering students to get interested to deepen learning the AD approach. Rehearsing and deepening the AD theory behind the BK method through easy and practical examples is the way to open the discussion with and among all the students. The WTB and ELT&IA engineering students had the most potential to learn and use the AD approach, but initially had the least trust in its added value for solving their problems. The Enhanced BK Approach (BK+) made it possible to detect this and dedicate more attention to these students, influencing their preconceptions on the AD approach. The contextualization of the AD theory into the mechanical and electrical domains, and its concretization through measurable examples, involved and captured the attention of the WTB and ELT&IA engineering students in the most efficient manner. This brought all the HU engineering students to the same starting level of acquaintance and confidence, contributing to an overall positive effect on learning conceptualizing.

7.2 Weaknesses of the Enhanced BK Approach (BK+)

The BK+ Approach utilizes electro-mechanical analogies and is specifically tailored to WTB and ELT&IA engineering students in the context of this research. It is important to note that the applicability of this approach may be limited to students with similar study backgrounds and cannot be generalized to other disciplines. Additionally, the influence of introductory courses on the assimilation of AD principles could not be fully examined in this exploratory research. It is essential to acknowledge that this research, conducted within the HU context, is not intended to provide conclusive findings. The small sample size ($N = 25$) and the constraints on repeating the case studies in the same context or other learning environments prevent the generalization of results. Furthermore, the absence of comparative studies hinders the ability to establish the actual effectiveness of the approach. To gather more robust evidence and draw more accurate conclusions, a deeper understanding of the AD theory and more extensive engagement with the HU engineering students would have been beneficial.

8 Conclusion

Explaining the AD principles behind the BK method enables us to detect what unites and distinguishes the HU engineering students in using their existing knowledge to learn and use the BK method. The TBK engineering students showed immediate eagerness and preparedness to assimilate the BK method, while the WTB and ELT&IA engineering students required more explanation and addressed preconceptions about its utility. The WTB engineering students were the most receptive to mathematical explanations, while both the TBK and ELT&IA students had moderate abilities in this aspect. Although the TBK engineering students achieved a better overall understanding of the AD approach and its added value for learning conceptualization to better organize the EDP (Engineering Design Process), both the WTB and ELT&IA students lagged. Preconceptions of the WTB and ELT&IA HU engineering students about the utility of the BK method were addressed, enabling them to finally perceive the BK method as worthwhile and relevant to their learning goals.

In short, the study's final evaluation shows that the BK method had the greatest impact on WTB engineering students, leading to significant improvements in assimilation skills. ELT&IA students also made progress, while TBK students remained consistent.

This successful integration of disciplinary knowledge and AD principles through the BK method promoted learning and competence among HU engineering students. Additionally, no significant performance differences were found between WTB and ELT&IA students, highlighting the positive impact of the enhanced BK method on assimilating the AD approach. Overall, the BK method effectively met the learning goals and brought all students to a similar level of preparation.

9 Further Improvements

This research holds significance as it represents the first study focused on teaching AD to vocational engineering students. However, it is important to acknowledge the limitations of the study, such as its narrow scope and small sample size. Further research is warranted to enhance our understanding in this area. This could involve expanding the participant pool, implementing pre and post-tests, and conducting statistical analyses to assess the instructional strategy's impact. To gain deeper in-sights, providing students with more time for reflection and a better understanding of the AD theory would be beneficial. Additionally, establishing a control group and a test group at the onset of the study, conducting interviews throughout with recorded answers, and involving an assistant to facilitate data collection would enhance the evaluation process. Moreover, to better prepare engineering students for the AD approach, it is recommended to review the engineering programs at the HU. One suggestion would be to integrate the BK+ Approach method into the Systems Engineering subject, ensuring its inclusion for all engineering students starting from their first year.

References

1. Ullman, D.G.: The Mechanical Design Process. McGraw-Hill Science/Engineering/Math, New York (2015)
2. Puik, E.: Application of axiomatic design for agile product development. MATEC Web Conf. **223**, 01004 (2018). https://doi.org/10.1051/mateconf/201822301004
3. Fredriksson, B.: Holistic system engineering in product development, The Saab-Scania Griffin, p. 95 (1994)
4. Stojcevski, A.: Learning to solve 'design problems' in engineering education. Washington Accord International Engineering Alliance (2014). http://www.Ieagreements.org/assets/Uploads/Documents/History/25YearsWashingtonAccord-A5booklet-FINAL.pdf. Accessed 05 Mar 2017
5. Woline, T., VanDerZanden, A.M.: Student perceptions of problem-solving skills and evaluation of a web-based environment for case-study work in landscape horticulture. HortTechnology **20**(4), 820–824 (2010). https://doi.org/10.21273/HORTTECH.20.4.820
6. Puik, E., van Duijn, J., Ceglarek, D.: Guidelines for application of the constituent roadmap of product design based on axiomatic design. MATEC Web Conf. **127**, 01013 (2017). https://doi.org/10.1051/matecconf/201712701013
7. Vosniadou, S., Brewer, W.F.: The concept of the earth's shape: A study of conceptual change in childhood (Center for the Study of Reading Technical report No. 467) (1989)

8. Rumelhart, D., Norman, D.: Accretion, tuning and restructuring: Three modes of learning. In: Cotton, J.W., Klatzky, R. (eds.), Semantic Factors in Cognition. Erlbaum. As cited in Semantic Scholar Homepage (1978). https://www.semanticscholar.org/paper/Accretion%2C-tuning-and-restructuring%3A-Three-modes-of-Rumelhart-Norman/569c9194c7fad64d5fc15deb5ea65ccb0ddd015e. Accessed 05 Mar 2017
9. Dagobert Soergel's Homepage (n.d.). UBLIS571DS-06. 1a-1Reading10RumelhartAccrtionTuningAndRestructuring.pdf. Retrieved from https://www.dsoergel.com/UBLIS571DS-06.1a-1Reading10RumelhartAccretionTuningAndRestructuring.pdf. Accessed 05 Mar 2017
10. Rumelhart, D., Norman, D.: Analogical processes in learning. In: Anderson, J.R. (ed.) Cognitive Skills and Their Acquisition. Erlbaum, Hillsdale, NJ (1981)
11. Worth, R.: Quantitative problem solving in science: cognitive factors and directions for practice. Chin. Univ. Educ. J. **26**(1), 13–29 (1998)
12. Sweller, J.: Cognitive load during problem solving: effects on learning. Cogn. Sci. **12**, 257–285 (1988). Cognitive Load. http://www.instructionaldesign.org/theories/cognitive-load/. Accessed 05 Mar 2017
13. Derry, S.J.: Cognitive schema theory in the constructivist debate. Educ. Psychol. **31**(3–4), 163–174 (1996)
14. Anderson, J.R.: The architecture of cognition, vol. 5, pp. 231–255. Psychology Press, London (1996)

Connecting the V-Model and Axiomatic Design; An Analysis How Systems Engineering Methodologies Relate

Erik Puik(✉) and Michiel van Osch

Fontys University of Applied Sciences Eindhoven, Eindhoven, The Netherlands
erik.puik@fontys.nl

Abstract. V-Model and Axiomatic Design are two proven systems engineering methodologies to support the design process of complex systems. Both methodologies originated in the late seventies. The V-Model is the more successful of the two, partly due to its simplicity and straightforwardness. AD is more powerful but also more complex. AD has a clear way of requirements management in domains that represent the customer needs, system functionality, physical execution, and manufacturing processes. Both models are generally seen as independent from each other. Also, in teaching courses they are considered two separate methodologies. In this paper it is investigated what the overleap of the V-Model and Axiomatic Design is, and if there can be defined a continuous learning path for (professional) teaching from V-Model to Axiomatic Design. The overlap appears to be quite substantial, the execution process of the V-Model can be seamlessly mapped on the process of Axiomatic Design, making the V-Model mainly a subset of Axiomatic Design. Axiomatic Design contributes to the V-Model. It improves understanding of the V-Model because Axiomatic Design clarifies artifacts that are not consciously embedded in the v-model but are still present.

Keywords: V-Model · V-Modell XT · Axiomatic Design · Systems Engineering

1 Introduction

The V-Model originated as a systems-engineering tool to support the process of software development [1, 2]. Over time it has been transformed to optimize its functionality for product development in general. Nowadays, the V-Model is the most widely applied systems-engineering tool that, among other things, makes part of the INCOSE and NASA systems-development methods [3, 4].

Axiomatic Design (AD) is a design methodology that assists the designer during the various stages and activities of the design process [5]. The goal is to enable the designer being able to make the right decisions along the product development process. AD strongly focusses on the knowledge of the designer as it enables understanding of the difficulties in a product design [6]. Unfortunate design decisions will be recognized in the early stage when the Independence Axiom is applied. This enables appropriate measures before errors escalate and delay a design project.

E. Puik et al. (Eds.): ICAD 2023, LNNS 849, pp. 207–220, 2024.
https://doi.org/10.1007/978-3-031-49920-3_14

Both methods, V-Model and AD, are valuable to the knowledge base of the modern engineer and are often part of a modern curriculum for engineering students, however, they are considered to be different systems engineering methodologies. This paper examines whether this is a correct assumption, how the two methodologies relate further, and if AD, being the more complex but also more extensive methodology, might be considered to be a more complete version of the V-Model.

This paper is structured as follows. Section 2 gives an overview of current literature. Section 3 defines the exact research scope of this paper. The research methodology is described in Sect. 4, and in Sect. 5 the actual comparison of V-Model and AD takes place. The discussion in Sect. 6 reflects on the results of that comparison.

2 Background

There is a reasonable amount of literature about the comparison of V-model and AD. A search for papers on scholar.google.com, that have 'V-Model' in the title, delivers 965 results (same result as 'V Model'). 'Axiomatic Design' in the title gives 1710 results. Both 'V-Model' and 'Axiomatic Design' gives only a single hit (Tarenskeen et al., about IT architecture [7]).

When search words 'V-Model' and 'Axiomatic Design' are applied, that not necessarily in the title but in the whole document, 320 results are found. These 320 documents have been used as a starting point. Most of these papers are only moderately relevant since they just mention the V-Model or AD without focussing on the meaning of the systems engineering process. About 30 papers go beyond just naming them and indeed compare V-Model and AD. These papers have been categorized in Table 1.

Table 1. Literature on V-Model and AD in relation to this research

Issues Considered	Addressed By	Related Work
Papers that compare properties of AD and V-Model in some way	Reference or relevant work about V-Model	Rook [1], V-Modell 97 [8], Forsberg & Mooz [9], V-Modell XT [10], Gausemeier & Moehringer [11, 12], Graessler & Henze [13]
	Reference work AD	Suh [4], Suh [5], Suh [13]
	Focusing on a specific design problem	Thomas et al. [15], Ognjanovic et al. [16], Rolli et al. [17], Fardelas [18],
	Investigate similarities/differences of V-Model and AD	Tarenskeen et al. [7], Mlambo et al. [19], Balkhair [20]

(*continued*)

Table 1. (*continued*)

Issues Considered	Addressed By	Related Work
	Specific comparison of V-Model and AD	Suh & Do [21], Do & Suh [22], Chung & Suh [23], Chung [24], Puik et al. [25], Malaek et al. [26], Puik & Ceglarek [27], Thomas & Mantri [28, 29], Xinyu [30]
	Investigate whether AD can be an extension of V-Model	*Covered in this paper*

From these papers, the strengths and limitations of the V-Model and AD were inventoried. These strengths and limitations are shown in Table 2.

Table 2. Comparison of strengths of V-Model and AD

	V-Model	AD
Strengths	• Supports system decomposition by breaking down complex systems in smaller, better manageable systems • It enables 'Gating' between hierarchical layers to prevent changing requirements to affect parallel activities • Functional requirements are fed forward to the later stages where modular subparts, parts, and systems are integrated • It structures integration and testing in reverse order of decomposition	• Systematic approach: A structured approach to design that ensures all necessary functions and constraints are met • Design optimization: Eliminates conflicts between functional FRs resulting in a more efficient design • Encourages interdisciplinary collaboration among engineers and designers from different fields
Limitations	• The V-Model causes quite some overhead in smaller projects • Poor support for requirements management like the definition of Functional Requirements (FRs) and Design Parameters (DPs) • Developed for software, no support for manufacturing processes	• AD can be complex and time-consuming, especially when design matrices are growing • As any structured approach, AD may limit creativity • Selection of design parameters and functional requirements may involve subjective judgments

This comparison indicates that both models have similar strengths like the structured approach and enforcement of a sound requirements definition. There also seem to be joint weaknesses, e.g. the overhead caused for their application, especially when managing larger projects. In any case, a closer comparison of the two methods seems worthwhile.

3 Scope of Research (GAP)

The V-Model is one of the most common and easy to understand SE-models. When teaching methodologies for Systems Engineering (or Engineering Design) the V-Model is typically the first methodology to address. It is considerably more widespread, than AD, mainly due to its adoption by governments and military organizations that have invested heavily in its development, documentation, and dissemination.

AD was developed by Nam Suh at the Massachusetts Institute of Technology, recorded in several books, and mainly adopted by academia. The latter is mainly caused by the relatively steep learning curve of AD that requires a somewhat more in-depth understanding of the design process.

3.1 Current Situation

In the current situation, V-Model and AD are typically taught as different system engineering methodologies. The relationship between V-Model and AD remains minor, they are two separate methods.

3.2 Desired Situation

Ideally, AD is an extension of the V model. Students who understand the V-Model can expand their body of knowledge and skills with AD. AD has an extended, and possibly steeper learning curve, but also delivers more added value. The two methods reinforce each other, and novice product developers use the right method for their problem cases.

3.3 Key Limitations (Research GAP)

i. V-Model and AD are not recognized as extensions of each other. They are seen as independent methods.
ii. Because they are not considered extensions of each other, there is no growth path from V-Model to AD.
iii. Many novice product developers stop expanding their knowledge after the V-Model and remain deprived of AD.

Given these key limitations, the main research question of this paper was defined as:

> *How can we bridge the gap between the V-Model and AD when teaching systems engineering methods so that students can logically and intuitively progress from V-Model to AD?*

4 Research Methodology

4.1 Methodology for This Research

The investigation initially focuses on the similarities between V-Model and AD. Subsequently, it is examined whether the two methodologies can be placed in line with each other. Based on apparent similarities, it is investigated in which both models have unique properties. Thus, an impression is obtained of:

- The overlap of both methods.
- Unique features of the V-Model.
- Unique properties of AD.

4.2 Chronological Comparison of Activities in the V-Model and AD

During the design process, 'the designer' will be guided through the development process. Note that 'the designer' may be single or a group of designers of different professional disciplines. In either way, V-Model and AD will advise the designer when to apply one of the following activities:

- Functional specification of the product to be developed.
- Decomposition of project, product, systems, and sub-systems.
- Integration of subsystems to products and services.
- Testing of the parts, subsystems, product, and factory- and site-acceptance tests.

The research methodology that will be applied here is to monitor the proposed activities of both models in chronological order and compare them afterwards. The aim is to find similarities in the execution of the models in the area of the above subjects. In this way, congruence and deviations between the models become clear.

At first, the V-Model will be chronologically analyzed and secondly this will be done for AD. Afterwards we will reflect on these results which provides insight into how V-Model and AD support the designer.

4.3 How this Methodology Addresses the Key Limitations (Research GAP)

This research methodology will address the in paragraph 3.3 mentioned key limitations because:

i. If sensible overlap between both methods is found it indicates that V-Model and AD should not be seen as independent models.
ii. The overlapping segment of the methodologies can perhaps serve as a connecting factor with which a continuous development path can be realized.
iii. A continuous development path is challenging for students. In enables them to look beyond the V-Model and develop a broader view on systems engineering.

5 Case: Investigation of the Commonalities and Differences Between V-Model and AD

There is no such thing as a single V-model. Although the basic concept is similar, many different versions have been developed over the years. In this research, the German V-Modell XT will be applied for comparison with AD. V-Modell XT is a well thought out and meticulously documented version of the general V-Model [10]. It emphasizes the importance of testing in product and software development and ensures that testing activities are performed in parallel with development activities. The result is a more structured and systematic approach to development that prioritizes quality and reliability. The V-Modell XT will be applied in this paper as 'the Standard' for the V-Model.

5.1 Analysis of Sequential Operations of V-Modell XT

The first analysis is performed on the V-Modell XT. This model plots the hierarchical analysis vertically as a function of elapsed running time of the project. The horizontal timeline is dynamic, which means that time is not necessarily plotted linearly on the horizontal axis [31]. The hierarchical analysis consists of two parts. First, decomposition is applied, and a project is downwards broken down in subsequently products, systems, subsystems, and parts. Secondly, the direction is reversed, and decomposition replaced by upward integration. As such, subsystems, systems, and products are subsequently built together by the assembly of parts. The V-Modell XT and the axes as explained are shown in Fig. 1.

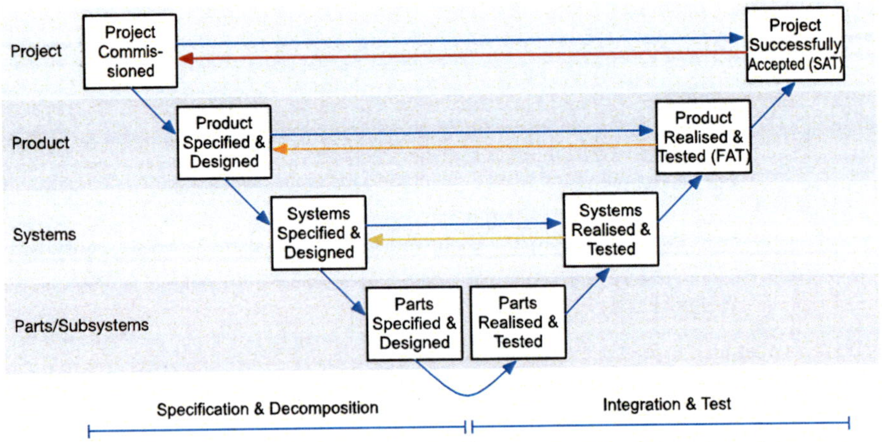

Fig. 1. V-Modell XT. The horizontal axis is a timeline that plots the order of activities to be executed according to the model. The vertical axis represents hierarchical decomposition in downward direction and integration upwards.

The V-Model is generally a gating model, which means that if a gate (a block in the schematic diagram) is completed it is not meant to be reopened and project choices made thus far are final. This enables different teams to work concurrently on subparts of the project in relative isolation Because gates that are closed are not reopened, they can be certain that there is no change of project context. Note that if the designers fail at completing a following gate in the project, which means that the project is stuck, escalation is inevitable. In this case, the gates have to be opened anyway and the project falls back to an earlier stage. In this case all parallel teams in that branch need to be informed about the changes and reconsider their work.

The arrows from left to right indicate that specifications that were defined in the conceptual phase of the project (also called 'left leg') are forwarded to the integration stages which guarantees that tests are performed in the right context. The arrows in opposite direction should ideally not be used because that would indicate that integration and testing has failed, and the project needs to be repaired. The yellow arrow indicated a minor failure (bottom arrow pointing to the left), the orange arrow (middle pointing left) is more serious, and the red (upper pointing left) arrow would indicate a disastrous failure. An exception to this rule would be the case that an iterative development cycle is foreseen. The first iteration, with limited functionality, would lead to the development of the total system in a second iteration. Also, in this situation the timeline continues from left to right, but there are two or more 'Vs' chronologically placed behind each other.

5.2 Analysis of Sequential Operations of AD

The analysis of sequential execution steps for AD is more complex than that of the V-Modell XT. Firstly, AD applies a more advanced way of system specification compared to the V-Model. Functional Requirements (FRs), Design Parameters (DPs), and Process Variables (PVs) are carefully brought together in harmony by optimization of the 'Design Structure Matrix'- and 'Process Structure Matrix'. To complete this, AD uses a procedure called zigzagging [5] and reverse zigzagging [25], The zigzagging process is shown in Fig. 2.

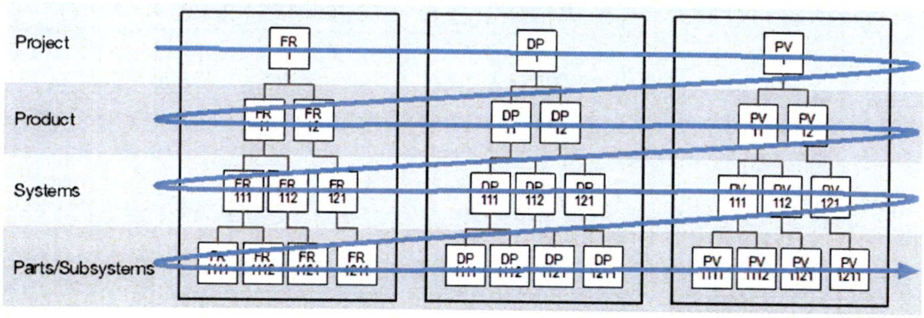

Fig. 2. Zigzagging through the domains in AD.

The vertical axis shows the level of decomposition and is very comparable to that of the V-Model. The hierarchical level on this axis can in principle like the V-Model be chosen freely. In this case the levels are copied from the V-Modell XT.

As shown in Fig. 2, zigzagging starts at the highest hierarchical in the physical domain level and 'zigs' right to enforce alignment of FR_1, DP_1, and PV_1. After completion of that level, it 'zags' back to the next lower hierarchical level to align respectively FR_{11}, DP_{11}, PV_{11}, and FR_{12}, DP_{12}, and PV_{12}. This process is repeated three times till all four levels have been determined. AD may be seen as a gating model but with some flexibility. If the process of zigzagging gets stuck unexpectedly, AD has defined rules how to act. This is usually done by escalating the hierarchical level one step upwards, fix the problem in the design and restart zigzagging from that point [5].

The process of zigzagging is combined the completion of the Design- and Process Structure Matrices of AD (|A| & |B|) and if necessary, the decoupling of these matrices.

Deviating from the V-Model, AD does not plot the timeline on the horizontal axis. Time moves along the path of the arrow that characterizes the zigzag process but here again it is not a linear relationship since the number of FRs, DPs, and PVs gets larger at the bottom of the hierarchy (dynamic timeline).

When arrived at the bottom the process of zigzagging stops and reverse zigzagging is started. The Structure Matrices |A| and |B| should be decoupled at this point.

Successively, the path of reverse zigzagging is exact opposite of the initial zigzagging process. This means that reduction of the information content, or more practically put, making the relations between PVs, DPs, and FRs robust, starts with the relation between PV_{1211} and DP_{1211} followed by that of DP_{1211} and FR_{1211}. Note that zigzagging always uses the order FR → DP → PV and reverse zigzagging PV → DP → FR (Fig. 3).

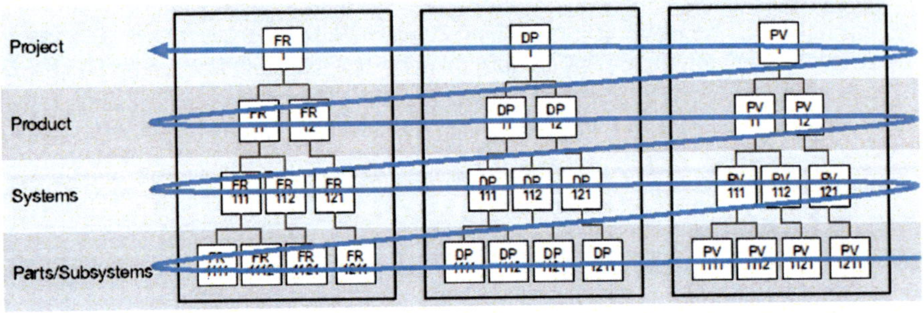

Fig. 3. Reverse zigzagging is exact opposite of zigzagging.

5.3 Comparison of Sequential Operations of V-Modell XT and AD

When the activities of the zigzagging process in AD are plotted chronologically, as shown in Fig. 4, a pattern emerges that resembles the left leg of the V-Modell XT.

And if reverse zigzagging is added, this pattern emerges further and adds the right leg of the V-Modell XT. This is shown in Fig. 5.

Specifications for testing in the right leg were defined during the conceptual definition when the process of zigzagging (down) took place and are used again for testing during the process of reversed zigzagging (up). These specifications are forwarded from the left to the right leg. Note that these specifications do not only concern the FRs but also the DPs and the PVs. These specifications are indicated with the triple blue arrows.

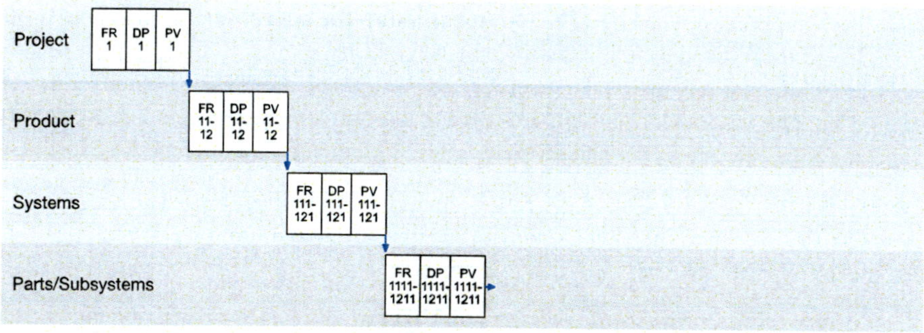

Fig. 4. Zigzagging through the domains in AD.

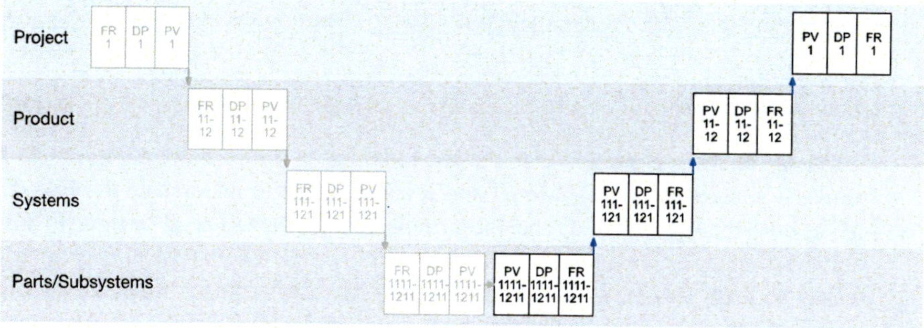

Fig. 5. Reverse zigzagging through the domains in AD added to Fig. 4.

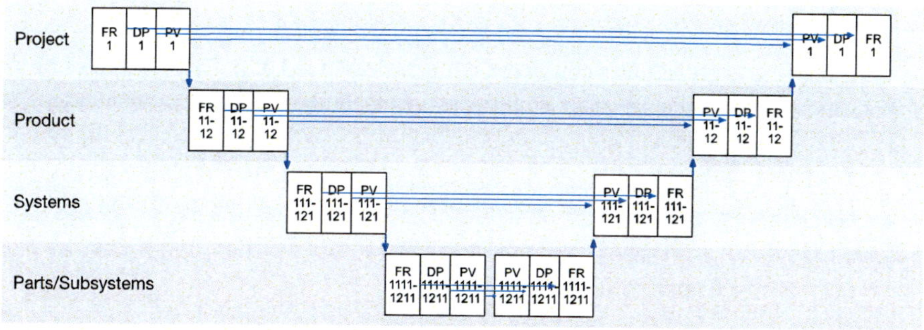

Fig. 6. Like the V-Model, the requirements are forwarded to the integration stages in AD

6 Discussion

6.1 Strengths

If the chronological steps of V-Model and AD, as analysed in Sect. 5 are lined up, V-Model and AD show strong resemblance. Both guide the designer in the exact same way through the design process. Both models also provide insight in project hierarchy and both support forwarding of requirements from the conceptual stage of the project (left leg of the V-Model) to the testing stages (right leg).

Note that AD uses a considerably more sophisticated way to specify FRs, DPs, and PVs. The V-Model, originally invented to support software developments does focus on the FRs, however, DPs and PVs are not really embedded in the model. Note also, when applying the V-Model, the independence of the FRs and the DPs is not automatically maintained, as AD dictates when satisfying the Independence Axiom. The Information Axiom, not explicitly investigated here, seems to work for the V-Model comparably to AD because testing the rigidity of the system does also reduce the information content of the system. This would require more investigations for exhaustive understanding. It may be concluded that AD is indeed the more comprehensive methodology compared to V-Modell XT.

The V-Model is a gating model. AD is basically the same. When the process of zigzagging fails, the decomposition returns to the previous hierarchical level [5].

6.2 Weaknesses

AD may be more complete than the V-Model but is it also more complex. Its learning curve is steeper than the V-Model and it takes longer to understand the FRs, DPs, and PVs, how they are related, and the principle of Information in design. In this sense, the more complex model is the more powerful model. The V-Model is simpler but also easier to learn and may offer sufficient performance in most situations. With its more straightforward structure it has proven to perform for large projects and because of that it was embraced by NASA and INCOSE [3, 4].

6.3 Limitations

In this research AD was compared with the V-Modell XT. This version of the V-Model was selected because of its soundness in performance and documentation. The more traditional implementations of the model may not match AD as well as the comparison in this paper. But even if the blocks in both models do not match completely, the underlying approaches are quite comparable. Hierarchy and sequentiality are similarly addressed.

6.4 Other Considerations

As described in Sect. 2, some comparisons between V-Model and AD have been made before. A specifically important one was described by Suh and Do [21, 22] and is shown in Fig. 7. In this version of the V-Model, which is a modified version of El-Haiks interpretation from the perspective of AD [33], is shown that is applied for object-oriented software programming. It is peculiar that FRs and DPs are written underneath each other ('Define FRs' and 'Map to DPs'). This implies that functionality, system design, and realization are developed as decomposition evolves while going down in the V-Model. This is inconsistent with the findings of this paper (Fig. 6) as it appears that FRs, DPs, and PVs are always grouped at the same hierarchical level. This difference is seen in more versions of the V-Model, e.g. the version that Boeing applies as basis for

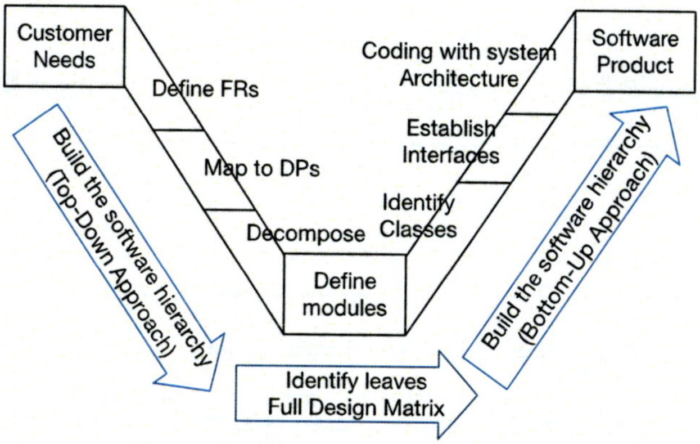

Fig. 7. V-Model and AD according to Do and Suh [22].

its 'Diamond Model' [34]. However, many versions as inventoried by Graessler et al. [35] comply to the findings in this paper.

These differences seem not infrequently the consequence of a lack of unambiguous definition of the different types of requirements. The fact that, in many of these models, there is a single moment of definition of requirements at the top of the left leg of the V, is at the expense of the quality of the system's analysis in the rest of the V-Model. As a remedy for this problem, an interesting approach is given by Graessler [36] that promotes continuous requirements elicitation and management through the whole V-Model. This is shown in Fig. 8.

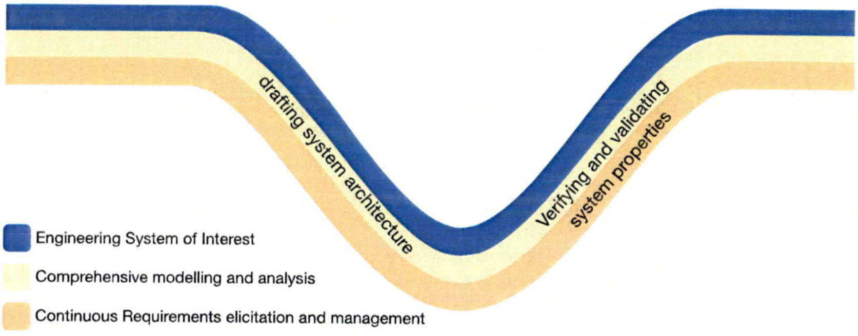

Fig. 8. Enhanced V-Model according to Graessler [36].

This version of the V-Model shows great overlap with the requirements management and zigzagging processes of AD.

6.5 Future Work

As explained in Sect. 6.1, this research could be expanded by an investigation of the role of the axioms in AD and how their functionality is embedded in the V-Model.

7 Conclusion

In the project, the V-Model was compared to AD by successively following the activities proposed by both models in chronological order. It appears that the V-Modell XT and Axiomatic Design have a lot in common. Strengths of the V-Model is its simplicity that makes it suitable for larger projects in larger organisations. Drawback of the V-Model is the unambiguous definition of the different types of specifications. This is where Axiomatic Design comes forward quite strongly. The latter is the more comprehensive of the two methodologies, especially considering the requirements management, with the clear definitions of FRs, DPs, and PVs, and the process of zigzagging. The higher level of complexity of Axiomatic Design is its weaker point. For educational purposes, the V-Model and Axiomatic Design can be taught as matching methodologies that enforce each other.

Acknowledgements. This research was supported by the Fontys University of Applied Sciences Eindhoven, and the project 'NXT GEN SE, funded by the Dutch government.

References

1. Boehm, B.W.: Verifying and validating software requirements and design specifications. IEEE Softw. **1**(1), 75–88 (1984)
2. Rook, P.: Controlling Software Projects. Softw. Eng. J. **7**, 1–16 (1986)
3. Walden, D., et al.: INCOSE Systems Engineering Handbook: A Guide for System Life Cycle Processes and Activities, 4th Edn. INCOSE (2015)
4. NASA: NASA Systems Engineering Handbook - NASA SP-2016–6105 Rev2: NASA (2016)
5. Suh, N.P.: The Principles of Design. Oxford Series on Advanced Manufacturing (1990)
6. Suh, Nam P.: Complexity Theory and Applications. Oxford University Press, New York (2005). https://doi.org/10.1093/oso/9780195178760.001.0001
7. Tarenskeen, D., Bakker, R., Joosten, S.: Applying the V model and axiomatic design in the domain of IT architecture practice. Procedia CIRP **34**, 263–268 (2015). https://doi.org/10.1016/j.procir.2015.07.035
8. Dröschel, W., Heuser, W., Midderhoff, R.: Inkrementelle und objektorientierte Vorgehensweisen mit dem V-Modell 1997 (1st edn.). De Gruyter Oldenbourg (1997)
9. Forsberg, K., Mooz, H.: System Engineering for Faster, Cheaper, Better. Center for Systems Management, Inc. (1998)
10. Höhn, R., Höppner, S.: Das V-Modell XT: Grundlagen Methodik und Anwendungen. Springer International Publishing, Heidelberg (2008). https://doi.org/10.1007/978-3-540-30250-6
11. Gausemeier, J., Moehringer, S.: VDI 2206- a new guideline for the design of mechatronic systems. IFAC Proc. Vol. **35**(2), 785–790 (2002). https://doi.org/10.1016/S1474-6670(17)34035-1
12. Gausemeier, J., Moehringer, S.: VDI 2206- a new guideline for the design of mechatronic systems. IFAC Proc. Vol. **35**(2), 785–790 (2002). https://doi.org/10.1016/S1474-6670(17)34035-1
13. Graessler, I., Hentze, J.: The new V-model of VDI 2206 and its validation. Automatisierungstechnik **68**(5), 312–324 (2020). https://doi.org/10.1515/auto-2020-0015
14. Suh, N.P.: Axiomatic Design: Advances and Applications. Oxford University Press, Oxford (2001)

15. Thomas, J.D., Lee, T., Suh, N.P.: A function-based framework for understanding biological systems. Ann. Rev. Biophys. Biomol. Struct. **33**(1), 75–93 (2004). https://doi.org/10.1146/annurev.biophys.33.110502.132654
16. Ognjanović, M., Ristić, M., Živković, P.: Reliability for design of planetary gear drive units. Meccanica **49**(4), 829–841 (2014). https://doi.org/10.1007/s11012-013-9830-8
17. Rolli, F., Pecoraro, F., Luzi, D., Ricci, F.L., Pourabbas, E., Parretti, C.: Clinical risk evaluation of medical device software: an axiomatic design-based methodology. MATEC Web Conf. **301**, 00005 (2019). https://doi.org/10.1051/matecconf/201930100005
18. Fardelas, G.: Ship Design through Axiomatic Design Approach, Sustainable Engineering Principles and Artificial Intelligence Methods. Hellenic Naval Academy, 2008; Department of Mechanical Engineering & Engineering and Management Massachusetts Institute of Technology (2021)
19. Mlambo, P., et al.: Methodologies for mechatronic systems design: attributes and popularity. Zimbabwe J. Sci. Technol. **12**, 117–127 (2018)
20. Balkhair, H.: Development of a Multi-Layer V-Model Design Process and Computational Tools for Mechatronic Conceptual Design. The University of British Columbia; Faculty of Mechanical Engineering (2019)
21. Suh, N.P., Do, S.-H.: Axiomatic design of software systems. CIRP Ann. **49**(1), 95–100 (2000). https://doi.org/10.1016/S0007-8506(07)62904-7
22. Do, S.H., Suh, N.P.: Object-oriented software design with axiomatic design. In: First International Conference on Axiomatic Design, Institute for Axiomatic Design (2000)
23. Chung, J., Suh, N.P.: Computer Aided Geometric Topology and Shape Design Within Axiomatic Design Framework. Institute for Axiomatic Design, Institute for Axiomatic Design, p. 11 (2002)
24. Chung, J. Functional Requirements to Shape Generation in CAD. Mechanical Engineering in Partial Fulfillment of the Requirements for the Degree of Doctor of Philosophy in Mechanical Engineering at the Department of Mechanicat Engineering, Massachusetts Institute of Technology (2002)
25. Puik, E.C.N., et al.: A method for indexing axiomatic independence applied to reconfigurable manufacturing systems. In: Worcester Polytech, The Seventh International Conference on Axiomatic Design, Institute for Axiomatic Design, p. 9 (2013)
26. Malaek, S.M.B., et al.: A new systems engineering model based on the principles of axiomatic design. J. Indus. Intell. Inf. **3**(2), 1–9 (2015)
27. Puik, E.C.N., Ceglarek, D.: Axiomatic Product Design in Three Stages: A Constituent Roadmap That Visualises the Status of the Design Process by Tracking the Knowledge of the Designer. Imece2015–50993, vol. V015T19A003, ASME, p. 10 (2015)
28. Thomas, J., Mantri, P.: Axiomatic design/design patterns mashup: part 1 (theory)'. Procedia CIRP **34**, 269 (2015). https://doi.org/10.1016/j.procir.2015.08.010
29. Thomas, J., Mantri, P.: Axiomatic design/design patterns mashup: part 2 (cyber security). Procedia CIRP. **34**, 276 (2015). https://doi.org/10.1016/j.procir.2015.08.011
30. Xinyu, Y.: Development of Spine Design Process and Implementation of Axiomatic Design Theory for Cyber Physical System Design Analysis. University of Strathclyde; Department of Design, Manufacture and Engineering Management (2016)
31. Scheithauer, D., Forsberg, K.: V-Model Views. INCOSE Int. Symp. **23**(1), 502–516 (2014)
32. Puik, E.: Risk adjusted, concurrent development of microsystems and reconfigurable manufacturing systems. University of Warwick (2017)
33. El-Haik, B.: The Integration of Axiomatic Design in the Engineering Design Process (1999)

34. Seal, D.: The System Engineering 'V'--Is it Still Relevant in the Digital Age? Global Product Data Interoperability Summit (2018)
35. Graessler, I., Hentze, J., Bruckmann, T.: V-models for interdisciplinary systems engineering. In: Society, T.D. (Ed.), International Design Conference - Design 2028 (2018)
36. Graessler, I.: A new v-model for interdisciplinary product engineering. In: DS 1992: Proceedings of the DESIGN 2018 15th International Design Conference, vol. 2018, pp. 747–756 (2017)

Axiomatic Design and Health

A Combined Axiomatic Design-MCDA Method for Selecting Medical Systems Operating on a Common Telemedicine Platform

Chiara Parretti[1(✉)], Fernando Rolli[1], Riccardo Barbieri[2], Gabriele Arcidiacono[1], and Paolo Citti[1]

[1] Department of Engineering Science, Guglielmo Marconi University, via Plinio 44, 00193 Rome, Italy
c.parretti@unimarconi.it

[2] Dipartimento Ingegneria Industriale, Università Degli Studi di Firenze, via di Santa Marta 3, Firenze, Italy

Abstract. The Covid-19 pandemic has triggered a significant push towards the digitalization of the Italian healthcare system. The National Recovery and Resilience Plan (PNRR) has designed a Digital Health national platform that is implemented based on microservices. However, the technological heterogeneity of healthcare companies poses difficulties in using a common healthcare platform in terms of interoperability. The first issue is selecting, for each health protocol, the basket of medical systems to be adopted, which must be compatible with this infrastructure and appropriate for the operating context. In this article, the authors propose a methodology to select healthcare systems based on axiomatic design and MCDA techniques. The expected result is to identify, in the first phase, the set of functionally acceptable solutions, and in the second phase, to select the most suitable basket based on evaluation criteria that are not necessarily functional.

Keywords: Axiomatic design · information entropy · Healthcare digital transition

1 Introduction

The National Recovery and Resilience Plan (PNRR) aims to develop a national platform for Digital Health (Fig. 1) based on micro-services, which will be made available to the various Italian regions. This initiative, funded by the European Union, aims to provide basic tools to all Italian healthcare companies to ensure that essential healthcare treatment levels, as guaranteed by the Constitution, are met. It should be noted that in Italy, the healthcare system is of the "universal type", which means that it is the responsibility of the public sector and managed by the regions. This has resulted in a diversification of the levels of service offered, with each region having specific systems within the scope of what is permitted by the Ministry of Health. This technological heterogeneity poses a critical issue in the use of a common health platform, as it presents challenges for the interoperability between different systems. Moreover, a study conducted on around

E. Puik et al. (Eds.): ICAD 2023, LNNS 849, pp. 223–233, 2024.
https://doi.org/10.1007/978-3-031-49920-3_15

800 healthcare professionals from different healthcare companies has shown that only 3% of the interviewees use Telemedicine systems, particularly for consultation activities with colleagues. The collected data shows that only 18% of users have received dedicated training, highlighting the need for significant investment in technology and skills development to ensure the widespread diffusion of digital medicine. Therefore, a robust methodology is necessary to assist decision-makers in selecting the most suitable medical systems for the platform to meet the needs of patients and healthcare professionals. The methodology involves two parts: identifying admissible solutions, i.e. the compositions of medical devices that can be used for a specific health protocol, and selecting the most suitable solution for the operative context. The independence axiom constitutes a powerful tool to identify admissible baskets. However, for the second problem category, the information axiom may not be enough, and other selection tools, such as multi-decision analysis techniques criterion (MCDA), are required. The authors propose using AHP to define the relative importance of individual selection criteria, while the decision-making process is carried out based on the information entropy concept.

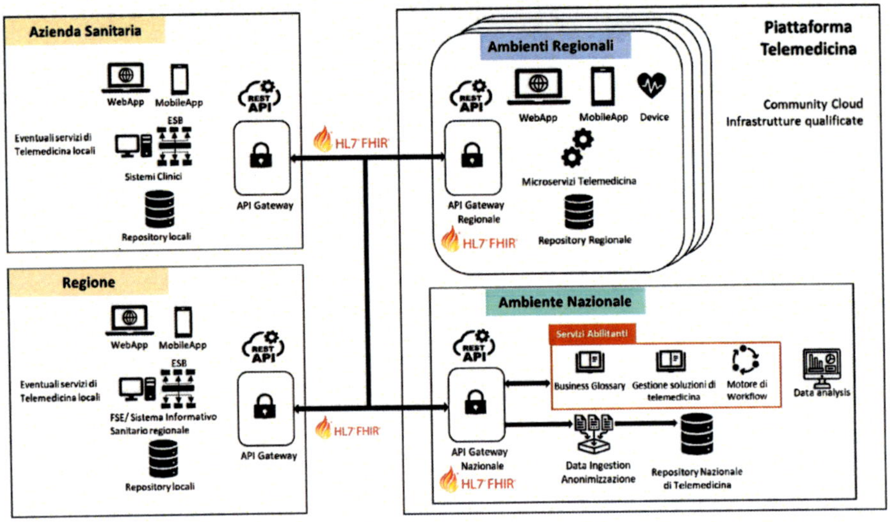

Fig. 1. Digital Health national platform block diagram

2 Devices Basket

The activation of the national telemedicine platform enables interoperability with the medical systems and devices used by healthcare companies in the Italian regions [1]. However, there are critical issues related to the heterogeneity of these systems in terms of technology and operating methods. Unlike in other countries, the Italian health service does not have a single contracting station for equipment and medical support services purchase [2]. Each spending center, which coincides with the single local health authority (ASL), is responsible for autonomously proceeding with the market acquisition of necessary instruments, even for the systems that the Ministry of Health has expressed a favorable preventive opinion about [3]. This has led to the use of heterogeneous technological devices, which can make integration on a unitary telemedicine platform extremely difficult. A telemedicine system may require the performance of several functions, such as the measurement of numerous vital parameters and the completion of inter-therapeutic medical interventions at home [4, 5]. This involves the use of various instruments, which may be technologically incompatible with each other or functionally redundant in joint use [6]. Therefore, it is necessary to define a set of appliable devices, called a "basket," for a specific therapeutic protocol. The device choice is up to the specialist doctor who follows the patient, but it may be appropriate to predefine the possible baskets based on axiomatic design. Starting from the user requirements required by healthcare professionals, it is possible to identify the functional requirements to be met [8]. The definition of the functional requirements allows the construction of a set of device baskets necessary to activate a specific health protocol on the telemedicine platform using axiomatic design. The axiom of independence guarantees the logical coherence of the use of the devices in joint form, while the axiom of information allows the selection of the least complex basket [9]. Regarding the evaluation of system complexity, authors propose a reformulation of the information axiom that extends the evaluation of the complexity of a system beyond the evaluation of the functional requirements [10, 11]. Authors suggest that the non-functional elements of a system should also be considered in the overall assessment. Functional requirements represent what the system must do, and all design methodologies are based on a detailed analysis of the systems functional requirements to be implemented [8]. However, there is no rigorous evaluation mechanism for non-functional elements in the design process [10]. To overcome this limitation, authors have placed emphasis on the opportunity to integrate non-functional elements characterizing the system to be designed into a formal process [11]. In this study, authors classify the non-functional elements of a system according to what is defined in the field of software engineering [14, 15]. The aim is to highlight that the non-functional elements of a system themselves do not constitute a single set of characteristics, but in order to better estimate the complexity of a basket, it is necessary to categorize these elements into homogeneous groups.

Therefore, by analogy with software systems, it will be possible to introduce the following classification of the non-functional elements that can characterize a system:

- Non-functional requirements (NFR);
- Project requirements and constraints (PRC).
- Non-functional requirements represent specific properties associated with the system. They can be divided into three further subcategories [14, 15]:
- Quality Requirements (QRs). Represent the quality characteristics of the device (Performance, Reliability, Safety, Maintainability, Functional suitability, ...);
- System Environment Requirements (SER). Describe the operating context of the system in terms of number and type of users, type of application environment and access methods.
- Technical Requirements (TR). Describe the technologies and technical standards, to which the device must refer.
- Process Domain (Usability, Compatibility and Portability)

Project Requirements and Constraints (PRC) refer to requirements and constraints that do not directly affect the operational management of the system [15]. They pertain to activities such as coordination, training, and the expertise level of personnel using the equipment. Conceptually, the selection process for the proposed robust basket of devices involves integrating non-functional system elements into axiomatic design, as shown in Fig. 2.

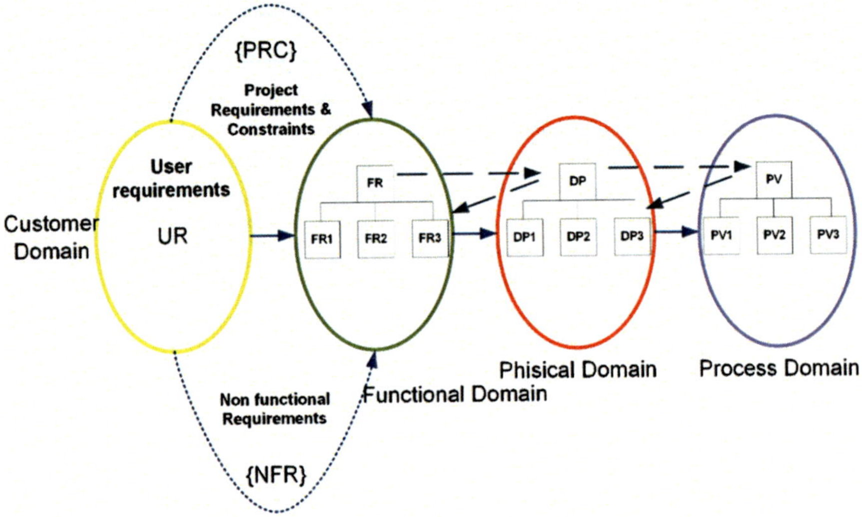

Fig. 2. General Block Diagram forn an AD based method for robust devices basket selection

3 Medical Devices Basket Complexity Evaluation as an Information Axiom Extension

The AD standard approach to evaluating system complexity considers only the functional requirements [11–13]. Essentially, it aims to identify the design solution that satisfies the same functional requirements with the least amount of information content [16]. However, in this study, the authors suggest a new definition of system complexity that takes into account non-functional aspects of basket valuation. To achieve this, they propose reformulating the information axiom, as illustrated in Fig. 3.

This generalization involves the assessment of admissible sets of medical devices, denoted as "Baskets" (B), which satisfy the independence axiom, across different categories of non-functional elements in the system. This can be achieved by constructing a specific relationship matrix for each of these categories (as shown in Table 1), where the admissible baskets are listed along the rows and the non-functional elements, which serve as evaluation criteria, are listed along the columns. The elements (a_{ij}) of the matrix indicate the impact of the j-th non-functional element on the corresponding basket B_i.

From a conceptual standpoint, the creation of relation matrices does not completely resolve the issue [10, 11]. The relationship matrices, as currently formulated, can present

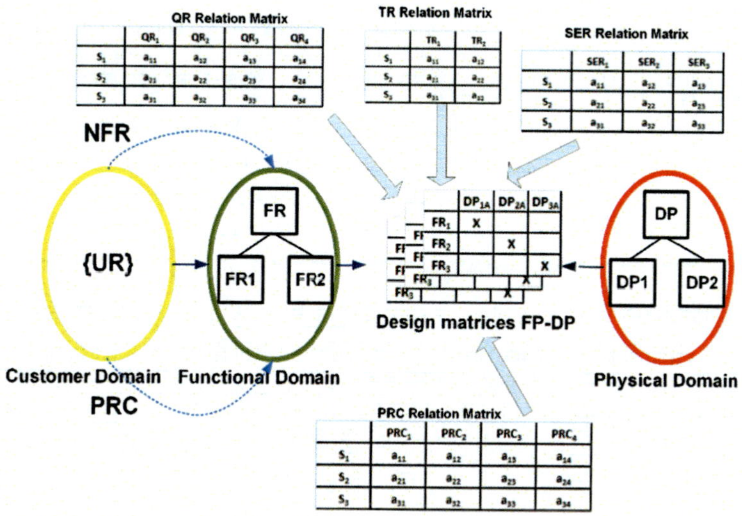

Fig. 3. Information axiom redefinition based on the system non-functional element's introduction

Table 1. Relation Matrix

	NFR1	NFR2	NFR3	NFR4
B_1	a_{11}	a_{12}	a_{13}	a_{14}
B_2	a_{21}	a_{22}	a_{23}	a_{24}
B_3	a_{31}	a_{32}	a_{33}	a_{34}

uncertain situations where it is not possible to select a unique basket. Additionally, the problem of determining the specific weight or coefficient of each non-functional element of the relation matrix persists [10]. This evaluation involves assigning equal importance to each non-functional element in the most appropriate basket selection of the medical protocol. However, this assumption of equivalence is unrealistic. To address these limitations, it is possible to use a particular type of multi-criteria evaluation, formulated based on subjective value judgments that healthcare professionals can assign to device baskets, identified through the axiom of independence [10, 11]. This type of evaluation allows for the creation of a second table, called a comparison table (Table 2), which assigns weights to the evaluation elements present in a relationship table (Table 1).

This new matrix will no longer present indeterminacy situations because the impact of each non-functional element is weighted by a specific weight. At this point, the problem becomes determining the weighting coefficients to be attributed to the selection criteria. For this purpose, it is possible to use the information entropy concept, as reported by Pourabbas et al. [10]. This concept provides the analytical tools to determine, based on the value judgments a_{ij} distributions, reported in the matrix in Table 1, an estimate of the coefficients W_j [17], where $0 < W_j < 1$ [12]. The value of W_j will be greater the more the judgments distribution attributed to the baskets constituting the comparison matrix will

Table 2. Comparison Matrix

	NFR_1	NFR_2	NFR_3	NFR_4
	(W_{QR1})	(W_{QR2})	(W_{QR3})	(W_{QR4})
B_1	a_{11}	a_{12}	a_{13}	a_{14}
B_2	a_{21}	a_{22}	a_{23}	a_{24}
B_3	a_{31}	a_{32}	a_{33}	a_{34}

present strongly discordant evaluations with respect to the j-th non-functional element impact. This implies that the j-th non-functional element carries greater weight than the others do. Conversely, value judgment distributions with low variability will result in low evaluation coefficients, i.e. closer to 0. Information entropy is defined by Shannon [18], and in this context provides a tool for evaluating the variability of value judgments [19]. The methodology can be applied by a team of specialists (professional medical personnel), following a set of precise rules. These rules are necessary to avoid situations of cognitive bias, which can arise when evaluations are based on subjective value judgments. Tversky and Kahneman [20] demonstrated in the field of cognitive psychology that even expert professionals may be susceptible to distorting phenomena when making value judgments. The human mind can assign logically coherent judgments only when two alternatives are compared [21]. In light of this cognitive evidence, Saaty [22] introduced the analytic hierarchy process (AHP) methodology, according to which the value judgments that specialist physicians attribute must be formulated as comparisons between only two baskets at a time. This rule involves redefining comparison matrices in terms of comparing baskets for each non-functional item to be evaluated.

Table 3 represents an example of a comparison matrix according to the AHP approach rules. In this case, the comparison between the same baskets B_i gives the value 1. Instead, if the comparison between the baskets B_i and B_j is given the value a_{ij}, the comparison between B_j and B_j is given the value $1/a_{ij}$. These rules of value judgments attribution make it possible to minimize the cognitive bias phenomenon.

Table 3. Comparison Matrix for solution alternate for any specific non-functional requirement [22].

QRs	B_1	B_2	B_3	B_4
B_1	1	a_{12}	a_{13}	a_{14}
B_2	$1/a_{12}$	1	a_{23}	a_{24}
B_3	$1/a_{13}$	$1/a_{23}$	1	a_{34}
B_4	$1/a_{14}$	$1/a_{24}$	$1/a_{34}$	1

Furthermore, these judgments must be made on a specific scale of values base (Table 4) [11, 22]. This also makes it possible to provide a measure classification that can be associated with the comparison between different baskets.

Table 4. Evaluation score matrix

Values	S_i vs. S_j level of importance
1	i and j have same importance
3	i moderately more important j
5	i more important than j
2, 4	Intermediate importance levels

4 Robust Basket Selection

The rules introduced in the previous paragraph allow selecting the robust basket of medical devices based on the estimate of the weighting coefficients associated with the non-functional elements' characteristic of the specific health protocol. These weighting coefficients are estimated considering, for each non-functional evaluation element, the relative comparison matrix (Table 3), on which an information entropy generalization is applied.

The information entropy H(x) of a discrete probability distribution p(x) is a positive function defined according to the following formula [18]:

$$H(x) = -\sum\nolimits_{x}^{X} p(x) logp(x) \tag{1}$$

where X represent a set of instances x.

In order to apply Eq. 1 to a comparison matrix (Table 3) it is necessary to proceed with the matrix normalization [10]. This operation is performed by replacing in the matrix of Table 3, the evaluation judgments a_{ij}, as defined by the expert evaluators, by the corresponding normalized elements A_{ij}, obtainable as follows:

$$A_{ij} = \frac{a_{ij}}{\sqrt{\sum_{j=1}^{4} a_{ij}^2}} \tag{2}$$

So, from the comparison matrix Table 3, it is obtained the normalized comparison matrix A. Based on this new matrix, Eq. 1 become:

$$H(A_i) = \sum\nolimits_{j=1}^{4} H(A_{ij}) = -\sum\nolimits_{j=1}^{4} A_{ij} logA_{ij} = W_i^{NFE_s} \tag{3}$$

where $W_i^{NFE_s}$ represents the weighting coefficient associated with the i-th row of the comparison matrix relating to the non-functional element NFE_s.

Repeating the calculation for each i-th row of the normalized comparison matrix A, the following vector of weighting coefficients [10] is obtained:

$$W^{NFE_s} = \begin{bmatrix} W_1^{NFE_s} \\ W_2^{NFE_s} \\ W_3^{NFE_s} \\ W_3^{NFE_s} \end{bmatrix} \tag{4}$$

These weighting coefficients can be put together in an overall comparison matrix, such as the one shown in Table 5. It brings together all the weighting coefficients calculated for the non-functional elements considered.

Table 5. Weighting matrix of non-functional items

	NFE_1	NFE_2	–	NFE_m
B_1	$W_1^{NFE_1}$	$W_1^{NFE_2}$	–	$W_1^{NFE_m}$
B_2	$W_2^{NFE_1}$	$W_2^{NFE_2}$	–	$W_2^{NFE_m}$
–	–	–	–	–
B_n	$W_n^{NFE_1}$	$W_n^{NFE_n}$	–	$W_n^{NFE_m}$

He B_{i*} robust basket will be the solution with the highest $W_i^{NFE_m}$ parameters sum within the B set of admissible This algorithm enables the definition of a ranking among n allowable baskets (B_i) while considering the non-functional elements specified in the operational context. However, as illustrated in Fig. 2, non-functional elements can pertain to various categories, and therefore, evaluating them simultaneously is inappropriate. To address this, a specific extension of the aforementioned method can be utilized, wherein structured hierarchical evaluation is employed. The Analytic Hierarchy Process (AHP) method can support this approach, and it enables the use of the entropy criterion through a decision tree (refer to [22, 23]). By evaluating the weighting coefficients (W_j) associated with sub-criteria, it is possible to perform successive aggregations, as shown in Fig. 4. This criterion facilitates the determination of weighting coefficients for various sub-criteria (QR, TR, SER) and allows the creation of Non-Functional Requirement (NFR) comparison matrices that encompass the three functional requirement categories. Additionally, it is possible to combine NFRs and Performance-Related Characteristics (PRCs) to generate an overall comparison matrix that enables the selection of the robust Si* solution based on a hierarchical application of information entropy. This generalization enables the separation of elements with distinct characteristics into homogeneous subsets to achieve a more accurate evaluation (refer to [23]).

$$S_{i*} = Max_{i=1}^{n}\left(\sum\nolimits_{j=1}^{m} W_i^{NFE_m}\right) \tag{5}$$

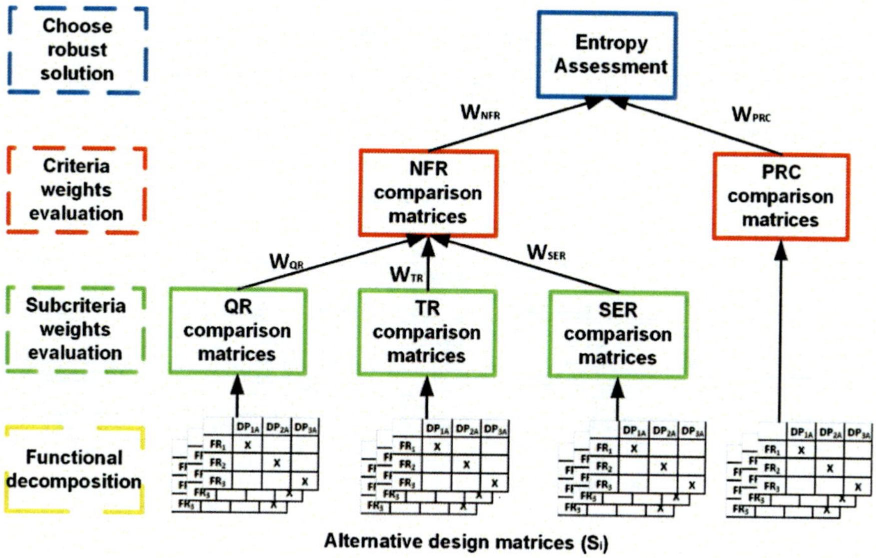

Fig. 4. Information entropy method to evaluate a robust basket. A method generalization

5 Conclusions

A national platform for delivering microservices in telemedicine represents a significant opportunity to enhance the Italian healthcare system. However, it also presents considerable challenges to local health companies, in terms of not only transitioning to digital services but also regarding interoperability, usability, and safety. The proposed approach offers several advantages in this context. Firstly, the use of axiomatic design as a tool for basket composition enables the identification of device collections that meet all functional requirements for both patients and healthcare professionals for specific healthcare protocols. Axiomatic design helps avoid combinations of technological or functional incompatibilities and reduces the duplication of redundant functions. This is particularly relevant given the technological heterogeneity of devices used in Italian healthcare companies, which may cause incompatibilities during technological integration with the telemedicine platform. To address this issue, the proposed approach involves a reformulation of the information axiom to redefine the concept of complexity using a wider set of criteria that include non-functional characteristics of devices. These criteria may include the level of interoperability, usability, portability, security, and confidentiality of processed data. The concept of information entropy is used to estimate the relevance of these non-functional elements based on value judgments attributed to the adoption of specific device baskets by teams of specialist doctors. However, selecting these instruments is the responsibility of the specialists, who may find it challenging to formulate comprehensive judgments. In these cases, subjective judgments are formulated through a comparison of two alternative solutions against a well-defined scale of values, reducing the possibility of cognitive bias. The information axiom reformulation can also be adopted to include economic aspects in the decision-making process. In this case, the complexity of the system will have an economic dimension, which is also relevant given the needs of spending review.

References

1. Piano Nazionale di Ripresa e Resilienza, Missione 6: Salute Componente 1 (M6C1): Reti di prossimità, strutture e telemedicina per l'assistenza sanitaria territoriale, ALLEGATO – "INDICAZIONI METODOLOGICHE PER LA PERIMETRAZIONE DELLE PROPOSTE DI PPP PER LA PIATTAFORMA NAZIONALE DI TELEMEDICINA. https://www.agenas.gov.it/images/agenas/In%20primo%20piano/telemedicina/Allegato_Indicazioni_Metodologiche_di_Perimetrazione_04.05.2022.pdf
2. Matt, D.T., Arcidiacono, G., Rauch, E.: Applying lean to healthcare delivery processes-a case-based research. Int. J. Adv. Sci. Eng. Inf. Technol. **8**(1), 123–133 (2018)
3. Gazzetta Ufficiale della Repubblica Italiana, DECRETO LEGISLATIVO 7 dicembre 1993, n. 517, GU Serie Generale n. 293 del 15–12–1993 - Suppl. Ordinario n. 113
4. Parretti, C., Pourabbas, E., Rolli, F., Pecoraro, F., Citti, P., Giorgetti, A.: Robust design of web services supporting the home administration of drug infusion in pediatric oncology. In: Proceedings of the 13th International Conference on Axiomatic Design (ICAD) (2019)
5. Parretti, C., Rolli, F., Pourabbas, E., Citti, P.: Axiomatic selection of health and social care web services on the basis of use cases. In: Proceedings of the 12th International Conference on Axiomatic Design (ICAD) (2018)
6. Khayal, I.S., Farid, A.M.: Healthcare system design. In: Suh, N.P., Cavique, M., Foley, J.T. (eds.) Design Engineering and Science, pp. 543–562 Springer, Cham (2021). https://doi.org/10.1007/978-3-030-49232-8_19
7. Ministero della Salute, "Banca dati dei dispositivi medici", https://www.salute.gov.it/interrogazioneDispositivi/RicercaDispositiviServlet?action=ACTION_MASCHERA
8. Liu, A.: Problem definition. In: Suh, N.P, Cavique M., Foley J.T. (eds.), Design Engineering and Science, pp. 167–189. Springer, Cham (2021). https://doi.org/10.1007/978-3-030-49232-8_5
9. Suh, N.P.: Axiomatic Design - Advances and Applications, Oxford University Press, Oxford (2001)
10. Pourabbas E., Parretti C., Rolli F., Pecoraro F.: Entropy-based assessment of nonfunctional requirements in axiomatic design IEEE Access **9**, 156831–156845 (2021)
11. Mabrok, M., Ryan, M., Efatmaneshnik, M.: Integrating nonfunctional requirements into axiomatic design methodology. IEEE Syst. J. **11**, 2204–2214 (2015)
12. Puik, E.C.N.: Complexity in axiomatic design, In: Suh, N.P, Cavique M., Foley J.T. (eds.), Design Engineering and Science. Springer, Cham (2021). https://doi.org/10.1007/978-3-030-49232-8_12
13. Arcidiacono, G., Placidoli, P.: Reality and illusion in virtual studios: axiomatic design applied to television recording. Procedia CIRP **34**, 137–142 (2015)
14. The COSMIC Functional Size Measurement Method, Guideline on Non-Functional & Project Requirements (2015)
15. Buglione, L.: The next frontier: measuring and evaluating the non-functional productivity Metr. Views, IFPUG Newsl. **6**(2), 11–14 (2012)
16. Fradinho, J., Gonçalves-Coelho, A.M.: The information axiom and robust solution. In: Suh, N.P, Cavique M., Foley J.T. (eds.), Design Engineering and Science. Springer, Cham (2021). https://doi.org/10.1007/978-3-030-49232-8_11
17. Venkata Rao R.: Decision Making in the Manufacturing Environment. Using Graph Theory and Fuzzy Multiple Attribute Decision Making Methods, Springer, London (2006). https://doi.org/10.1007/978-1-84628-819-7

Applying Axiomatic Design to Risk Assessment

Kenji Iino(✉) and Masayuki Nakao

The University of Tokyo, Tokyo 113-0033, Japan
kiino@esc.u-tokyo.ac.jp

Abstract. As the complexity of our society increase, managing risk is an important task in maintaining healthy and sustainable societies. Risk assessment is a widely used method to identify events associated with high probability of occurrence and severe consequences. Although simple and easy to understand, risk assessment is over-simplified with a single estimate of occurrence probability and a single severity evaluation of each event. Axiomatic design (AD) decomposes an overall functional requirement (FR) into a set of element FRs. Although an uncoupled design is ideal with each FR mapping to a single design parameter (DP), a design, in many cases, is coupled and one or more single FRs may depend on multiple DPs. This study shows how we can apply the design equation formulation of AD to risk assessment by identifying the failure probability of each DP, estimating the severity of not meeting each FR, and then calculating the risk associated with each FR. The method identifies improvement on which DP is most effective in reducing the overall system risk. Our analysis also shows how unwanted design interference can cause unexpected serious negative consequences. The designer is encouraged to apply AD to recognize design interference and spend efforts in removing them before production.

Keywords: Risk assessment · Axiomatic Design · Design record graph

1 Introduction

Risk assessment [1] is a popular tool for managers and business groups for identifying risks that exist in their business environment with probabilities of occurrence that are too large with severe consequences. Risk is typically calculated with the equation:

$$\text{(Risk)} = \text{(Probability of occurrence)} \times \text{(Severity of consequence)} \tag{1}$$

without definite rules for what numbers to give to the two terms in the right-hand side of the equation. Thus, there is no established quantification guidelines for risk.

A risk assessment session often proceeds with a table on the side to aid the person or group of people performing the task (performer). Figure 1 shows a typical risk assessment table.

With this table on the side, the performer lists out foreseeable events that are unwanted. The performer then makes two independent evaluations for each event; one is how likely the event takes place, and the other, what are the consequences of the event.

E. Puik et al. (Eds.): ICAD 2023, LNNS 849, pp. 261–274, 2024.
https://doi.org/10.1007/978-3-031-49920-3_17

		Likelihood			
		Very Likely	Likely	Unlikely	Very Unlikely
Consequence	Fatal	Critical	Critical	High	Moderate
	Major Injury	Critical	High	Moderate	Low
	Minor Injury	High	Moderate	Low	Low
	Injury Negligible	Moderate	Low	Low	Very Low

Fig. 1. Risk Assessment Table

These evaluations have no rules for making them and thus, relies on the gut feeling of the performer.

If an event is "Very Likely" to happen and with the consequence of "Fatality," the risk assessment table says that the event is critical, and the business must take some measures to lower the likelihood of the event happening or lessen the severity of the consequence before proceeding further with their operations. Even if the likelihood is only "Likely," the table stills says that it is critical. In contrast, if the event leads only to "Minor Injuries" with a likelihood of "Unlikely," the event has low priority in terms of requiring some actions.

Performing risk assessment, therefore, is simple and easy for those maybe not so well equipped with mathematical proficiency in linear algebra. It yet gives a nice visual representation of analysis using a table. In fact, however, performing risk assessment is far better that not carrying out any risk analysis. Note for this analysis, that each unwanted event is handled independently. It proceeds with an implicit assumption that event A has no effect on the likelihood of event B.

This table is often called "risk assessment matrix" from its regular appearance with all cells filled with quantities. Some practices of risk assessment assign numbers to likelihood and consequence and each cell shows the result of multiplying the two evaluations. If the likelihood and consequence quantifiers are larger with higher probability of occurrence and more serious outcome, respectively, larger products in the cells indicate events that businesses have to act on.

This risk assessment table showing the level of risk for each event, however, is not like a matrix used in linear algebra. Such mathematical matrices indicate mapping from one linear space to another. In terms of Axiomatic Design (AD) [2], the matrix that maps a design parameter (DP) to a functional requirement (FR) vector is called the design matrix (DM) and AD will examine a DM to tell the quality of the design it represents. The first axiom of AD drives designs to be uncoupled, i.e., each FR is realized by a single DP and the other DPs has no effect on the performance of FR. In this case, the DM is diagonal like the following Eq. (2) shows.

$$\begin{Bmatrix} FR_1 \\ FR_2 \\ \vdots \\ FR_n \end{Bmatrix} = \begin{bmatrix} X & 0 & \cdots & 0 \\ 0 & X & \cdots & 0 \\ \vdots & \vdots & \ddots & \vdots \\ 0 & 0 & \cdots & X \end{bmatrix} \begin{Bmatrix} DP_1 \\ DP_2 \\ \vdots \\ DP_n \end{Bmatrix} \quad (2)$$

The cells in a risk assessment table are independent from each other. If we look at the quantities, or levels in a single row or column they increase or decrease monotonically simply because the headers likelihood and consequence are arranged monotonic. A quantity or level in a cell has nothing to do with those in the adjacent ones.

From its nature of evaluating each event independently, risk assessment does not evaluate effects of an event on others. Many risk assessments evaluate natural hazard events like earthquake, tsunami, storms, or fires, or risks with our daily work like, falling from a ladder, power outages, or a sudden leave or sickness of a worker. Evaluating levels of consequences of these events independently is somewhat acceptable, however, we know that many unwanted events can trigger occurrence of other events, like an earthquake causing tsunami or sickness leading to the worker falling from a ladder.

This paper proposes a more rigorous approach to risk assessment by applying design equation (DE) from AD to correlate DPs to FRs. Our method starts with a design equation with DPs having their probabilities of failure and the DE leads to finding likelihood of each FR failing. Each FR has its own severity of consequence and evaluating this consequence remains the same with conventional risk assessment. In other words, this paper shows a systematic approach to evaluate probability of a FR failure using AD.

2 Smart Phone Design and Its Risk Assessment

2.1 Smart Phone Design

To explain the concept of applying AD to risk assessment, we discuss an existing design of a smart phone and how we propose analyzing the risks associated with the FRs they have.

Figure 2 shows the back and front of a smart phone (iPhone 11) with its parts identified [3]. Some of the parts inside are guessed, and probably there are more parts, especially IC Chips for purposes.

Figure 3 shows the DPs we identified for our AD analysis. Note that we bundled CPU, Memory, AD converter, and DA converter into PC board. The DP nodes with dark gray indicate parts that are on the rear side or inside the unit and are invisible from the front. The DP nodes with light gray color are parts arranged on the sides.

2.2 Design Record Graph

A Design Record Graph (DRG) [4] starts with the overall FR for a product, iteratively subdivides the FR into smaller FRs until a set of FRs that are no longer practical for further subdivision. The elements in this FR set are called functional elements (FEs). Then an FE maps to one or more physical elements (PEs). Multiple PEs bundle to define small assemblies which next combine into larger assemblies. The binding continues until the product is defined. Figure 4 shows the DRG for iPhone 11 in Fig. 3.

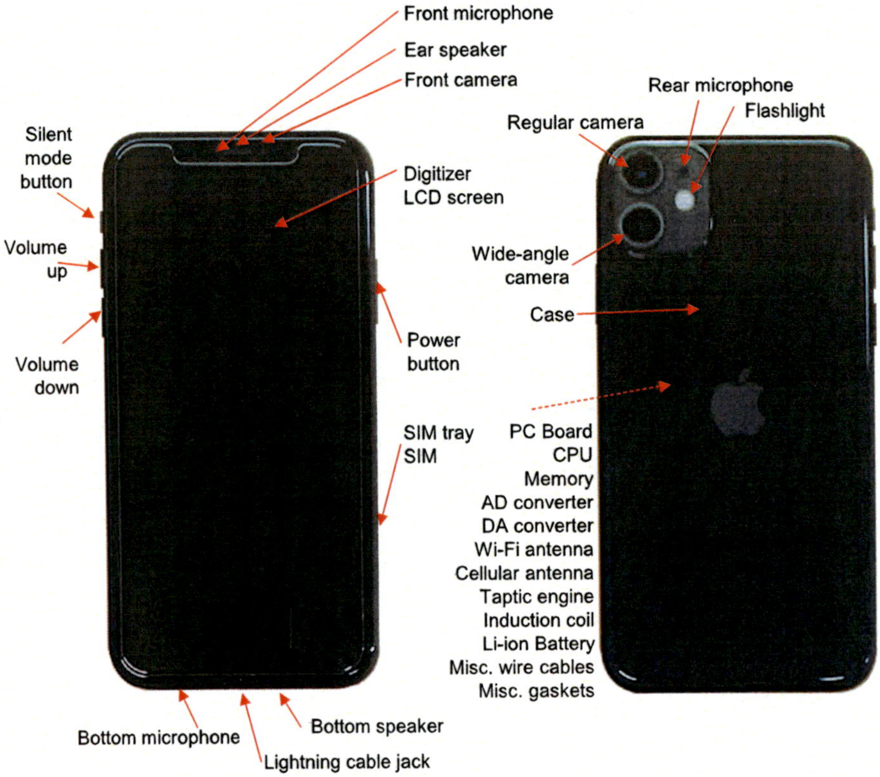

Fig. 2. iPhone 11 and its parts

In the next Subsect. 2.3, we map this DRG into a DE in AD. AD's first Independence Axiom tells the designer to keep independence of FRs [2]. Ideally one FR maps to a single DP. When we interpret this axiom to DRG, it means to have a ladder like mapping from functional space to physical space. Our experience with novice designers like graduate school students, suggests that working with the graphical interface of a DRG to establish a ladder like mapping is easier than to work directly with the DM. Figure 5 takes the third to the ninth elemental FRs from the top of Fig. 4 and their corresponding 8 elemental DPs and works out FRs rephrasing to reach a one-to-one correspondence.

Axiom 2, the Independence Axiom of AD, guides the designer in reaching a better design by aiming at minimizing the information content of the design when there are multiple design options. This approach minimizes the risk in not meeting the desired FRs [5]. In our paper, we show how risks of DPs affect the FRs with AD and we reach a maintenance scheme to turn unacceptable risks with FRs into reasonable ranges.

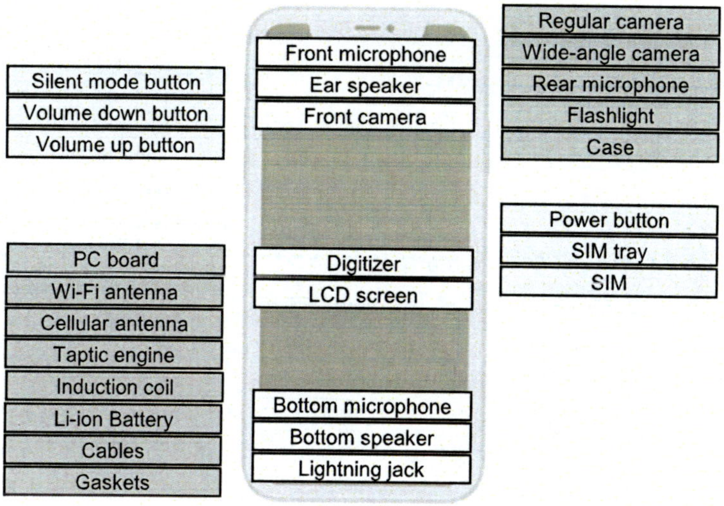

Fig. 3. DPs of iPhone 11 in our analysis

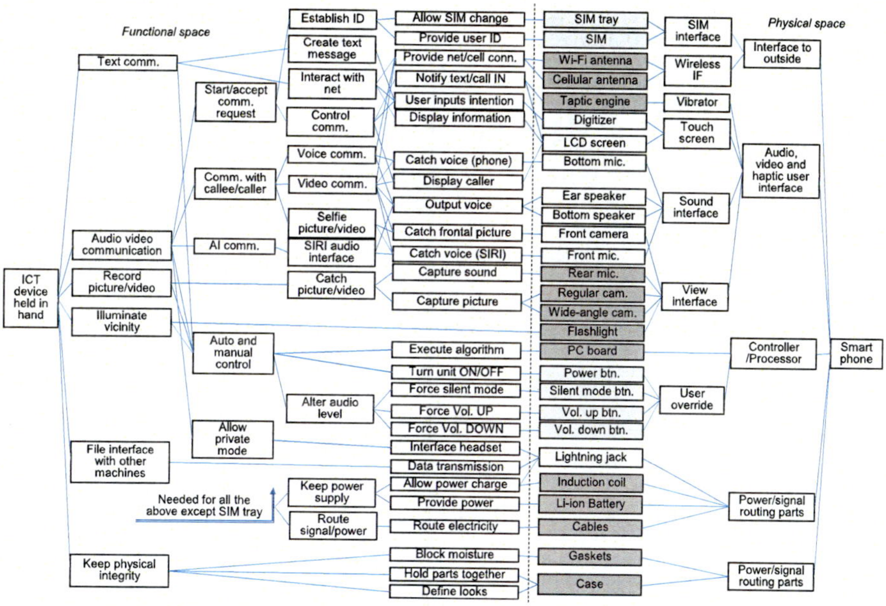

Fig. 4. DRG of iPhone 11

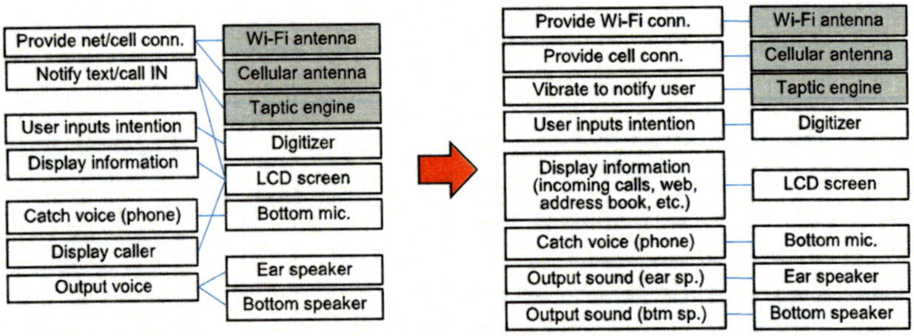

Fig. 5. Rephrasing FRs for independence

2.3 Design Equation

FEs in a DRG correspond to FR elements in AD, and PEs to DP elements. We can then write the DE for an iPhone 11 as follows:

$$\left\{\begin{array}{c} \text{Allow SIM change} \\ \text{Provide user ID} \\ \text{Provide net/cell conn.} \\ \text{Notify text/call IN} \\ \text{User inputs intention} \\ \text{Display information} \\ \text{Display caller} \\ \text{Catch voice (phone)} \\ \text{Output voice} \\ \text{Catch frontal picture} \\ \text{Catch voice (SIRI)} \\ \text{Capture sound} \\ \text{Capture picture} \\ \text{Illuminate vicinity} \\ \text{Execute algorithm} \\ \text{Turn unit ON/OFF} \\ \text{Force silent mode} \\ \text{Force Vol. UP} \\ \text{Force Vol. DOWN} \\ \text{Interface headset} \\ \text{Data transmission} \\ \text{Allow power charge} \\ \text{Provide power} \\ \text{Route electricity} \\ \text{Block moisture} \\ \text{Hold parts together} \\ \text{Define looks} \end{array}\right\} = \mathbf{DM} \left\{\begin{array}{c} \text{SIM tray} \\ \text{SIM} \\ \text{Wi-Fi antenna} \\ \text{Cellular antenna} \\ \text{Taptic engine} \\ \text{Digitizer} \\ \text{LCD screen} \\ \text{Bottom mic.} \\ \text{Ear speaker} \\ \text{Bottom speaker} \\ \text{Front camera} \\ \text{Front mic} \\ \text{Rear mic} \\ \text{Regular cam} \\ \text{Wide-angle cam} \\ \text{Flashlight} \\ \text{PC board} \\ \text{Power btn} \\ \text{Silent mode btn} \\ \text{Vol.up btn} \\ \text{Vol.down btn} \\ \text{Lightning jack} \\ \text{Li-ion Battery} \\ \text{Cables} \\ \text{Gaskets} \\ \text{Case} \end{array}\right\} \tag{3}$$

where Eq. (4) below shows the elements of DM in Eq. (3).

$$
\mathbf{DM} = \begin{bmatrix}
9&0&0&0&0&0&0&0&0&0&0&0&0&0&0&0&0&0&0&0&0&0&0&0&0&0&1\\
0&9&0&0&0&0&0&0&0&0&0&0&0&0&0&0&0&0&0&0&0&0&0&9&9&0&0\\
0&0&3&9&0&0&0&0&0&0&0&0&0&0&0&0&0&0&0&0&0&0&0&9&9&0&1\\
0&0&0&0&9&0&3&0&0&0&0&0&0&0&0&0&0&0&0&0&0&0&0&9&9&0&1\\
0&0&0&0&0&9&0&0&0&0&0&0&0&0&0&0&0&0&0&0&0&0&0&9&9&0&1\\
0&0&0&0&0&0&9&0&0&0&0&0&0&0&0&0&0&0&0&0&0&0&0&9&9&0&1\\
0&0&0&0&0&0&9&0&0&0&0&0&0&0&0&0&0&0&0&0&0&0&0&9&9&0&1\\
0&0&0&0&0&0&0&9&0&0&0&0&0&0&0&0&0&0&0&0&0&0&0&9&9&0&1\\
0&0&0&0&0&0&0&0&9&3&0&0&0&0&0&0&0&0&0&0&0&0&0&9&9&0&1\\
0&0&0&0&0&0&0&0&0&0&9&0&0&0&0&0&0&0&0&0&0&0&0&9&9&0&1\\
0&0&0&0&0&0&0&0&0&0&0&9&0&0&0&0&0&0&0&0&0&0&0&9&9&0&1\\
0&0&0&0&0&0&0&0&0&0&0&0&9&0&0&0&0&0&0&0&0&0&0&9&9&0&1\\
0&0&0&0&0&0&0&0&0&0&0&0&0&9&3&0&0&0&0&0&0&0&0&9&9&0&1\\
0&0&0&0&0&0&0&0&0&0&0&0&0&0&0&9&0&0&0&0&0&0&0&9&9&0&1\\
0&0&0&0&0&0&0&0&0&0&0&0&0&0&0&0&9&0&0&0&0&0&0&9&9&0&0\\
0&0&0&0&0&0&0&0&0&0&0&0&0&0&0&0&0&9&0&0&0&0&0&9&9&0&1\\
0&0&0&0&0&0&0&0&0&0&0&0&0&0&0&0&0&0&9&0&0&0&0&9&9&0&1\\
0&0&0&0&0&0&0&0&0&0&0&0&0&0&0&0&0&0&0&9&0&0&0&9&9&0&1\\
0&0&0&0&0&0&0&0&0&0&0&0&0&0&0&0&0&0&0&0&9&0&0&9&9&0&1\\
0&0&0&0&0&0&0&0&0&0&0&0&0&0&0&0&0&0&0&0&0&9&0&9&9&0&0\\
0&0&0&0&0&0&0&0&0&0&0&0&0&0&0&0&0&0&0&0&0&9&0&9&9&0&0\\
0&0&0&0&0&0&0&0&0&0&0&0&0&0&0&0&0&0&0&0&0&3&3&9&9&0&1\\
0&0&0&0&0&0&0&0&0&0&0&0&0&0&0&0&0&0&0&0&0&0&0&9&9&0&9\\
0&0&0&0&0&0&0&0&0&0&0&0&0&0&0&0&0&0&0&0&0&0&0&9&9&0&1\\
0&0&0&0&0&0&0&0&0&0&0&0&0&0&0&0&0&0&0&0&0&0&0&0&0&9&9\\
0&0&0&0&0&0&0&0&0&0&0&0&0&0&0&0&0&0&0&0&0&0&0&0&0&0&9\\
0&0&0&0&0&0&3&0&0&0&0&0&0&0&0&0&0&0&0&0&0&0&0&0&0&0&9
\end{bmatrix} \quad (4)
$$

The matrix in Eq. (4) has its non-zero elements in white background so they are easier to identify among the zero elements shaded in gray. Otherwise, the matrix would appear like a quiz of finding non-zero entries among all the zeros. Note that the DM in Eq. (4) is not diagonal. We produced this DM for Fig. 4 instead of working in the independence FR rephrasing in Fig. 5. Pursuing an uncoupled design, however, is not the point of this study.

Also, it is noteworthy that the Lithium-ion battery and Cables serve all electricity driven FRs. The Case also serves almost all FRs in a weak manner other than the FRs of blocking moisture, holding parts together, defining looks, and providing power with link strengths of 9. The lithium-ion battery's integrity strongly depends the case stiffness because bending or impact on the battery can cause it to smoke or start a fire.

Separating out DPs of the lightning jack for its three functions is difficult for the following reason. A lightning connector has 8 pins as Fig. 6 shows [6]. The pin configuration on the other side is in a flipped sequence so that you do not have to worry about connecting a lightning jack and a connector in a reverse manner.

When used for charging the Lithium-ion battery, we first thought only the two pins, Power and Ground, are in use. When we measured lighting cable pins connected to a USB AC power adaptor, the Power pin measured 5V relative to the Ground pin, however, the Lane 0 positive pin also measured 1V relative to the Ground pin. Maybe, it is the way of telling the lightning jack that the user intends to charge the battery. All other 5 pins measured 0V relative to the Ground pin. When in other modes of transferring signals for

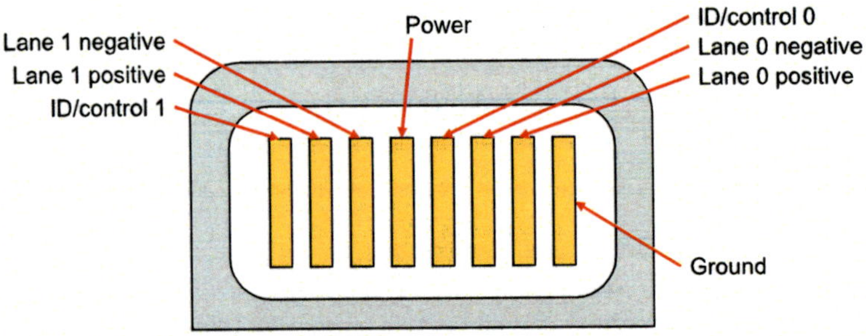

Fig. 6. Configuration of 8 pins with a lightning connector

the headset or data communication, the Ground pin must be used for identifying status of the other signal pins. If we want to define DPs for each FR mode, like in Fig. 5, we will define different DPs of "Lightning jack in charge mode," "Lightning jack in sound communication mode," and "Lightning jack in data transfer mode." But again, aiming for an uncoupled design is not the point of this study.

Note that the non-zero elements of the DM in Eq. (4) are 1, 3 or 9. We followed the convention used in Reference [7]. Directly quoting the quantification scheme for quantities 1, 3, and 9 with an additional condition for assigning 0,

0 = no link: the function is not affected at all failing contribution of the component;
1 = weak link: the function is lightly degraded failing contribution of the component;
3 = middle link: the function is only reduced failing contribution of the component;
9 = strong link: the function is completely annulled failing contribution of the component;

Which quantity to assign to each FR-DP relation was at the discretion of the authors.

3 Risk Analysis

3.1 Failure Probabilities of DPs

For our risk analysis, we start with estimating the failure frequencies of the DPs. The quantity we adopt is the annual failure rate (AFR), i.e., average number of failures per year [8]. It relates to mean time before failure (MTBF) as follows [8]:

$$AFR = 1/MTBFyears = 8760/MTBFhours \tag{5}$$

Table 1 shows the AFR values we assigned to our DPs in Eq. (3). We found most of these values by applying Eq. (5) to MTBF on the net for similar products, except, we assigned 10^{-4} to moving mechanical parts and 10^{-5} to stable parts. We based the value 0.33333(1/3), we assigned to lithium-ion batteries, on a net article [9] claiming 2 to 3 years of life for lithium-ion batteries. Table 1 shows these AFR values without solid references in gray background.

Other lithium-ion battery articles exist with different life spans up to 6 years. One of the authors, Iino, has been using an iPhone 11 with the first photograph taken in

Table 1. AFR values of DPs

Design Parameter	AFR
SIM tray	0.00010
SIM	0.00001
Wi-Fi antenna	0.00876
Cellular antenna	0.00876
Taptic engine	0.00548
Digitizer	0.00833
LCD screen	0.01667
Bottom microphone	0.01667
Ear speaker	0.00024
Bottom speaker	0.00024
Front camera	0.00253
Front microphone	0.01667
Rear microphone	0.01667
Regular camera	0.00253
Wide-angle camera	0.00253
Flashlight	0.00012
PC board	0.00001
Power button	0.00006
Silent mode button	0.00006
Volume up button	0.00006
Volume down button	0.00006
Lightning jack	0.00010
Induction coil	0.00833
Li-ion Battery	0.33333
Cables	0.00001
Gaskets	0.00001
Case	0.00001

November, 2016, thus, he has been using this unit for over 6 years with the original battery it came with. He also has an iPhone 5 in use since 2013. It only served as a phone unit for about two years and then has been left untouched until recently for looking up meaning of words at night. The original battery for this unit is also in service.

The AFR for smartphone parts largely depend on how the owner use the phone. Most academic users communicate via email on their PCs. Both authors of this paper receive hundreds of mails and reply to about 10% of them. Communications on the smartphone are, for both authors, twenty or less a day. Iino looked up his average screen time with his current iPhone 11 which showed 28 min. This is largely less than 4 h and 23 min for an average US mobile user [10]. We calculated the AFR for liquid crystal display with the 4.5 h of average screen time, and 100,000 h of life [11] that gives an MTBF of 100,000 / (4.5*365) = 60 years, and 0.01667 is 1/60 from Eq. (3).

Another note is that we did not include the more common problem of causing damage to smartphones, that is, dropping them on ground or on floor [12].

3.2 Functional Significance and Its Failure Rate

With the AFR estimates in the previous section, we can calculate annual failure rate of each FR (FFR) with:

$$\mathrm{FFR}_i = \sum_{j=1}^{m} E_{ij} \cdot \mathrm{AFR}_j \tag{6}$$

where FFR_i is the functional failure rate of the *i*-th FR, E_{ij} is DM element of the *i*-th row and *j*-th column and AFR_i is the AFR of the *j*-th DP.

Table 2 shows the resulting calculation results in the yellow column. Note that we also divided the results with 9 to keep the results less than 1. This is equivalent to using link strength values of (0, 1/9, 1/3, 1) instead of (0, 1, 3, 9) at the end of Sect. 2.3. The numbers 1, 3, and 9 are good for assigning quantities that correspond to human feelings, however, in calculating probabilities, keeping them less than 1.0 is more intuitive.

Table 2, in its second column from the left, shows the significance value Si the authors assigned to each FR, e.g., establishing connection to the internet via Wi-Fi or cellular network is essential to a smartphone, as well as the user commanding the smartphone what to do via the digitizer screen. On the other hand, illuminating the vicinity with the LED mounted on the back side is a convenient function, however, not essential to the smartphone. Here, we used the quantities 1, 3, and 9 again with a larger number assigned to a more important function. The third column from the left in Table 2 is the product of the significance of *i*-th FR (S_i) and its functional failure rate (FFR_i). We call this quantity Significant Functional Failure Rate (SFFR_i) expressed in the following equation.

$$\mathrm{SFFR}_i = \mathrm{S}_i \cdot \mathrm{FFR}_i / 9 \tag{7}$$

Note that all the electricity related functions are strongly affected by the lithium-ion battery AFR. We highlighted SFFR values of 0.3 or higher in Table 2 with pink background. They turn out to be smartphone functions that depend on power from the lithium-ion battery with user significance values of 9. Note the division by 9 in Eq. (7) is, again, normalizing the largest intuitive significance of 9.

Table 2. Significance and Functional Failure Rates of Smartphone FRs

	Significance of Function	SFFRi: Significant Functional Failure Rate AFR(Li-ion)=0.33333	FFRi: Functional Failure Rate AFR(Li-ion)=0.00010	SIM tray	SIM	Wi-Fi antenna	Cellular antenna	Taptic engine	Digitizer	LCD screen	Bottom mic.	Ear speaker	Bottom speaker	Front camera	Front mic.	Rear mic.	Regular cam.	Wide-angle cam.	Flashlight	PC board	Power button	Silent mode button	Vol.up button	Vol.down button	Lightning jack	Induction coil	Li-ion Battery	Cables	Gaskets	Case
Allow SIM change	3	0.00003	0.00010	9	0	0	0	0	0	0	0	0	0	0	0	0	0	0	0	0	0	0	0	0	0	0	0	0	0	1
Provide user ID	3	0.11112	0.00012	0	9	0	0	0	0	0	0	0	0	0	0	0	0	0	0	0	0	0	0	0	0	0	9	9	0	0
Provide net/cell conn.	9	0.34502	0.01179	0	0	3	9	0	0	0	0	0	0	0	0	0	0	0	0	0	0	0	0	0	0	0	9	9	0	1
Notify text/call IN	3	0.11479	0.01115	0	0	0	0	9	0	3	0	0	0	0	0	0	0	0	0	0	0	0	0	0	0	0	9	9	0	1
User inputs intention	9	0.34167	0.00844	0	0	0	0	0	9	0	0	0	0	0	0	0	0	0	0	0	0	0	0	0	0	0	9	9	0	1
Display information	3	0.11667	0.01678	0	0	0	0	0	0	9	0	0	0	0	0	0	0	0	0	0	0	0	0	0	0	0	9	9	0	1
Display caller	3	0.11667	0.01678	0	0	0	0	0	0	9	0	0	0	0	0	0	0	0	0	0	0	0	0	0	0	0	9	9	0	1
Catch voice (phone)	3	0.11667	0.01678	0	0	0	0	0	0	0	9	0	0	0	0	0	0	0	0	0	0	0	0	0	0	0	9	9	0	1
Output voice	3	0.11122	0.00043	0	0	0	0	0	0	0	0	9	3	0	0	0	0	0	0	0	0	0	0	0	0	0	9	9	0	1
Catch frontal picture	3	0.11196	0.00264	0	0	0	0	0	0	0	0	0	0	9	0	0	0	0	0	0	0	0	0	0	0	0	9	9	0	1
Catch voice (SIRI)	3	0.11667	0.01678	0	0	0	0	0	0	0	0	0	0	0	9	0	0	0	0	0	0	0	0	0	0	0	9	9	0	1
Capture sound	3	0.11667	0.01678	0	0	0	0	0	0	0	0	0	0	0	0	9	0	0	0	0	0	0	0	0	0	0	9	9	0	1
Capture picture	3	0.11224	0.00348	0	0	0	0	0	0	0	0	0	0	0	0	0	9	3	0	0	0	0	0	0	0	0	9	9	0	1
Illuminate vicinity	1	0.03705	0.00023	0	0	0	0	0	0	0	0	0	0	0	0	0	0	0	9	0	0	0	0	0	0	0	9	9	0	1
Execute algorithm	9	0.33335	0.00012	0	0	0	0	0	0	0	0	0	0	0	0	0	0	0	0	9	0	0	0	0	0	0	9	9	0	0
Turn unit ON/OFF	9	0.33340	0.00017	0	0	0	0	0	0	0	0	0	0	0	0	0	0	0	0	0	9	0	0	0	0	0	9	9	0	1
Force silent mode	3	0.11113	0.00017	0	0	0	0	0	0	0	0	0	0	0	0	0	0	0	0	0	0	9	0	0	0	0	9	9	0	1
Force Vol. UP	3	0.11113	0.00017	0	0	0	0	0	0	0	0	0	0	0	0	0	0	0	0	0	0	0	9	0	0	0	9	9	0	1
Force Vol. DOWN	3	0.11113	0.00017	0	0	0	0	0	0	0	0	0	0	0	0	0	0	0	0	0	0	0	0	9	0	0	9	9	0	1
Interface headset	3	0.11115	0.00021	0	0	0	0	0	0	0	0	0	0	0	0	0	0	0	0	0	0	0	0	0	9	0	9	9	0	0
Data transmission	3	0.11115	0.00021	0	0	0	0	0	0	0	0	0	0	0	0	0	0	0	0	0	0	0	0	0	9	0	9	9	0	0
Allow power charge	9	0.33615	0.00292	0	0	0	0	0	0	0	0	0	0	0	0	0	0	0	0	0	0	0	0	0	3	3	9	9	0	1
Provide power	9	0.33334	0.00012	0	0	0	0	0	0	0	0	0	0	0	0	0	0	0	0	0	0	0	0	0	0	0	9	9	0	9
Route electricity	9	0.33334	0.00011	0	0	0	0	0	0	0	0	0	0	0	0	0	0	0	0	0	0	0	0	0	0	0	9	9	0	1
Block moisture	3	0.00000	0.00002	0	0	0	0	0	0	0	0	0	0	0	0	0	0	0	0	0	0	0	0	0	0	0	0	0	9	9
Hold parts together	9	0.00001	0.00001	0	0	0	0	0	0	0	0	0	0	0	0	0	0	0	0	0	0	0	0	0	0	0	0	0	0	9
Define looks	3	0.00186	0.00557	0	0	0	0	0	0	3	0	0	0	0	0	0	0	0	0	0	0	0	0	0	0	0	0	0	0	9

4 Discussion

4.1 Lithium-Ion Battery Life

The 2 to 3 years of life for lithium-ion batteries neglects the fact that they are replaceable upon them reaching the end of their lives. Many users have experience with swollen lithium-ion batteries [13]. Pushing on a bulged lithium battery can lead to smoke or fire, thus, if we feel that our smartphones have grown fatter, they no longer stay flat on a flat surface, or the batteries start to drain quickly, we turn them in for service to have their batteries replaced.

If the user replaces a swollen battery, the life is elongated to 6 years from 3 years, and repeating this replacement can keep the life of the battery function of supplying power indefinite. Table 3 in its right two columns shows the SFFR and FFR when the AFR of lithium-ion battery is set to 0.00010, a value that assumes timely battery replacement by the user. Note that some of the FRs with significance of 9 gain high reliability, e.g., "Executing algorithm" or "Turning unit ON/OFF." The FRs of "Providing net/cell connection" and "User inputs intention" with significance 9 keep relatively high SFFR among the FRs due to their respective reliance on "Cellular antenna" and "Digitizer."

Table 3. Significance and Functional Failure Rates of Smartphone FRs with AFR of Li-ion battery at 0.33333 and at 0.00010

	Significance of Function	SFFRi: Significant Functional Failure Rate AFR(Li-ion)=0.33333	FFRi: Functional Failure Rate AFR(Li-ion)=0.33333	SFFRi: Significant Functional Failure Rate AFR(Li-ion)=0.00010	FFRi: Functional Failure Rate AFR(Li-ion)=0.00010
Allow SIM change	3	0.00003	0.00010	0.00003	0.00010
Provide user ID	3	0.11112	0.33335	0.00004	0.00012
Provide net/cell conn.	9	0.34502	0.34502	0.01179	0.01179
Notify text/call IN	3	0.11479	0.34438	0.00372	0.01115
User inputs intention	9	0.34167	0.34167	0.00844	0.00844
Display information	3	0.11667	0.35001	0.00559	0.01678
Display caller	3	0.11667	0.35001	0.00559	0.01678
Catch voice (phone)	3	0.11667	0.35001	0.00559	0.01678
Output voice	3	0.11122	0.33366	0.00014	0.00043
Catch frontal picture	3	0.11196	0.33587	0.00088	0.00264
Catch voice (SIRI)	3	0.11667	0.35001	0.00559	0.01678
Capture sound	3	0.11667	0.35001	0.00559	0.01678
Capture picture	3	0.11224	0.33671	0.00116	0.00348
Illuminate vicinity	1	0.03705	0.33346	0.00003	0.00023
Execute algorithm	9	0.33335	0.33335	0.00012	0.00012
Turn unit ON/OFF	9	0.33340	0.33340	0.00017	0.00017
Force silent mode	3	0.11113	0.33340	0.00006	0.00017
Force Vol. UP	3	0.11113	0.33340	0.00006	0.00017
Force Vol. DOWN	3	0.11113	0.33340	0.00006	0.00017
Interface headset	3	0.11115	0.33344	0.00007	0.00021
Data transmission	3	0.11115	0.33344	0.00007	0.00021
Allow power charge	9	0.33615	0.33615	0.00292	0.00292
Provide power	9	0.33334	0.33334	0.00012	0.00012
Route electricity	9	0.33334	0.33334	0.00011	0.00011
Block moisture	3	0.00000	0.00001	0.00001	0.00002
Hold parts together	9	0.00001	0.00001	0.00001	0.00001
Define looks	3	0.00186	0.00557	0.00186	0.00557

When AFR of the lithium-ion battery drops low with the user intervention, AFR values of other DPs of antennas, digitizer, and microphones turn relatively high.

5 Conclusion

Risk assessment of the type shown in Fig. 1 has a set of events and consequences that are identical and, thus, assumes each consequence depends only on its corresponding event and is independent from all other events. Such risk assessment, therefore, may be misleading, however, practicing one is far better than not performing any risk analysis.

The graphical interface of a DRG makes the zig-zagging easier in trying to figure out an uncoupled set of FRs for a design.

When armed with an algebraic tool of matrix computation, like AD, we can carry out risk analysis in a more systematic way. In the case of a smartphone, the power source of lithium-ion battery and cables that route the power and electrical signals influence almost all functions of the unit and without them, a smartphone turns into just a shiny block.

In our case of smartphone analysis, we identified its weak point of lithium-ion battery, however, with the user deciding to replace an old battery, severity of unacceptable risks drop.

Upon analyzing risks associated with more catastrophic events like earthquakes, tsunami, or tornadoes, we cannot intervene with the events themselves and thus, all we can do to reduce severity of consequences is to work on elements of the matrix, i.e., the link strength of 1, 3, and 9 and make the links weaker. For example, if we have a building that is very likely to collapse with a magnitude 7 earthquake (a strong link 9), we can reinforce its structure, so it is unlikely to collapse in case of such an event (a weak link 1).

References

1. Risk Assessment, U.S. Department of Homeland Security. https://www.ready.gov/risk-assessment. Accessed May 18 2022
2. Suh, N.P.: Axiomatic Design, Advances and Applications. Oxford University Press, Oxford (2001)
3. iPhone 11 technical specification, Apple. https://www.apple.com/iphone-11/specs/. Accessed 20 May 2022
4. Iino, K., Nakao, M.: Design record graph and axiomatic design for creative design education, PROCIR-D-16–00377. In: Proceedings of ICAD (2016)
5. Pallaver, M., Do, S.: Failure mode analysis as an implementation of axiom 2 in the axiomatic design functional decomposition process. In: Proceedings of ICAD 2013. https://www.axiomaticdesign.com/wp-content/uploads/18-pallaver-and-do-eproceedings.pdf. Accessed 30 Mar 2023
6. Shipyard, Can I use a lightning charger with a IPad 2, Apple Community. https://discussions.apple.com/thread/7588489. Accessed 26 June 2022.
7. Arcidiacono, G., Citti, P., Fontana, V., Martelli, T.: Reliability improvement of car sliding door using axiomatic approach. In: Proceedings of ICAD (2004)

8. Reliability Calculations and Statistics, Springer. https://link.springer.com/content/pdf/bbm:978-3-540-34582-4/1.pdf
9. What Is The Typical Lifespan Of Lithium-Ion Battery? Choose Solar Australia. https://www.choosesolar.com.au/typical-lifespan-lithium-ion-battery/. Accessed 29 Jun 2022
10. How mobile users spend their time on their smartphones in 2022, Insider Intelligence. https://www.insiderintelligence.com/insights/mobile-users-smartphone-usage. Accessed 29 Jun 2022
11. LCD Modules for Industrial Applications with an MTBF of 100,000 Hours, Crystal Display Systems. https://crystal-display.com/lcd-modules-for-industrial-applications-with-an-mtbf-of-100000-hours/. Accessed 29 Jun 2022
12. SWNS, Americans drop their phones at least four times a week, November 15, 2019. https://nypost.com/2019/11/15/americans-drop-their-phones-at-least-four-times-a-week/
13. Velazco, C.: Here's why your old rechargeable batteries ballooned in size, 18 April 2022, The Washington Post https://www.washingtonpost.com/technology/2022/04/18/lithium-ion-battery-swelling-why/

Axiomatic Design and The Diamond Model: Towards Defining the Role of Axiomatic Design in the Age of Data Driven Development

Michiel van Osch(✉) and Erik Puik

Fontys University Eindhoven, Rondom 1, 5612 AP Eindhoven, The Netherlands
{michiel.vanosch,e.puik}@fontys.nl

Abstract. In 2018 Boeing defined the Diamond model as a framework for model-based engineering. The Diamond model can be seen as an engineering model that unifies Product Lifecycle Management, Model Based Engineering and Digital Twinning. In essence, the Diamond model is a model for data driven development. Axiomatic Design is another model for systems engineering. In axiomatic design requirements are expressed in four design domains and two important design principles are enforced. In this paper we investigate how several qualities of axiomatic design are being applied in data driven development, using the diamond model as starting point. The goal is to investigate how AD can be beneficial to modern day design practice.

Keywords: Axiomatic Design · Diamond Model · Digital Twinning · Model-based engineering Data Driven Development · Product Lifecycle Management

1 Introduction

The Unified Modeling Language UML [1] has been a software modeling standard since 1987 and has been used for decades as a graphical language to represent software designs. At the time of the development of SysML [2] the term Model Based Engineering (MBE) [3] was created. In the last decade MBE has become a common method for engineering large scale (complex) systems. In recent years digital twinning [4] has also become popular. A digital twin is a virtual model of a system. A digital twin is more than a simulation because it uses real-time data from the physical system as input and can also provide real time input to the physical system (and hereby control the physical system). Product Lifecycle Management (PLM) is an engineering methodology where all aspects of a product life are part of the engineering process, from conception to retirement. Data is shared in all phases of the development and use through digital models. INCOSE has identified MBE, digital twinning and digitization as key methodologies required for systems engineering in the future in its vision for 2035 [5]. To unify MBE with physical system development and digital twinning in one model, Boeing defined the diamond model in 2018 [6, 7]. The Diamond model is a visual representation of the MBE and PLM process. The models are used for digital twinning purposes via a so called digital thread which is the infrastructure for data sharing between physical system and model(s).

E. Puik et al. (Eds.): ICAD 2023, LNNS 849, pp. 275–284, 2024.
https://doi.org/10.1007/978-3-031-49920-3_18

Axiomatic design is another method for engineering systems [8]. It presents a different view on systems engineering compared to the popular V-model. This paper investigate how these principles of AD are being applied in MBE and PLM. This research is performed by relating four key principles of AD to the diamond model.

- Separating a design in four domains
- decomposition
- relation between times in different domains
- zigzagging between domains

The goal of this research is to see whether applying AD next to MBE and PLM can have added value in engineering complex systems, which is considered future work.

This paper is organized as follows. In Sect. 2 background information is given on the V-Model, MBE, digital twinning, and the diamond model. In Sect. 3 the use of the AD domains and the use decomposition is investigated. In Sect. 4 the relation between domains and the principle of zigzagging between domains is investigated. In Sect. 5 the results of this research are discussed. In Sect. 6 the conclusions of this investigation are summarized.

2 Background Information

2.1 The V-Model

The V-Model has been a commonly used process for engineering a system for decades. Many different definitions exist of this model. In this paper the definition of the V-Model according to the VDI 2206 guideline is used as reference [9].

In the VDI model, development is started by defining the requirements. After these requirements are defined the system design is made. The system specification describes the requirements of the system from a system perspective, e.g., "the system should be able to carry goods packed in boxes". Via a creative process, different concepts of the system are generated that lead to a high-level system design (system architecture). In this architecture the different system components are already distinguished but not specified. Detailed design consists of defining the requirements for these individual components (component specification) and creating the detailed design of each component. In the VDI definition of the V-Model this is called domain-specific design. After this step the system is built and integrated. First each component is assembled and tested. When all tests are passed the components are integrated end the integration of components is tested. After this step the system is tested against the system requirements. The difference between integration testing and system testing is that in integration testing it is only tested whether components are connected as required while in system testing it is tested that the system functions according to the system specification. Finally the system is validated by the customer, on the customer site. When testing fails at some level and a redesign needs to be made the process falls back to making adaptations in the corresponding design phase and following the steps of the model again. Tests for each phase are defined in the design phases to guarantee the system is testable (Fig. 1).

In theory this is a very rigid model. In practice often, a design is made iteratively by moving from requirements down to domain specific design and (e.g. because it turns

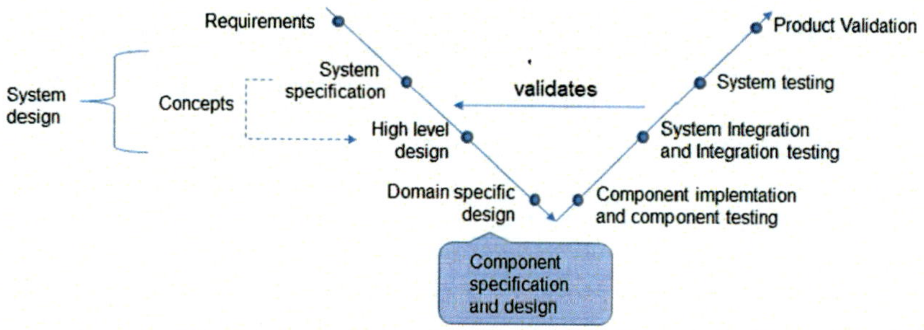

Fig. 1. The V-Model.

out a requirement could not be satisfied within certain hidden constraints) the system architecture and even requirements are adapted and a new version of the design is made.

2.2 Model Based Engineering

In classical engineering different teams are working mostly separately and specifications are made on paper. Verification is done by having design reviews. The system is tested based on test plans and test cases defined on paper. In model-based engineering (MBE) virtual models are made throughout the engineering process. Engineers from different disciplines collaborate to design and redesign the system through a shared digital environment. The customer is involved in the process giving feedback on virtual models and simulations of the design. The system is tested by test cases generated automatically from the models and correctness is validated against the models. Tool support is used to validate e.g., whether a system architecture is complete with respect to the system specification. In the model-based engineering definition used in this paper models are graphical representations of a system on different levels such as UML models, SysML models, but also cad designs and simulations.

2.3 The Diamond Model

In 2018 the diamond model (Fig. 2) was first presented by Daniel Seal of Boeing during the Global Product Data Interoperability Summit (GPDIS) [10]. Boeing was looking for a methodology for MBE represented by something as recognizable as the systems engineering V. The Diamond Model in essence adds a mirrored V shape on top of the regular V resulting in a diamond shape representing the design phases for the virtual system. This makes the Diamond model applicable for both MBE as well as digital twin development. The connection between the top half and bottom half of the diamond is the so-called digital thread, which is the data-driven architecture that links together information during the product life cycle [11]. The Diamond Model also incorporates the modeling and virtual realization of the production system, which is a deviation from the V-Model.

Within the diamond model, the virtual models on the left side specify the physical system and the right-hand side allows validation of the physical system. The top side

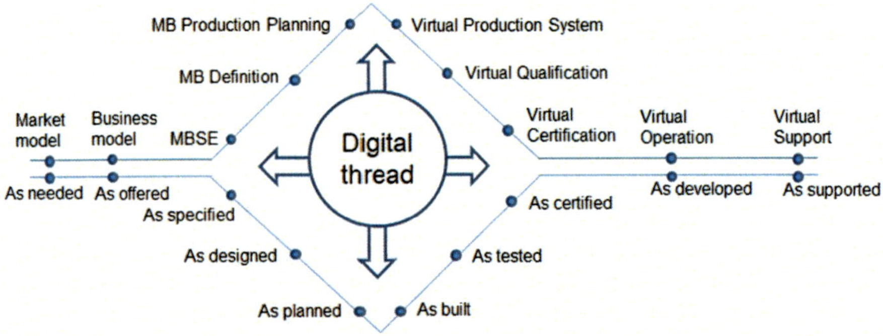

Fig. 2. The Diamond Model as defined by Boeing.

and the bottom side not necessarily have to stay in sync. On the top side it is allowed to move faster through phases and go back to earlier phases when redesign is needed.

In the top part of the diamond, the market model defines the customer needs and the business model defines the business needs. MBSE stands for Model Based Systems Engineering and defines the system specification (requirements) and concept of operations (concepts); and the Model Based Definition stands for the high level (system) design and the detailed design. The MB Production Planning defines how the system is built. The virtual production system is the digital twin of the actual production system. Virtual Qualification allows to partly validate the quality of a solution in a virtual environment so that less qualification needs to be done on the real system. Virtual Certification allows to validate whether the system meets the proper standards in a virtual environment. Virtual operation allows monitoring the system while it is in operation. Virtual support allows to trouble shoot a system virtually when problems occur.

The bottom part of the diamond defines the PLM process. In this process the different phases are defined by their milestones. Compared to the V-Model, it not only expresses the design phases from problem definition to finished product, but it also defines product operation phases. On the left-hand side, 'As needed' marks that the user requirements are defined. 'As offered' marks that the scope of the project is defined, and product will be delivered as agreed with the customer. Note that it may be that not all initial customer needs will be satisfied by the offer. 'As specified' marks the system requirements are specified for the physical system. 'As designed' marks that the physical system is designed. 'As planned' marks that that the production environment is ready, and building can start. On the right-hand side 'As build' marks the physical system has been built, 'As tested' marks the physical system functions as tested, and 'As certified' marks the physical system is certified as required to the applicable regulations. 'As develops' marks that the system functions as desired and 'As supported' marks that the system is supported as how it is supposed to be supported.

Vertically and horizontally the design stages, virtual system and physical system milestones are linked. E.g., production is supposed to be planned as designed in the virtual production system and the system is build using that production system.

3 Inside Domains

3.1 Domains

AD separates the design into four domains: (1) Customer attributes (CA) that define the needs of the customer; (2) Functional Requirements (FR) that define the behavioral requirements of the system; (3) Design Parameters (DP) that define the physical attributes of the system; and (4) Process Variables (PV) that define how the system is made.

Research in linking AD domains to MBE is scarce. Previously, Farid defined a mapping between AD domains and SysML diagrams [12]. Wang et al. map their MBSE design methodologies to AD [13], in doing so creating a new AD domain called the behavioral domain defining Behavioral Entities (BE) between FR and DP.

By taking the diamond model as starting point a link can be made as well (see Table 1) between AD and MBE by looking at the definitions:

Table 1. A mapping between MBE and AD domains.

MBE	AD
Market Model + Business Model	Customer Attributes
System Specification (part of MBSE)	Functional Requirements
MB Definition	Design Parameters
MB Production Planning	Process Variables

The As-X milestones in PLM where originally described by Airbus in 2013 [14, 15]. However, in these papers a clear definition of the products of each milestone is not given. A definition of the main products belonging to several design milestones is provided by Pinqué et al. [16]. In this article the product belonging to the As specified milestone is structured (system) requirements. The deliverable belonging to the As designed milestone is the engineering Bill of Materials (eBOM), describing all the (physical) parts of the system. The deliverable belonging to the As planned milestone is a manufacturing BOM (mBOM) describing how each part is produced. Table 2 shows the mapping from PLM milestones to AD domains. Furthermore, the deliverable belonging to the as needed and as offered milestone are the customer/market requirements [17].

Table 2. A mapping between PLM milestones and AD domains.

PLM milestone	PLM product	AD
As needed + As offered	Customer/market Requirements	Customer Attributes
As specified	Technical requirements	Functional Requirements
As designed	engineering BOM	Design Parameters
As planned	manufacturing BOM	Process Variables

3.2 Decomposition

An important principle of Axiomatic Design is decomposition. Items like requirements are decomposed into sub requirements which are further decomposed until every item has “bite-size” solutions. This brings structure in the design to be made and helps in validating that the design is complete. Decomposition is used both in MBE and PLM as well. In MBE a SysML requirements diagram has a tree structure. In SysML a block diagram is hierarchical. It is possible to define blocks within blocks.

In PLM customer requirements are decomposed [17]. Technical requirements and BOMs are structured hierarchically as well and called by this name [16]. MB Production planning is also hierarchical as it is directly linked to the product structure. Each component and subcomponent need to be produced in some way. Table 3 gives an overview on how decomposition is implemented in MBE, PLM, and AD.

Table 3. Decomposition in MBE, PLM, and AD.

MBE	PLM	AD
SysML requirements diagram	Structured customer requirements	Customer attributes
SysML requirements diagram	Structured technical requirements	Functional requirements
SysML e.g. block diagrams	Structured eBOM	Design parameters
Hierarchical production planning	Structured mBOM	Process variables

4 Between Domains

4.1 Relation Between Items in Domains

In Axiomatic Design Items in different domains are linked as well. Items within a tree structure in one domain link to the items on the same hierarchical level in another domain. This structure gives insight in the completeness of the design. This relation is expressed via design matrices. A design matrix describes the mapping between items in different domains in a matrix form. In MBE, SysML offers the concept of traceability matrices to validate whether all requirements are satisfied in the next step of the design. This concept is very similar to the design matrices of AD. The difference between design matrices and traceability matrices lies in the purpose. Design matrices in AD are used to optimize the design and create decoupled designs. These are designs where ideally one functional requirement is linked to one design parameter.

Also, in PLM items in each stage are linked. In PLM these are called linking structures. E.g., All items in an eBOM link to an item in an mBOM (Fig. 3).

$$\begin{Bmatrix} FR1 \\ FR2 \end{Bmatrix} = \begin{bmatrix} X & 0 \\ 0 & X \end{bmatrix} \begin{Bmatrix} DP1 \\ DP2 \end{Bmatrix}$$

	Satisfied by	
	Block 1	Block 2
Req1	X	
Req2		X

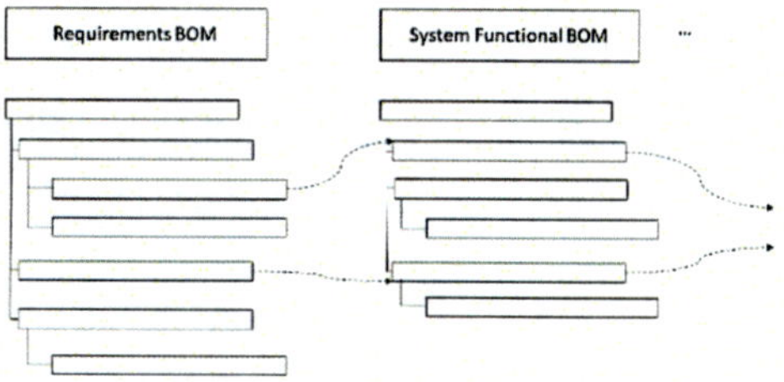

Fig. 3. AD design matrix (top left), SysML traceability matrix (bottom left), and PLM linking structures [18] (right)

4.2 Zigzagging

An Axiomatic Design is made by zigzagging between domains. In Axiomatic Design the Domains are ordered from left to right. Deriving items from left to right is called Zig- and deriving items from right to left called Zag. For instance, during design functional requirements lead to new design parameters (Zig) and process variables lead to new design parameters as well (Zag) (Fig. 4).

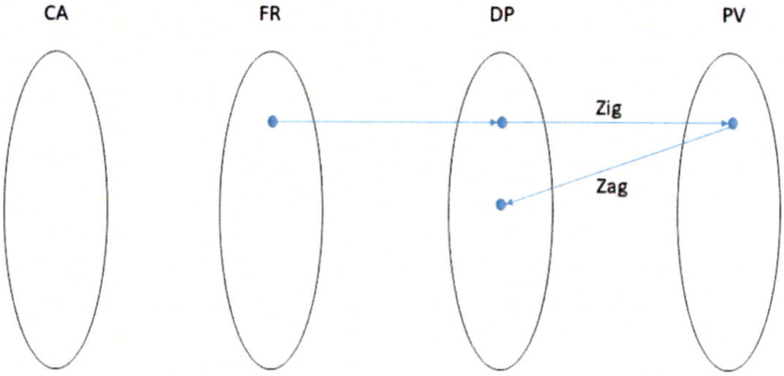

Fig. 4. Zig and Zag in AD

The process of Zigzagging is not made explicit in the Diamond Model. However, in engineering practice it is applied often. Iterative design can be seen as zigzagging. In MBE a credo is "fail fast, fail often". Designers are encouraged to make rough models in all domains and improve upon them to come up with the final design. MBE tools allow continuous tweaking of models in every stage of the design and through version management the history is maintained so designers can go back to a previous version.

A case can be made that Zigzagging is already embedded inside the V model. Figure 5 presents a mapping of the V-Model to AD. The Customer needs in our V-Model directly map to customer attributes in axiomatic design. The System specification maps to system level functional requirements in AD (FR). Through the concepts the system design is acquired (DP). The system design defines the physical system itself and is therefore in our view a representation of the design parameters. The domain specific design leads to new requirements (FR') and from those requirements a component design is defined (DP').

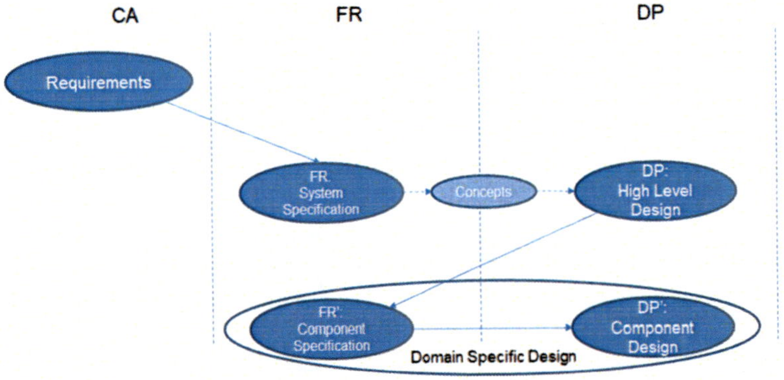

Fig. 5. A mapping of the V-Model to AD

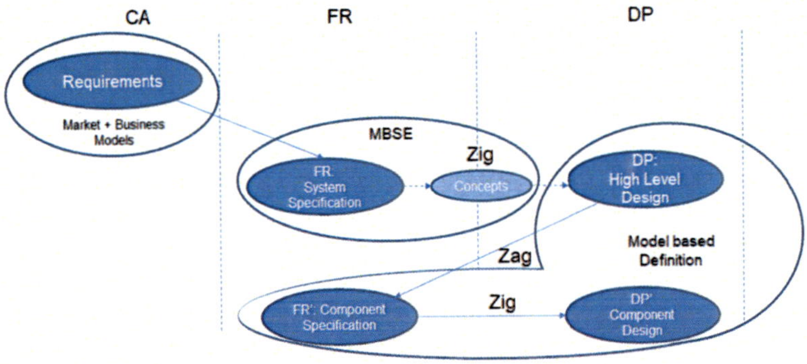

Fig. 6. A mapping of the V-Model, AD, and the Diamond model

The MBE phases in the diamond model can now directly be mapped into this representation of the V-Model. The customer attributes correspond to the market models and business models of the MBE Diamond. The System specification and concepts correspond directly to MBSE. High level design, component specification and component design correspond directly to MBD (Fig. 6).

The workflow in the Diamond model is not clearly defined. The top half and the bottom half of the diamond model are not synchronized. Several iterations can be made in the top half before on the bottom half the respective BOMs are generated. After that, when the design is revised the BOMs are updated as well. However, a hard case that the principle of zigzagging occurs in the bottom side of the Diamond model cannot be made.

5 Discussion

This paper analyses several attributes of AD and whether they are also applied in MBE and PLM processes. One important feature of AD has not been investigated, namely applying the two design axioms of creating uncoupled designs and limiting the information in the design. It will be interesting to investigate in which way in MBE and PLM these axioms can be enforced in making "good" designs. A second issue that has not been investigated yet is how MBE and PLM ensure completeness of a design. How does the designer know the design is finished? This is also a challenge in AD. A third issue that has not been investigated yet is how constraints (like cost, weight, etc.) are managed in MBE and PLM, as it is also an issue in AD design.

When all of these are investigated a clear picture arises on what aspects of AD are applied in practice. This may answer the question of how applying AD can still strengthen modern engineering practice. For instance, BOM structures are very similar to the design domains in AD. Would looking at the BOM structures in PLM and creating design matrices of BOMs help in creating decoupled model-based designs? Or could it be beneficial to have MBE inform an AD and vice versa, and generate the BOMs from the AD?

6 Conclusions

MBE, PLM, and digital engineering more and more becomes (if not already is) the preferred way of engineering complex systems. The Diamond model is an elegant model for linking MBE to PLM. AD is a different design methodology. In this paper several attributes of AD were compared to how they are applied in MBE and PLM, using the Diamond model as reference. For the most part the principles of AD are also applied in some form in MBE and PLM. The goal of this research is to define the role of AD in future engineering practice, which is ongoing research.

Acknowledgements. This research was part of the project "Digital Twin Academy" supported by the EU through the Interreg Euregio Maas-Rhein program, the province of Noord-Limburg (the Netherlands) and the province of Noord-Brabant (the Netherlands).

References

1. ISO/IEC 19505: Information Technology - Object Management Group Unified Modeling Language (OMG UML), ISO (2017)
2. ISO/IEC 19514: Information Technology – Object Management Group Systems Modeling Language (OMG SysML), ISO (2017)
3. INCOSE: INCOSE Systems Engineering Vision 2020, INCOSE-TP-2004-004-02 (2017)
4. INCOSE: SEBoK. https://www.sebokwiki.org/wiki/Digital_Engineering. Accessed 25 Jan 2023
5. Stoewer, H., Nichols, D.: Building the systems engineering workforce of the future, White paper (2022)
6. Seal, D.: The Model-Based Engineering (MBE) Diamond, A Framework for Digital Transformation, at INCOSE International Workshop 2020, Presentation (2020)
7. Seal, D.: The Model-Based Engineering (MBE) Diamond, A Framework for Digital Transformation. https://www.youtube.com/watch?v=yDzF-K1i1sg. Accessed 25 Jan 2023
8. Suh, N.P., Bell, A., Grossard, D.: On an axiomatic approach to manufacturing and manufacturing systems. J. Manuf. Sci. Eng. **100**(2), 127–130 (1978)

9. Gausemeier, J., Moehringer, S.: New guideline VDI 2206 – a flexible procedure model for the design of mechatronics systems. In: ICED 03, International Conference on Engineering Design (2003)
10. Seal, D.: The systems engineering "V" – is it still relevant in the digital age? At Global Product DATA Interoperability Summit (GPDIS) 2018, Presentation (2018)
11. Singh, V., Willcox, E.: Engineering design with digital thread. In: AIAA/ASCE/AHS/ASC Structures, Structural Dynamics, and Materials Conference (2018)
12. Farid, A.M.: An engineering systems introduction to axiomatic design. In: Farid, A.M., Suh, Nam P. (eds.) Axiomatic Design in Large Systems, pp. 3–47. Springer, Cham (2016). https://doi.org/10.1007/978-3-319-32388-6_1
13. Wang, H., Li, H., Tang, C., Zhang, X., Wen, X.: Unified design approach for systems engineering by integrating model-based systems design with axiomatic design. Syst. Eng. J. Int. Council Syst. Eng. **23**(1), 49–64 (2020)
14. Mas, F., Rios, A., Menéndez, J.L., Oliva, M.: A process oriented approach to modeling the conceptual design of aircraft assembly lines. Int. J. Adv. Manuf. Technol. **67**, 771–794 (2013)
15. Mas, F., Menéndez, J.L., Oliva, M., Gomez A., Rios, J.: Collaborative engineering paradigm applied to the aerospace industry. In: IFIP Advances in Information and Communication Technology (2013)
16. Pinqué, R., Rivest, L., Segonds, F., Veron, P.: An illustrated glossary of ambiguous PLM terms used in discrete manufacturing. Int. J. Prod. Lifecycle Manage. **8**(2), 142–171 (2015)
17. Siemens: Customer Needs Management: Leveraging Requirements Management to Deliver Products that Customers Will Buy, White-paper. https://www.siemens.com/teamcenter. Accessed 23 June 2014
18. Bitzer, M., Vielhaber, M., Kaspar, J.: Product Lifecycle Management – How to Adapt PLM to Support Changing Product Development Processes in Industry? NordDesign (2016)

Axiomatic Design Meets Model-Based Systems Engineering (MBSE)

David S. Cochran(✉), Joseph Smith, and John Fitch

Purdue Fort Wayne, Fort Wayne, IN 46805, USA
cochrand@pfw.edu

Abstract. The be able to design sustainable systems relies on the ability to associate and communicate design information to meet customer needs. Model-Based Systems Engineering (MBSE) provides a means to integrate the classes of information that Axiomatic Design (AD) uses to establish a hybrid (best-of both) design approach. The strength of AD is that it offers axioms to make effective design decisions. MBSE can associate all sources of information for the Systems Engineering (SE) lifecycle. AD offers design veracity. MBSE manages information such as the logic and thought process required in design decision making, how design decisions influence risk, and the results of test plans to verify and validate a design. To advance AD and MBSE, an information metamodel is required to integrate information classes used by AD and MBSE. The benefit of integrating AD with MBSE is first that an overall design decision pattern may be used to kick-start any new product or system design. The Functional Requirements (FRs) of a design may then be associated with the decision pattern. Second, MBSE characterizes use cases very well. Customer Needs (CNs) and FRs may then be tied to the description of how a product or service is used. The information metamodel also ties risk to a chosen solution during FMEA. To address procedural errors described by Thompson, MBSE manages the identified need to distinguish between architectural FR – PS coupling, and performance FRm – DP coupling. When combined, AD-MBSE enables lifecycle design information integration, the ability to express new design viewpoints, and a single source of truth for a System of Interest. The information metamodel has been tested on the classical water faucet design illustration and the skin graft work by Gabela and Suh.

Keywords: system · systems engineering · systems architecture · axiomatic design · modeling · viewpoints

1 Background

1.1 Axiomatic Design

Axiomatic Design represents the requirements derivation process as occurring across four domains [1]:

- Customer Needs (CN)
- Functional Requirements (FR)

E. Puik et al. (Eds.): ICAD 2023, LNNS 849, pp. 285–297, 2024.
https://doi.org/10.1007/978-3-031-49920-3_19

- Design Parameters (DP) also known as Physical Solution (PS)
- Process Variables (PV)

As used in product development, AD seeks to maintain the independence of the FRs through the selection of DPs, i.e., to eliminate coupling between FRs that results from a chosen design solution. This simple and powerful insight has generated improvements to product/system designs across multiple industries and technology domains.

1.2 Observations Regarding AD Practice

Despite its well-published successes, AD practice varies widely across its broad global community due to a lack of standards in taxonomy, for example, practitioners may write FRs in a variety of styles and may often confuse FRs with other types of requirements [2, 3]. Some FRs are stated as quantitative parameters to be improved, e.g., '*decrease infection probability*' (for an artificial skin graft product and grafting process) and others are stated as pure functions, e.g., '*prevent infection.*'

AD practice also varies widely with respect to how AD is applied. For instance, AD may be used to develop new products and systems, or as a language to express design innovations and inventions that others have developed without applying the rigor of AD's Axioms 1 and 2.

Thompson attributes difficulties in defining FRs to the limitations of AD's requirements classification scheme. For example, Thompson describes that there are only two categories for which requirements information can be captured; Functional Requirements and constraints. Thompson goes on to point out that additional information exists, but AD has no formal way of capturing the remaining information [2]. The additional design information (i.e., non-functional attributes of the system of interest and design goals (selection/optimization criteria) is unlikely to be effectively considered in the design process for two reasons. First, this additional design information may be deemed unimportant or unnecessary if the classification scheme does not explicitly provide a method or taxonomy for capturing the information. Second, the AD practitioner may choose to improvise by capturing additional design information as FRs. This improvision can lead to five common procedural errors when defining FRs as noted by Thompson [2]:

1. Mixing FRs with design parameters (DPs) - The designer confuses the "how" with the "what" and ultimately limits the design space by embedding the solution in the requirement.
2. Mixing FRs with other types of requirements – non-Functional Requirements, selection criteria, and optimization criteria are captured as FRs.
3. Mixing the FRs of the various stakeholders and of the system of interest – The designer captures FRs of the stakeholders vs. pure functions that the system must perform.
4. Mixing the FRs of the system of interest and of related systems – A poor definition of the system boundary can lead to the incorporation of FRs of related systems into the design of the system of interest. A mix of functions (FRs) that the system of interest must perform with functions that related systems must perform will convolute the functional definition of the system of interest.

5. Defining negative FRs – The FR reads as what the system should not do instead of what the system must do.

1.3 AD's Limitations and Systems Engineering Objectives in the Design Lifecycle

In addition to design procedural issues noted in Sect. 1.3, AD's viewpoint in assessing design acceptability based primarily on maintaining the independence of FRs, is missing numerous key elements of a full life cycle domain-independent systems engineering methodology. Systems engineering as discipline defines design requirements for each phase of a product or enterprise systems and derivatives to achieve the following design results.

- The first desired result is to provide a design synthesis engine that conceives full solutions to a problem by combining multiple system viewpoints and elements in novel ways.
- Second, systems engineering requires the expression of a model of the system physical architecture and the interfaces between system elements. Design decompositions must have system build instructions that explicitly define the interfaces between the constituent parts and the configuration of the assembled parts. The limitation with AD decompositions is that the information required to do integration of the constituent parts (PSs/DPs) is not defined or captured. For example, a list or pile of parts is powerless to deliver the required FRs and performance because it does not define how to build the system.
- Third, systems engineering provides a method to model the dependencies and interactions between system functions in terms of control flow and item flow which define the inputs/outputs of matter, energy, or information.
- Fourth, system engineering requires a design to maintain alignment and completeness, correctness, and consistency between the system physical, functional, mathematical (performance) and ECAD models. Every design decision that answers, "How will this FR be satisfied?" further elaborates each of these four models to create a digital thread.
- Fifth, systems engineering provides an integrated approach for identifying and addressing system failure modes and risks. What could go wrong and what should be done to mitigate it?
- Six, A complete design decision-making methodology that screens out infeasible solutions and identifies best-performing solutions, taking uncertainty into account while capturing and communicating decision rationale.
- Seven, A published and standardized information metamodel that provides a schema (classes and relationships) to capture design information. The metamodel enables suppliers to develop software tools that fully implement AD and to provide data interfaces with other engineering toolsets.
- Eight, A mechanism for accounting for emergent behaviors of a system that cannot be attributed to a single component in a system decomposition hierarchy.
- A technique for reliably and comprehensively identifying the derived requirements that flow from commitment to a specific design alternative in any design decision.

2 Model-Based Systems Engineering

This section discusses how the current practice of Model-Based Systems Engineering addresses the objectives of systems engineering described in Sect. 1.3, both procedural and gaps in capability, associated with AD theory and practice. The Systems Engineering Vision 2035 document, published in 2021 by the International Council on Systems Engineering (INCOSE), identifies five engineering practices that are deemed to be transitioning from a status of emerging to becoming Standard Practice in Systems Engineering [4].

- Product Line Engineering
- Agile Methods
- Design for Resiliency
- *Model-Based Systems Engineering (MBSE)*
- System of Systems

Referencing the INCOSE Systems Engineering Vision 2020 (published in 2007), the fourth edition of the INCOSE Systems Engineering Handbook defines MBSE as "the formalized application of modeling to support system requirements, design, analysis, verification, and validation activities beginning in the conceptual design phase and continuing throughout the development and later life cycle phases" [5]. MBSE, when contrasted with traditional document-based engineering approaches, is believed to offer significant benefits, including:

- Improved communication among stakeholders.
- Increased ability to manage system complexity.
- Improved product quality.
- Enhanced knowledge capture and reuse.
- Improved ability to teach and learn SE fundamentals.

The handbook provides an overview of leading MBSE methodologies, with details of two leading approaches, Function-Based Systems Engineering (FBSE) and Object-Oriented Systems Engineering Method (OOSEM). The handbook asserts that the "system model is the primary artifact of the SE process" without providing a definitive set of models that must be present for an engineering process to qualified as model based. Examples of system models, from ISO/IEC/IEEE Standard 15288 [6], include:

- Functional
- Behavioral
- Temporal
- Structural
- Mass
- Layout
- Network

Early proponents of MBSE created numerous innovations in modeling languages, visualizations and software tools that are well beyond the scope of this paper. ISO/IEC/IEEE 42010 recognized the diversity of modeling practices and representations by introducing the concept of architecture *viewpoints* and *views* [7].

The standard defines an architecture viewpoint as a "*work product establishing the conventions for the construction, interpretation and use of architecture views to frame specific system concerns.*" Viewpoints frame the architectural concerns of system stakeholders and define the notation and models used to capture and communicate each concern.

The standard defines an architecture view as a "*work product expressing the architecture of a system from the perspective of specific system concerns*". Views are instances of a viewpoint applied to a specific system. The relationship between viewpoints and views may be summarized as:

> *"A viewpoint is a way of looking at systems; a view is the result of applying a viewpoint to a particular system-of-interest."* [6]

For example, system stakeholders have a valid concern in desiring to understand and communicate:

- How a set of requirements and design goals drove the outcome of a design decision.
- How the chosen alternative creates new derived requirements that will impact the rest of the system design.

In response to this concern, an architectural viewpoint may be defined associated with Requirement-Decision-Requirement (R-D-R) Traceability that specifies a notation for visualizing this traceability thread in graphical or tabular form by including information consumed and created by the decision-making process:

- Requirements: Functional Requirements (FRs), Functional Requirement Measures (FRm's) and constraints.
- Evaluation Criteria: Factors used to evaluate the effectiveness of the alternatives.
- Decision: The fundamental question to be answered.
- Alternatives: Possible solutions to be evaluated.
- Risks & Opportunities: Ways that alternatives could fail or do better than expected.
- Mitigation and growth actions: Methods to reduce risks and grow opportunities.
- Derived Requirements: FRs, FRm's and constraints that are inherent consequences of a chosen alternative.

An instance of this viewpoint, the R-D-R Traceability *view* would be created when the R-D-R Traceability viewpoint specification is applied to a unique decision within a project. An example of such a view from the design of a cellular manufacturing system is shown in Fig. 1, below – adapted from [8].

Viewpoints and views provide a liberating structure that offers flexibility in capturing and visualizing system knowledge in a way that yields optimum insights for stakeholders across the system life cycle. These views and viewpoints break the pattern observed in document-centric engineering in which the method by which knowledge is stored and the method by which knowledge is communicated are coupled in the document artifact.

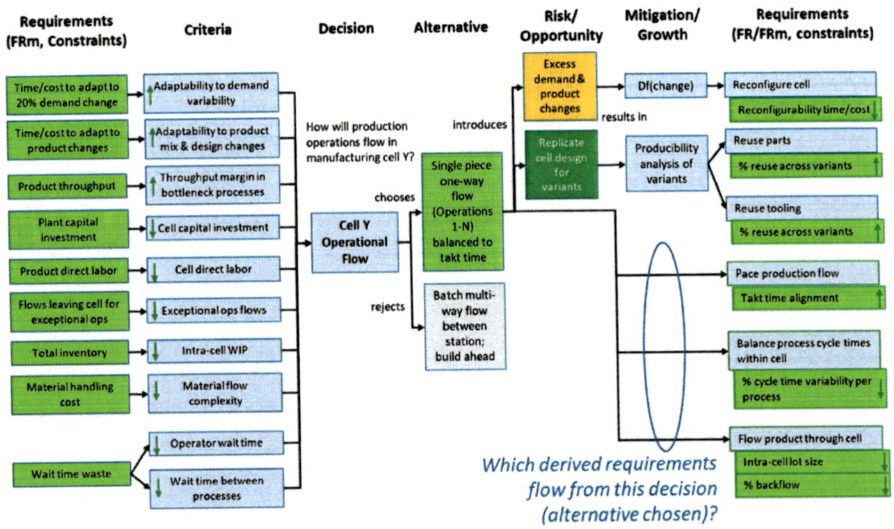

Fig. 1. Requirement-Decision-Requirement Traceability view for a Manufacturing System Design Decision.

3 Class-Based Information Metamodels

Views and viewpoints depend on precisely structured knowledge about the system and the thinking that drove its design. Knowledge structures (information metamodels) are typically described in terms of Entity classes, Relationships and Attributes (ERA) that support digital capture within a variety of database and modeling tools.

- Classes: The building blocks of engineering knowledge. *Requirements, decisions, components, risks, tasks,* etc.
- Relationships: Connections between entities that form the Digital Thread. *satisfies, performs, analyzes, results in, includes, couples,* etc.
- Attributes: "Fields" that precisely define characteristics of the entities with a class. *Priority* (of a decision), *Threshold* (value of a requirement), *Severity* (of a risk). *Selection Rationale* (of an alternative). *Target Date* (for completion of a task).

The demand for efficient information metamodels can only increase with increasing industry emphasis on Digital Engineering and the Digital Thread to address increasing system scale and complexity [9]. The engineering community needs ERA models that capture the essential information about the problem definition, design decision-making and solution descriptions and can populate an efficient set of system views that optimally engage stakeholders in various stages of the system life cycle. The goal of this set of views is collaborative thinking and fit-for-purpose insight. Different stakeholders need different views at different stages of product development. The ideal information metamodel for a specific development project includes the minimum number of entity classes, relationships, and attributes to support the highest value viewpoints and views. A compact representation of system knowledge helps to tame complexity, reduces ambiguity, enables filtering and navigation, and supports model integrity/validity checking.

Every engineering task populates instances of the SE information metamodel, i.e., new instances of each class, derivation/allocation/traceability relationships between these instances and attributes within them.

Viewpoints and views based on strict notations and modeling rules can be checked for completeness and consistency using automated or semi-automated techniques. Although such checks can detect structural defects within the model, these checks do not guarantee that the model represents the best possible system solution to the problem under analysis.

4 Development of the PFW AD/MBSE Information Metamodel

To illustrate the importance of a comprehensive, but efficient information metamodel, this paper highlights recent research on the Manufacturing System Design Decomposition, V10.0 (MSDD 10.0). MSDD 10.0 provides a design pattern for conceiving, evaluating, and creating sustainable manufacturing systems. The Purdue Fort Wayne (PFW) ERA model that is the basis for MSDD 10.0 is shown in Fig. 2, below – adapted from [8].

At the highest level of abstraction, the PFW information metamodel consists of three layers:

- Requirements Layer: Stakeholder needs, use cases and associated steps (actions), and formal requirements, including FR, FRm, and constraints.
- Decision Layer: The fundamental questions/issues that demand an answer/solution and the data that inform each decision.
- Solution Layer: Physical and logical architecture of the system based on design decisions.

These classes of knowledge and the relationships between them form the basis for multiple system viewpoints. Examples of the high value viewpoints include:

- Requirements hierarchy: Decomposition of FRs and associated FRm's for each FR.
- Functional Flow Block Diagram (FFBD): System functions, their dependencies (control flow) and the item flow between them.
- FRm Flowdown: Mathematical relationships between FRm's at each level of the FR decomposition.
- Decision Breakdown Structure: Hierarchy of design decisions with recommended alternatives.
- System Breakdown Structure: Elements (hardware/software components, people, facilities, or data) that make up the physical system.
- Physical Block Diagram: System elements and their interfaces.
- Digital Thread – Requirements-Decision-Requirements Traceability: See Fig. 1.

The ability to generate the desired views from these viewpoints demanded a significant extension to the modeling language (Vitech's Systems Metamodel) that was the basis for the MBSE tool (Vitech GENESYS) used as the software platform for this research initiative. Knowledge classes that were added include:

- Functional Requirement (FR)
- Functional Requirement Measure (FRm)
- Decision

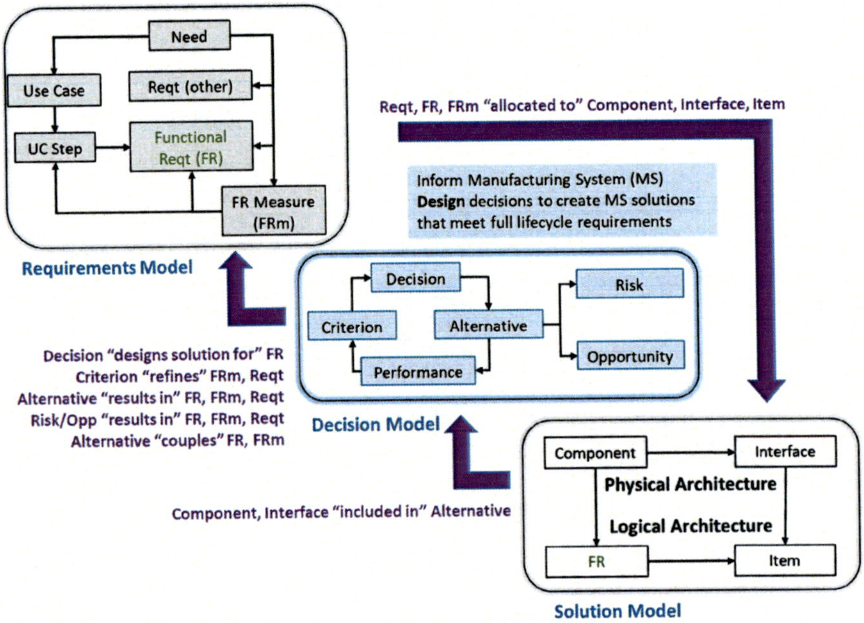

Fig. 2. Information Metamodel – Basis of the MSDD 10.0 Design Pattern

- Criterion
- Alternative
- Performance
- Risk Mitigation

Numerous relationships were added between these new classes and between the new classes and existing classes in Vitech's Systems Metamodel. The additional knowledge coverage and precision offered by new classes and relationships can enable substantial reduction of the procedural errors identified by Thompson and "fill" many of the methodology gaps observed in AD theory.

4.1 Functional Modeling Extensions

Relative to AD theory, the most significant extension in the PFW information metamodel has been forging a clear distinction between FRs and FRm's. Removing the FR-FRm ambiguity observed in AD practice directly addresses Thompson's error #2: *Mixing FRs with other types of requirements.*

The extended PFW information metamodel limits FRs to be "pure" functions that express the transformation of inputs into outputs. With this definition, an FR is always named using a Verb-Noun (Direct Object) syntax, e.g., *Prevent bacterial infection* or *Stimulate dermal cell growth* [10]. The dependency logic between FRs is captured using control flow notation to address such logical constructs as parallelism (potential concurrency of functions), exclusivity (OR branching between either FR1 or FR2), iteration, looping or replication.

In this model, FRm's are always associated with one and only one FR. FRm's act as adverbs, measures of performance, that specify "How well" the FR must be performed to meet stakeholder expectations. The FRm, *Probability of bacterial infection*, specifies the required (threshold) value of the *Prevent bacterial infection* FR. The FRm, *Dermal cell growth rate*, specifies the threshold value of the *Stimulate dermal cell growth* FR. A FR may (and most often will) be specified by multiple FRm's to address different aspects of "goodness" (stakeholder value) such as throughput, efficacy, output quality, efficiency of the transformation process, and consistency of outputs (e.g., robustness or repeatability) in the face of variable inputs.

The one-to-many FR-FRm cardinality noted above has a significant impact on AD decomposition views in which FRm's have been misidentified as FRs. It is inefficient to identify a single Physical Solution (PS) for each FRm, i.e., to turn each FRm into a separate "How will this level of performance be delivered?" decision. Most often a PS should be chosen as the method to satisfy a FR by meeting or exceeding the levels of performance demanded by multiple FRm's, all within relevant constraints, e.g., cost, time, resource consumption or other policy-driven limitations.

4.2 Functional Modeling of System Context

A functional model may be constructed for a system of interest that elaborates the role this system plays in broader context of its use cases or mission scenarios across the system life cycle. Such a top-level (contextual) functional model clarifies each functional interaction with stakeholders, e.g., users or maintainers, in terms of the inputs received from or outputs provided to these individuals/roles. Similarly, a functional model of the system context will explicitly define system functional interactions with external systems. As such, this model can directly resolve Thompson's error #3: *Mixing the FRs of the various stakeholders and of the system of interest* and error #4: *Mixing the FRs of the system of interest and of related systems.* "Mixing" is avoided by explicit allocation of all life cycle functions to the system of interest or to specific stakeholders or external interfacing systems in the broader system context.

4.3 Functional Modeling to Refine Decompositions

Precisely defining the control flow relationships between FRs and their inputs/outputs (item flow) clarifies the scope of each function – a process which can lead to discovery of missing or overlapping functions in the system model. Without an understanding of the functional model as a network of interacting functions, it is difficult to determine whether the intent of FR1: *Support wound healing* is completely and non-redundantly satisfied by child FRs 1.1 – 1.N, e.g., *Close wound, Form scar, Prevent bacterial infection, Aspirate wound, Prevent edema, Regenerate tissue, ...* Therefore, limiting the functional model notation to a decomposition hierarchy viewpoint increases the risk of errors and ambiguity in defining the FRs, with subsequent negative impacts on the remainder of the design process.

4.4 Functional Modeling to Distinguish Types of Coupling

Enforcing the FR-FRm distinction indirectly addresses Thompson's procedural error #1: *Mixing FRs with design parameters (DPs).* The AD viewpoint known as the Design Matrix makes sense as a mathematical construct intended to capture a design equation with coefficients that relate the value of each **FR** to the value of all DPs. However, such math is possible and meaningful only if the Design Matrix depicts FRm-DP relationships where both are FRm's and DPs are quantitative in nature and share units that can be related in the form of an equation. In common AD practice, the DPs have been redefined as Physical Solutions (PSs), not the quantitative measures of performance that are derived from these PSs. Despite shortfalls in the notation, AD practitioners have intuitively understood that there must be a mapping between:

- Functions and the physical solutions that deliver them: PS - > performs - > FR.
- The performance required (FRm's) for each function (FR) and the PSs that exhibit that performance, i.e., satisfy the FRm's by delivering required performance against a set of DPs.

By not distinguishing FRs-FRm's and morphing DPs into solutions, the definition of coupling that is visible in the Design Matrix has become ambiguous. It could be either:

- *Architectural coupling* where the item flow between functions (FRs) is such that changing how Function A is accomplished, with a different PS alternative, will have excessive ripple effort on numerous functions that send inputs to or receive outputs from Function A.
- *Performance coupling* where the ability to achieve the performance specified for Function A (in terms of any FRm) is affected by DPs associated with more than one PS.

These two types of coupling are very different mechanisms and have differing impacts on the success of a design. Architectural coupling drives up system complexity (think "spaghetti code" or a "Rube Goldberg" design) and the life cycle cost of design changes because of the number of shared inputs/outputs across all functions. Performance coupling creates a competition/tradeoff between measures of performance in which the improvement of one FRm that is valued by stakeholders results in the worsening of another valued FRm. It appears possible that a design with excessive architectural coupling may exhibit no performance coupling and vice versa. Further research is suggested to explore this hypothesis.

4.5 Decision Modeling Extensions

To satisfy a FR (by meeting the FRm's that specify the FR), there must always be a "*How will the system deliver Function X*?" decision. Without explicitly modeling the decision and the data that informs this choice, designers:

- Increase the risk of poor solutions, i.e., alternatives that are destined to fail and disappoint stakeholders.
- May often overlook novel solutions that might represent significant leaps in stakeholder value and satisfaction.

- Have difficulty in objectively evaluating a range of possible solutions.
- Have difficulty in capturing and communicating decision rationale to stakeholders to gain their commitment (in the form of goodwill and resources) to implementation.
- Fail to capture rejection rationale for non-viable alternatives, leading to second-guessing and fruitless rework.
- Greatly increases the cost of change, i.e., revisiting a decision when requirements change or assumptions/estimates concerning the solution are invalidated.

The traditional Design Decomposition/Map used in AD displays only the FR (or sometimes FRm's misclassified as FRs) and the chosen PS to satisfy the FR. The full decision logic behind that choice is not captured nor visualized in views that can prevent the potential decision failure modes listed above.

If a PS has been chosen to fulfill a FR, a decision has been made, but likely without sufficient preservation of decision rationale to ensure decision quality. Dependence on human memory is not a winning strategy for designing complex systems or in the face of potential staff turnover.

PFW's addition of a more thorough model of decision-making information resolves most of the concerns listed above:

- Decision: Frames the question to be answered and provides the context for all other decision analysis data.
- Criterion: Provides a method for clarifying the influence of any requirement (FR, FRm or constraint) or design goal in the context of a specific decision. Defines success and provides an objective way to evaluate solution alternatives.
- Alternative: Explicit definition of possible solutions, whether physical (combination of interacting components) or otherwise (range of use cases that a product will support, or the value proposition associated with each use case).
- Performance: Estimates of the effectiveness of each alternative against each criterion. When combined, this data populates an evaluation matrix or various graphical representations of the merits of the competing solutions.
- Risks and Opportunities: Potential tiebreakers between leading alternatives based on projections of what could go wrong or what could go better than expected.

Capture of the Decision-> *chooses* -> Alternative -> *results in* -> (Derived) Requirement thread makes explicit how the chosen solution *results in* the next level requirements. All requirements are derived from answers/solutions chosen in other, typically "upstream" decisions. The Alternative -> *results in* -> Requirement (FR, FRm, constraint) relationship specifies the source of next-level requirements and localizes the potential changes to the requirements that might occur if a different alternative must be chosen in the future. Similarly, the identification of risks and opportunities supports a derivation trace to requirements added for risk mitigation (by reducing likelihood or severity) or opportunity growth (by increasing likelihood and positive impact).

4.6 Architecture Modeling Extensions

AD identifies a DP or (PS) as the means to satisfy a FR. In practice, the definition provided for a DP/PS is too imprecise to specify how to build the system from physical or software components, facilities, or human tasks. PFW's extended information metamodel includes a PS -> *includes* -> Component relationship to overcome this ambiguity. Decisions (through alternatives chosen) are the source of solution architectures; more than one decision may contribute to defining the components and component-to-component interfaces required to satisfy the system requirements. Incorporation of this relationship resolves a loose end between AD and generally accepted systems engineering and systems architecture practices and viewpoints. When combined with the functional modeling approach discussed in Sect. 3.1, the PFW information metamodel provide a mechanism for aligning the physical architecture decomposition with the functional architecture decomposition at each branch of the decomposition hierarchy.

5 Conclusions

The PFW experience with an extended information metamodel supports the hypothesis that MBSE constructs are complementary with AD and can increase the overall value delivered during design. Highlights include:

- FR-FRm precision is needed to distinguish between performance and architectural coupling, both of which have differing impacts on system success and require different methods to resolve.
- Functional modeling, in the form of control and item flows, is a powerful tool to improve the completeness and quality of the FRs that drive AD.
- Functional modeling of the context of the system of interest can reduce confusion between system FRs and FRs allocated to stakeholders and external interfacing systems.
- Decision modeling extensions can improve decision quality and buy-in, provide a method for deriving next-level requirements in the AD zig-zagging process, and reduce the cost of managing change.
- Information metamodel extensions enable additional high value viewpoints that can improve stakeholder engagement across the system life cycle.

6 Future Research

Additional research is suggested to further investigate the benefits of combining the fundamental elements of AD with MBSE principles and practices. Examples of such research topics include:

- Confirmation of the distinction and independence between architectural and performance coupling implied by the FR-FRm distinction.
- Evaluation of various requirements classification schemes that are common to the systems engineering and product development communities.
- Prototyping and evaluation of viewpoints enabled by AD/MBSE; clarification of the applicability and benefits associated with new viewpoints.
- Methods for "keeping alive" multiple solution alternatives through multiple layers of design decomposition and the impact of such practices on viewpoints, numbering schemes, naming conventions, etc.

- Redefinition of AD's zig-zagging process to account for synchronization of the physical architecture model, functional architecture model and mathematical system performance model at each branch of the design decomposition.
- Formalizing and refining heuristics for defining derived requirements from a chosen solution alternative.
- Methods for incorporating state/mode and state/mode transition models into the design process.
- Methods for reducing the redundancy between system risks, failure modes and hazard analysis models.
- Investigation of the potential for simplifying the information metamodel by eliminating the CN-FR-DP-PV domains as requirement classes and replacing them as subclasses that express the context of FRs and FRm's. In general, separate class from context on all entities.

References

1. Suh, N.P.: The Principles of Design. Oxford University Press, Oxford (1990)
2. Thompson, M.K.: A classification of procedural errors in the definition of functional requirements in axiomatic design theory. In: Thompson, M.K. (ed.) Proceedings of the 7th International Conference on Axiomatic Design (Chapter ICAD-2013-16). ICAD (2013)
3. Kulak, O., Cebi, S., Kahraman, C.: Applications of axiomatic design principles: a literature review, Expert Syst. Appl. **37**, 6705–6717 (2010)
4. INCOSE: Systems engineering vision 2035 – engineering solutions for a better world (2021). https://www.incose.org/about-systems-engineering/se-vision-2035
5. INCOSE: Systems Engineering Handbook: A Guide for System Life Cycle Processes and Activities, Version 4.0. Wiley, New York (2015)
6. ISO/IEC/IEEE: Standard 15288. Systems and Software Engineering – System Life Cycle Processes. International Organization for Standardization, Geneva, Switzerland (2015)
7. ISO/IEC/IEEE: Standard 42010. Systems and Software Engineering – Recommended Practice for Architectural Descriptions of Software-Intensive Systems. International Organization for Standardization, Geneva, Switzerland (2011)
8. Cochran, D.S., Smith, J., Fitch, J.A.: MSDD 10.0: a design pattern for sustainable manufacturing systems. J. Prod. Manuf. Res. **10**, 964–989 (2022)
9. Rauch, E., Vickery, A.R.: Systematic analysis of needs and requirements for the design of smart manufacturing systems in SMEs. J. Comput. Des. Eng. **7**(2), 129–144 (2020)
10. Gebala, D., Suh, N.P.: An application of axiomatic design. Res. Eng. Des. Theory Appl. Concurr. Eng. **3**, 149–162 (1992)

Model-Based Systems Engineering in Smart Manufacturing - Future Trends Toward Sustainability

Ali Asghar Batalebl[1(✉)], Erwin Rauch[1], and David S. Cochran[2]

[1] Industrial Engineering and Automation (IEA), Faculty of Engineering, Free University of Bolzano-Bozen, Bolzano, Italy
Aliasghar.Batalebl@unibz.it

[2] Center of Excellence in Systems Engineering, Purdue University Fort Wayne, 2101 East Coliseum Blvd., Fort Wayne, IN 46805, USA

Abstract. Creating a more agile and productive industrial base using intelligent and emerging technologies relying on systems engineering is not only a progress key for any entity regardless of its size but also is the durability factor in the nowadays competitive industrial environment. Therefore, to promote the traditional document-based information exchange, serial design procedures, and single disciplinary analysis, systems of systems thinking would need to be expanded in organizations. This strategy should be implemented from the beginning of the project definition and stakeholder needs to the entire product development process throughout the V-diagram and persists throughout the product's operational life. Despite recently developed tools and significant growth and movement from the level of Industry 1.0 to 4.0 toward smart manufacturing, many researchers are still trying to push the boundaries of manufacturing. However, companies have faced many challenges in addressing such technology growth trends in their long-term enterprise strategy and design. Although the Model-based Systems Engineering (MBSE) tools by visual modeling of the communication of the information alleviate some difficulties for companies in many respects, bridging between systems-level decisions, design requirements, and sustainability dimensions through connecting MBSE and Multidisciplinary Systems Design Optimization (MSDO) can promise strategic advantages and innovation. Providing such a combined tool with visual modeling helps manufacturers trace the effect of their decisions and achieve sustainable manufacturing goals faster. By reviewing the application of MBSE in smart factories, this paper will provide future research fields for further development to enable sustainable innovation in manufacturing and factory design.

Keywords: Smart Manufacturing · Sustainable Manufacturing · Systems Thinking · Model-based Systems Engineering · Multidisciplinary Systems Design Optimization

E. Puik et al. (Eds.): ICAD 2023, LNNS 849, pp. 298–311, 2024.
https://doi.org/10.1007/978-3-031-49920-3_20

1 Introduction

Despite supply crosswinds and instability of the marketplace, the manufacturing industry strongly persists in surpassing the expectations of previous years [1]. Leading companies strive to create a digital environment that allows them to achieve dimensions of sustainability space (i.e. economic, environmental, and social sustainability) as much as possible through a concurrent procedure. On the other hand, the level of innovation maturity within factories has a remarkable impact on their competitiveness and profitability. Therefore, creating a more agile and productive industrial base using intelligent and emerging technologies relying on systems engineering is not only a progress key for any entity regardless of its size but also is the durability factor in the nowadays competitive industrial environment. As the level of digital transformation defines the level of innovation maturity companies have achieved, leaders should leverage digital technologies, adopt intelligent strategies for future products, and drive whenever possible toward sustainability [2, 3]. In this respect, this paper explores that Model-based Systems Engineering (MBSE) relying on systems of systems thinking strategy should be at the top of the agenda for many companies that try to survive and improve productivity. Therefore, the triangle of intelligent manufacturing should cover Innovation, Digitalization, and System Thinking, to companies keep pace with technology (Fig. 1).

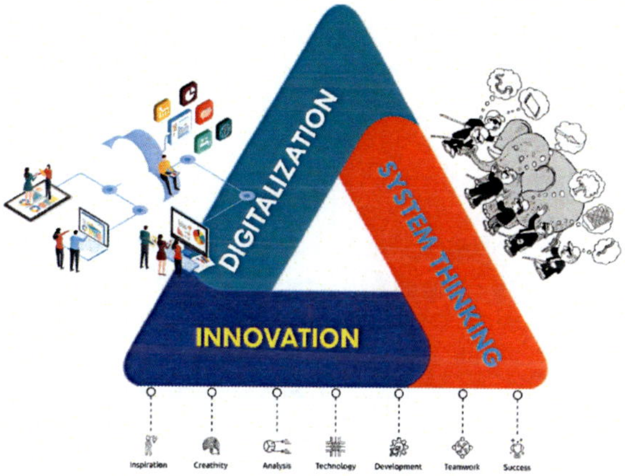

Fig. 1. The triangle of systems engineering induced intelligent manufacturing.

On the other hand, with the substantial increase in demand for personalized products, manufacturing architecture has become extremely complex both in terms of concept and structure. Designing such a factory deals with several internal and external collaborations at the system level as well as mechanical, electrical, automation, and other relevant fields at the sub-system level, which further reveals the need to consider a Multidisciplinary Systems Design and Optimization (MSDO) framework. In the last decades, lots of research addressed the topics of digital twins (DT) [4, 5], MBSE [6, 7], and MSDO [8, 9] separately. Also, several literature reviews have been done on each topic [10, 11]. Despite many followers in these fields, today we need all of them in one framework.

While MBSE is expanding in the manufacturing industry, new methodologies based on Systems Engineering (SE) concepts have been developed to adapt the manufacturing procedure to new demands, in which the system's architecture and requirements are followed concurrently through the product life cycle from design and development to manufacturing and retirement/replacement. These new methodologies which call the agile approach and rely on MBSE and MSDO have been trying to bridge the gap between mentioned critical subjects [12]. The ability of agile methodologies as practical improvement in engineering and other fields has been demonstrated in many companies [13].

However, one of the vital challenges in current manufacturing processes is that DT, MBSE, and MSDO are performed as three different activity streams, based on separate tools and requiring specific expertise. In the future industry should benefit from the capabilities of all three SE, DT, and MSDO methodologies in dealing with complex manufacturing problems.

To address these issues, in the following, a brief overview of key parts of this paper including the industrial revolution, MBSE, and MSDO presented. Then, in the discussion section, some research initiatives with a focus on bridging between MBSE and MSDO are highlighted. Finally, the paper is ended up with an outlook on future directions within manufacturing toward sustainability.

2 Design and Optimization Methodologies

The topic of optimality and productivity in the presence of variation and uncertainty that are inevitable parts of any manufacturing and assembly of complex real-world systems is not a new one. It goes back to Six Sigma and the reliability concepts in the early 1990s when William Smith, a reliability engineer at Motorola, proposed the concept of Six Sigma to alleviate the high failure rate of Motorola's products. After that, many companies like Motorola, General Electric, Allied Signal, Black and Decker, Honeywell, ABB, and Bombardier proclaimed that they had impressive business performance achieved through this strategy [14]. Design for Six Sigma (DFSS) behaves as a management strategy that helps companies provide an efficient roadmap to improve manufacturing procedures to eliminate defects in products, processes, and services. According to DFSS, many procedures such as DMAIC (i.e. Define, Measure, Analyze, Improve, and Control) or DMADV (comprising Define, Measure, Analyze, Design, and Verify) had emerged to help certify the final quality of the product [15]. The role and situation of considering DFSS and DMAIC/DMADV in the product life cycle are presented in Fig. 2.

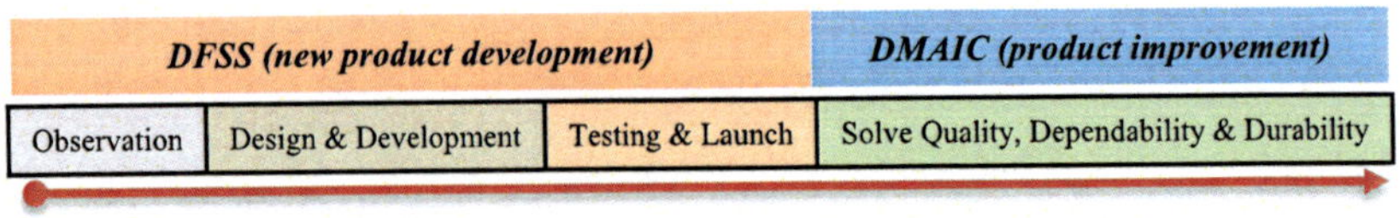

Fig. 2. DFSS and DMAIC/DMADV in the product life cycle

However, the traditional optimal design process which is based on a sequential approach although has its advantages, it does not include online interdisciplinary interactions and finally leads to local optimality and complexity in the decision-making as well as a gap between product design and prototype manufacturing [16]. Despite these challenges and as the traditional method is time-consuming with inevitable iterations on the whole design and development process, the engineering community had been needed a paradigm shift in design methodology for complex engineering systems. To overcome or at least alleviate those problems, new methodologies known as Concurrent Engineering (CE) and Multidisciplinary Design Optimization (MDO) had been developed which are relying on parallelization. A schematic comparison between the traditional and CE methods is illustrated in Fig. 3 [17]. CE aims to provide a balanced design through full and formal multi-disciplinary integration and optimization concurrently in all disciplines [18]. Also, one of the popular definitions of MDO is "*a methodology for the optimal design of complex engineering systems and subsystems that coherently exploits the synergism of mutually interacting phenomena using high fidelity analysis with formal optimization*" [19]. Publishing lots of literature in these fields demonstrates the successful application of CE and MDO on various engineering projects from design to manufacturing in the last decades [20–22].

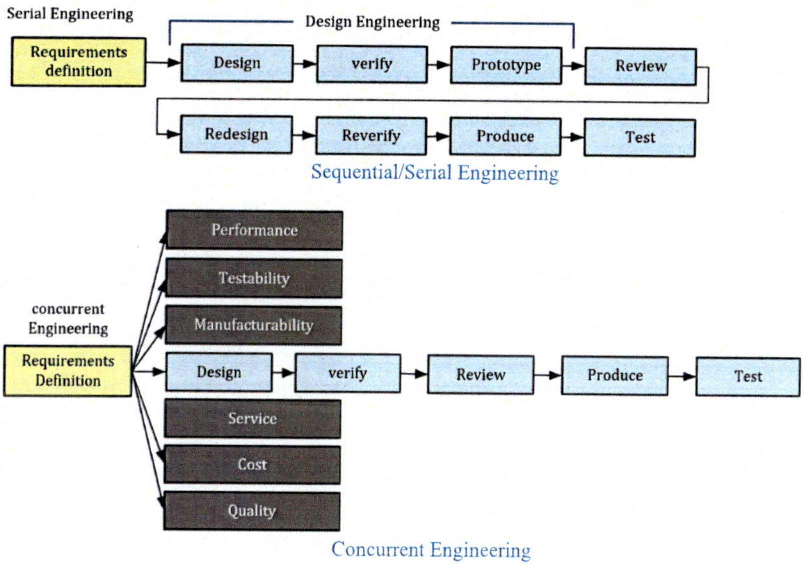

Fig. 3. Traditional Sequential Engineering versus Concurrent Engineering [17]

Furthermore, although real-world manufacturing mainly suffers from the various system and sub-system requirements, the curse of dimensionality regarding considering disciplines, and the multi-disciplinary nature of the involved disciplines, these issues may be intensified by considering different sources of uncertainties in the product lifecycle realization. The uncertainty sources can be divided into the following general categories: mission, design, manufacturing, and operation [23] (see Fig. 4). To alleviate such challenges, features like flexibility [24], modularity [25, 26], and automation [27] have been utilized within the manufacturing industry. Besides, Systems Modeling and

Simulation (SMS) through Uncertainty-based Design Optimization (UDO) methodologies like Robust Design Optimization (RDO) and Reliability-based Design Optimization (RBDO) are other major enablers for fulfilling system requirements and constraints in the presence of uncertainties. The RDO is a design methodology for achieving a product less sensitive to various uncertainties. Also, RBDO is a methodology to have an optimal product that fulfills a predetermined and acceptable level of failure [23].

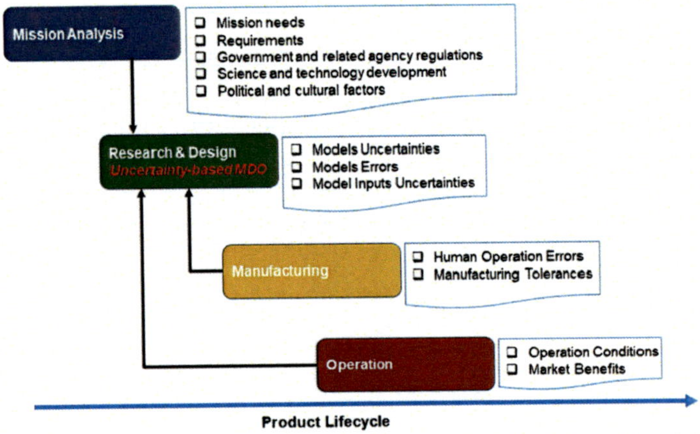

Fig. 4. Uncertainty sources in product life cycle [23]

The main challenges and future research in UDO fields of study have been addressed in [28]. By introducing computational burden as the main problem in applying UDO methods to real-world problems, new research fields like Surrogate-Assisted Optimization (SAO) and Evolution Control Strategies (ECS) as powerful paradigms have emerged over the last two decades [29–31].

Another design methodology that has been developed in parallel with the concepts of DFSS, CE, and MDO is Axiomatic Design (AD), which is based on deriving the Functional requirements (FRs) and related Design Parameters (DPs) [32]. DPs are the key solutions that have to logically satisfy the specified set of FRs. Although numerous research has been done on AD and its application in the design of manufacturing systems, some researchers are still working on both the theory and practical application aspects [33–37]. According to the basis of AD (Fig. 5), it models the interactions between FRs as what we want to achieve and the DPs as what physical implementation we choose to achieve the FRs [34].

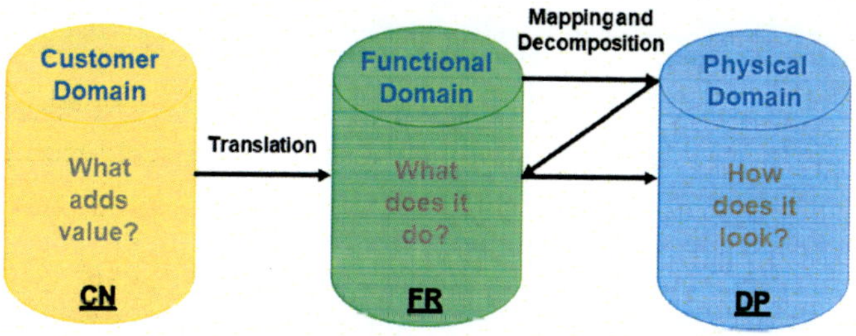

Fig. 5. The basis of AD [34]

3 Model-Based Systems Engineering

The traditional document-centric SE of real-world products always involves thousands of created and maintained documents meanwhile the product life cycle. Some of these documents include requirements specification, requirements traceability, design structure matrix, test scenarios and specifications, interface control documents, and so on. It is important to note that the information in these documents is not independent and in contrast, the change of information in any document needs to be traced and exchanged manually in all the other affected documents [38, 39].

In recent years, SE followed by Model-Based Systems Engineering (MBSE) has undergone major changes. The transition from traditional systems engineering to MBSE (i.e. document-centric to model-centric) is depicted in Fig. 6 [38]. As an alternative to the traditional document-based information exchange, MBSE has received more popularity within the industry. In MBSE, visual modeling of communication has made it easier to trace requirements and stakeholder needs. According to the *SMS_ThinkTank*™ [40], a global resource and leader in systems modeling and simulation, the best definition for MBSE is provided as follows: "*MBSE is the formalized application of modeling to support system requirements, design, analysis, verification, and validation activities*

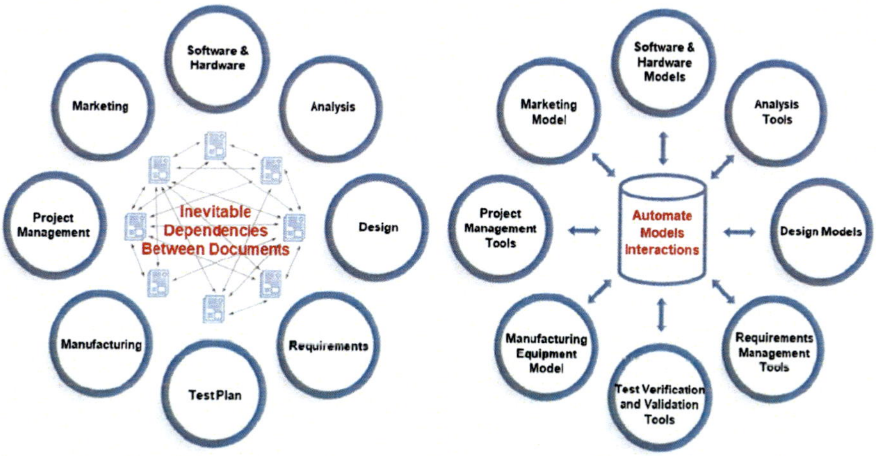

Fig. 6. Traditional SE (left) in comparison with MBSE (right) [38]

beginning in the conceptual design phase and continuing throughout development and later life cycle phases".

The International Council on Systems Engineering (INCOSE) has pointed out in its vision 2025 report that MBSE has provided a basis for profitable success in today and future industries [41]. In this way and relying on the traditional understanding of the lifecycle of a product or process (i.e. V-diagram), researchers develop a lot of frameworks based on the System Model Language (SysML). The Cameo [42], GENESYS [43], Modelica [44], and Capella [45] are some of the common MBSE tools. Therefore, the importance of diving into MBSE, whenever possible, is clear to the overall engineering community as asking for it. On the top, we have industrial companies like Boeing [46] and Airbus [47] which are pushing more and more MBSE and virtual integration as the way to interact with their suppliers in the future (Fig. 7).

While MBSE has progressively been used in industrial applications, many open issues still confine the execution of MBSE [48]. The teamwork nature of the MBSE, lack of knowledge of experts to work with relevant tools, information security, resistance to organizational culture change, and refrain from investing in new methods/software are some of these barriers. In any case, although companies are compelled to move in this direction, their steps depend on their organizational capabilities and are different for small to medium-sized enterprises.

4 Smart and Sustainable Manufacturing

The industry is undergoing an era of digital transformation. Since the dawn of the industrial age, despite recently developed tools and significant growth and movement from the level of Industry 1.0 to 4.0 toward smart manufacturing to achieve higher levels of innovation maturity, manufacturers have been evolving and adapting in response to new technological innovations and changing market demands. Also, many researchers are still trying to push the boundaries of manufacturing [49, 50].

During the last decade, the engineering community relying on Industry 4.0 technologies and specifically digital twin technology tries to connect systems and operations to achieve smart manufacturing. To attain this, virtual capabilities are required at many stages of the product life cycle. Therefore, the main transformative aspect of the digital twin is to position the DT in the SE life cycle by expanding the traditional understanding of the V-diagram from a sequential to an iterative view (like a W-diagram) at every stage based on a closed-loop process through including a specific virtual prototyping stage. The virtual stage is then used as the basis of DT in the second cycle (Fig. 8) [51].

Furthermore, both DT and the physical could be sustained by a linked MBSE tool, which supports data and workflow. Such a configuration guarantees bidirectional information transmission between the DT and the physical twin by serving MBSE as a digital thread [6, 51]. It is expected that digitalization become a distinguished capability within MBSE because of its four different levels of execution in the products life cycle (i.e. Pre-DT, DT, Adaptive DT, and Intelligent DT) while at the same time connecting cutting-edge technologies to MBSE push it toward new features in smart manufacturing to penetrate impressively in various industries.

On the other hand, in recent years, various sources forced the industry to move toward a new step of evolution, the step that sustainability is its core [52, 53]. It could

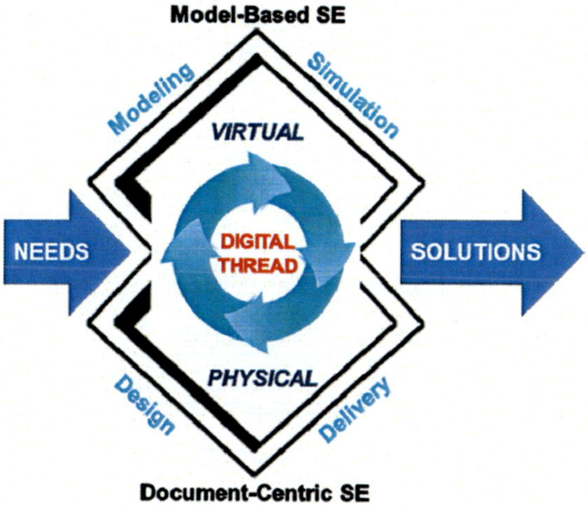

Boeing MBE Diamond [46]

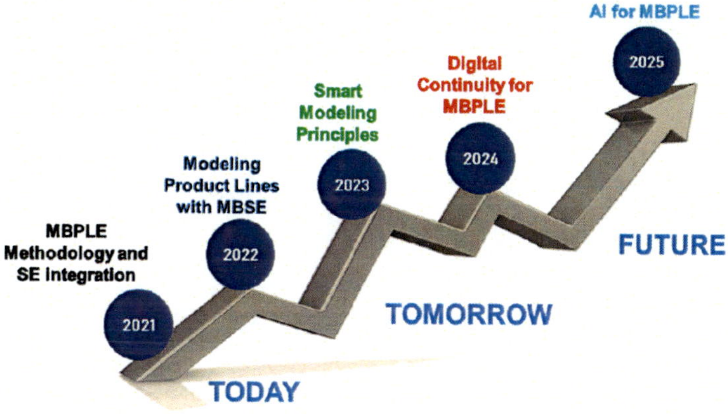

Airbus MBPLE Plan [47]

Fig. 7. Boeing and Airbus industries' MBSE and virtual integration Plan

be seen that this major factor with three dimensions of economic, environmental, and social sustainability (also known as Triple Bottom Line), not only is a multidisciplinary problem but also could be considered as a multi-objective optimization problem. When we consider different weights for the environmental, social, and economic, it deals with weak sustainability and aims to balance them. In contrast, strong sustainability focusing on the whole system dealt with the three subjects as nested and admits different weightings for the dimensions [54, 55]. Therefore, it is better to seek Pareto solutions in dealing with such problems to represent the best feasible design points that can be achieved (Fig. 9) [56]. It seems that sustainability is more of an organizational culture

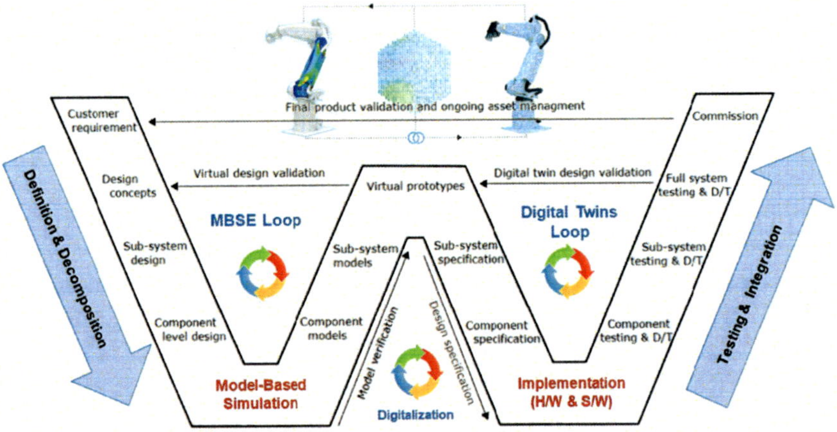

Fig. 8. Shifting from V-diagram to W-diagram toward digitalization [51]

than a structure or goal. Therefore, since sustainability is considered a major competition criterion between companies today, a reorientation of the manufacturing society is necessary, utilizing knowledge and values to generate notable changes.

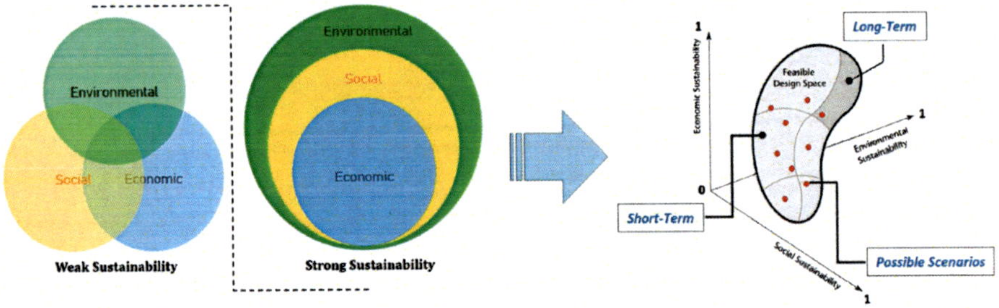

Fig. 9. Sustainability as a multi-objective optimization problem [55, 56]

5 Bridging the Gaps Toward Smart and Sustainable Manufacturing Systems Design

As aforementioned, by facing ever-increasing complexity in industrial systems and marketplace demands, and an uncertain environment, organizations have already begun transitioning from traditional SE to MBSE and digitalization to achieve agile procedures. Therefore, it is clear that this transition is no anymore a plus, it is a must. But, despite some successes, this shift is a challenging and time-consuming process. There is no straightforward and unique path to attain this. It depends on many factors, culture, facilities, maturity, experts' knowledge, level of communication and interactions, managers' and leaders' adoption strategy, and way of thinking.

Although different approaches like MSDO, MBSE, and DT have been taken and developed, it's time to bring them into an integrated framework. Currently, companies such as MathWorks [57] and GENESYS [58] are trying to provide the platform for this integration with the possibility of communicating different software on the MBSE platform. Furthermore, according to INCOSE Vision 2035 [59], a family of unified, integrated MBSE-SMS frameworks develops by 2035. They will leverage MSDO methodology and DTs and would fully integrate with the digital thread foundation to provide life cycle management systems.

For the practical integration of MBSE tools with MSDO and achieve sustainability in smart manufacturing, which is a multi-objective as well as multidisciplinary problem, the Free University of Bolzano and Purdue University Fort Wayne are starting a research project entitled "SFDD - Sustainable Factory Design Decomposition". Using MBSE approaches along with MSDO could alleviate difficulties in dealing with such multi-objective complex systems. MBSE is taking over the role of a formalized and digitally supported application of modeling to derive system requirements, evaluate system architectures, and analyze, verify, and validate design activities. Whereas MSDO focuses on numerical optimization (e.g. MATLAB-Simulink) for the design of systems that involve several disciplines or subsystems with multiple and interdisciplinary objectives and interactions due to sustainability goals. Providing such a combined tool relying upon visual modeling helps factory designers and stakeholders easier follow up on the effect of their decisions and achieve sustainable manufacturing goals more easily and faster.

6 Conclusion and Outlook

This paper proposed a general review of the field of systems engineering from the systems design and optimization view to digitalization and digital twin perspective. In this regard, after a brief introduction and illustration of the intelligent manufacturing triangle, the concept of Six Sigma and its procedures to increase reliability in the product life cycle is explained. In the next step, to find an alternative to traditional sequential design methodologies, we described the emerged CE and MDO approaches. To include different sources of uncertainties in design and attain feasible manufacturing and decrease the gap between design to practice, RDO and RBDO methods are explained. Meanwhile, SAO approaches based on machine learning and artificial intelligence have been developed to alleviate complexities with the computational burden of the mentioned design methodologies. Parallel to design and optimization, some research has been focused on methods like AD to work breakdown structure to clarify the problem definition in different levels of the system providing trees of information from stakeholder needs to requirements and physical solutions to find alternatives to make better decisions. With technology advancements and a competitive environment toward innovation and digitization, organization and Small and medium-sized enterprises have to change their thinking culture. MBSE is the master key and the best tool for the transition from traditional document-based information exchange space to digitalization in the least possible time.

Although different software has been developed in each era and now each is functional, reliable, and mature software separately, there is still a gap between their practical combination from the system of systems perspective and not a single-disciplinary view [60]. Therefore, as near-future research in the SFDD project, we will try to accelerate manufacturing factories' transition towards both profitable and ecologically and socially sustainable factories by combining SE, MBSE, and MDO. To achieve this goal, the research team will collect direct data regarding needs through semi-structured interviews asking users and stakeholders of factories (owner, manager, production engineers, associations, innovation clusters) and evaluate the relevance of collected data in focus group workshops. Afterward, AD will be used for translating these needs into technically sound functional requirements (FRs). Collected user needs containing non-solution-neutral data will undergo an AD reverse engineering approach for retrieving the underlying FRs. Candidate design parameters (physical solutions) (DPs) will be derived for each FR and metrics will be identified to make candidate solutions measurable and comparable. MBSE tools will be applied for supporting the modeling of requirements, design, analysis, verification, and validation. The full set of systems requirements and interactions will be evaluated afterward through MDO by establishing the mathematical model for each subsystem and using optimization algorithms to achieve finally an optimized design. Based on the Manufacturing System Design Decomposition (MSDD) approach [61] an evaluation tool will finally be developed to create a hands-on assessment tool evaluating the sustainability status of manufacturing companies and to guide factory and process designers in making their factories greener and socially sustainable.

Acknowledgments. This research is part of the "SFDD - Sustainable Factory Design Decomposition" project and has received funding from the Autonomous Province of Bolzano, (grant number TN221R).

References

1. 2023 Manufacturing Industry Outlook. https://www.deloitte.com/global/en/Industries/industrial-construction/analysis/gx-manufacturing-industry-outlook.html
2. Qi, Q., Tao, F., Zuo, Y., Zhao, D.: Digital twin service towards smart manufacturing. Procedia CIRP **72**, 237–242 (2018)
3. Kossiakoff, A., et al.: Systems Engineering Principles and Practice, vol. 83. Wiley, New York (2011)
4. Li, L., Lei, B., Mao, C.: Digital twin in smart manufacturing. J. Ind. Inf. Integr. **26**, 100289 (2022)
5. Shao, G., et al.: Digital twin for smart manufacturing: the simulation aspect. In: 2019 Winter Simulation Conference (WSC). IEEE (2019)
6. Madni, A.M., Madni, C.C., Lucero, S.D.: Leveraging digital twin technology in model-based systems engineering. Systems **7**(1), 7 (2019)
7. Cha, S., Vogel-Heuser, B., Fischer, J.: Analysis of metamodels for model-based production automation system engineering. IET Collaborative Intell. Manuf. **2**(2), 45–55 (2020)
8. Liu, G., Xiong, Y., Rosen, D.W.: Multidisciplinary design optimization in design for additive manufacturing. J. Comput. Des. Eng. **9**(1), 128–143 (2022)

9. Butkewitsch, S., Filho, H.M.: A systems engineering-driven decomposition approach for large-scale industrial decision-making processes. In: Smart Manufacturing, pp. 261–282. Elsevier (2020)
10. Liu, M., Fang, S., Dong, H., Cunzhi, X.: Review of digital twin about concepts, technologies, and industrial applications. J. Manuf. Syst. **58**, 346–361 (2021)
11. Son, Y.H., Kim, G.Y., Kim, H.C., Jun, C., Noh, S.D.: Past, present, and future research of digital twin for smart manufacturing. J. Comput. Des. Eng. **9**(1), 1–23 (2022)
12. Wu, Y., Zhou, L., Zheng, P., Sun, Y., Zhang, K.: A digital twin-based multidisciplinary collaborative design approach for complex engineering product development. Adv. Eng. Inform. **52**, 101635 (2022)
13. Komus, M., Ayelty, K.: Status quo agile 2019/2020. In: 4th International Survey Benefits and Challenges of (Scaled) Agile Approaches (2020)
14. Kumar, U.D., Crocker, J., Chitra, T., Saranga, H.: Reliability and Six Sigma Management. Springer, Heidelberg (2006). https://doi.org/10.1007/0-387-30256-5_10
15. Selvi, K., Majumdar, R.: Six sigma-overview of DMAIC and DMADV. Int. J. Innov. Sci. Mod. Eng. **2**(5), 16–19 (2014)
16. Walden, D., Roedler, G., Forsberg, K., Hamelin, R., Shortell, T.M.: Systems Engineering Handbook: A Guide for System Life Cycle Processes and Activities. Wiley, New York (2015)
17. Sofuoğlu, E.: Different Approaches to Concurrent Engineering. Research Gate Publication, New York (2011)
18. Sudhakar, K., Mujumdar, P.M., Isaacs, A.: Multidisciplinary Design Optimization – A Paradigm Shift in Design Methodology for Complex Engineering Systems. Centre for Aerospace Systems, Design & Engineering Department of Aerospace Engineering, IIT Bombay, 6 December 2002
19. Martins, J.R.R.A., Lambe, A.B.: Multidisciplinary design optimization: a survey of architectures. AIAA J. **51**(9), 2049–2075 (2013)
20. Stjepandić, J., Wognum, N., Verhagen, W.J.C.: Concurrent engineering in the 21st century. In: Concurrent Engineering in the 21st Century: Foundations, Developments, and Challenges (2015). https://doi.org/10.1007/978-3-319-13776-6
21. Ammar, R., Hammadi, M., Choley, J.-Y., Barkallah, M., Louati, J., Haddar, M.: The design and modeling of an optimized mechatronic system using a set based concurrent engineering. In: Fakhfakh, T., Karra, C., Bouaziz, S., Chaari, F., Haddar, M. (eds.) ICAV 2018. ACM, vol. 13, pp. 111–120. Springer, Cham (2019). https://doi.org/10.1007/978-3-319-94616-0_11
22. Yao, X., Moon, S.K., Bi, G.: Multidisciplinary design optimization to identify additive manufacturing resources in customized product development. J. Comput. Des. Eng. **4**(2), 131–142 (2017)
23. Yao, W., Chen, X., Luo, W., Van Tooren, M., Guo, J.: Review of uncertainty-based multidisciplinary design optimization methods for aerospace vehicles. Prog. Aerosp. Sci. **47**(6), 450–479 (2011)
24. Tolio, T.: Design of Flexible Production Systems. Springer, Heidelberg (2008). https://doi.org/10.1007/978-3-540-85414-2
25. Shaik, A.M., Rao, V.V.S., Rao, Ch.: Development of modular manufacturing systems—a review. Int. J. Adv. Manuf. Technol. **76**(5), 789–802 (2015)
26. Rauch, E., Matt, D.T., Dallasega, P.: Mobile on-site factories—scalable and distributed manufacturing systems for the construction industry. In: 2015 International Conference on Industrial Engineering and Operations Management (IEOM), pp. 1–10. IEEE (2015)
27. Lu, Y., Xun, X., Wang, L.: Smart manufacturing process and system automation–a critical review of the standards and envisioned scenarios. J. Manuf. Syst. **56**, 312–325 (2020)
28. Viana, F.A., Simpson, T.W., Balabanov, V., Toropov, V.: Special section on multidisciplinary design optimization: metamodeling in multidisciplinary design optimization: how far have we really come? AIAA J. **52**(4), 670–690 (2014)

29. Tenne, Y., Goh, C.-K. (eds.): Computational Intelligence in Expensive Optimization Problems. ALO, vol. 2. Springer, Heidelberg (2010). https://doi.org/10.1007/978-3-642-10701-6
30. Roshanian, J., Bataleblu, A.A., Ebrahimi, M.: A novel evolution control strategy for surrogate-assisted design optimization. Struct. Multidiscip. Optim. **58**(3), 1255–1273 (2018)
31. Bataleblu, A.A.: Computational intelligence and its applications in uncertainty-based design optimization. In: Bridge Optimization-Inspection and Condition Monitoring. IntechOpen (2019)
32. Kulak, O., Cebi, S., Kahraman, C.: Applications of axiomatic design principles: a literature review. Expert Syst. Appl. **37**(9), 6705–6717 (2010)
33. Brown, C.A., Rauch, E.: Axiomatic design for creativity, sustainability, and industry 4.0. MATEC Web Conf. **301**, 00016 (2019). EDP Sciences
34. Cochran, D.S., Hendricks, S., Barnes, J., Bi, Z.: Extension of manufacturing system design decomposition to implement manufacturing systems that are sustainable. J. Manuf. Sci. Eng. **138**(10) (2016)
35. Gualtieri, L., Rauch, E., et al.: Application of Axiomatic Design for the design of a safe collaborative human-robot assembly workplace. MATEC Web Conf. **223**, 01003 (2018). EDP Sciences
36. Rauch, E., Matt, D.T.: Artificial intelligence in design: a look into the future of axiomatic design. In: Suh, N.P., Cavique, M., Foley, J.T. (eds.) Design Engineering and Science, pp. 585–603. Springer, Cham (2021). https://doi.org/10.1007/978-3-030-49232-8_21
37. Park, S., Yoon, J., An, H., Park, J., Park, G.-J.: Integration of axiomatic design and design structure matrix for the modular design of automobile parts. Proc. Inst. Mech. Eng. Part B J. Eng. Manuf. **236**(3), 296–306 (2022)
38. Madni, A.M., Purohit, S.: Economic analysis of model-based systems engineering. Systems **7**(1), 12 (2019)
39. Purohit, S., Madni, A.M.: Towards making the business case for MBSE. In: Madni, A.M., et al. (eds.) Recent Trends and Advances in Model-Based Systems Engineering. Springer, Cham (2022). https://doi.org/10.1007/978-3-030-82083-1_28
40. Ladzinski, E.A., Tolle, D.: DIGITAL TWIN Its Role and Structure within a Modern Systems Engineering Approach. SMS_ThinkTank™ (2019)
41. INCOSE. A World in Motion, Systems Engineering Vision 2025 (2014)
42. No Magic. Cameo Systems Modeler User Guide. v. 18.1. No Magic (2015)
43. Vitech Corporation. GENESYS System Definition Guide. v. 6.0. Vitech Corporation, Blacksburg, VA (2018)
44. Modelica Association. Modelica - A Unified Object-Oriented Language for Systems Modeling. v. 3.4. Modelica Association (2017)
45. Roques P.: Systems Architecture Modeling with the Arcadia Method: A Practical Guide to Capella. Elsevier, New York (2017)
46. Seal, D.: The System Engineering 'V' – Is it Still Relevant in the Digital Age? Boeing Company, Global Product Data Interoperability Summit, Presentation (2018)
47. Ferrogalini, M., Forlingieri, M.: Transforming Airbus Through Product Line Engineering - Foundation and Real-Life Implementation, Airbus Company (2020)
48. Tschirner, C., et al.: Tailoring model-based systems engineering concepts for industrial application. In: 2015 Annual IEEE Systems Conference (SysCon) Proceedings. IEEE (2015)
49. Herrmann, F.: The smart factory and its risks. Systems **6**(4), 38 (2018)
50. Wagg, D.J., Worden, K., Barthorpe, R.J., Gardner, P.: Digital twins: state-of-the-art and future directions for modeling and simulation in engineering dynamics applications. ASCE-ASME J. Risk Uncertain. Eng. Syst. Part B Mech. Eng. **6**(3), 030901 (2020)

51. Schluse, M., Atorf, L., Rossmann, J.: Experimentable digital twins for model-based systems engineering and simulation-based development. In: Annual IEEE International Systems Conference (SYSCON), pp. 1–8. IEEE (2017)
52. Rauch, E., Cochran, D.S.: Sustainable introduction of Industry 4.0: a systematic literature review. In: Proceedings of the International Conference on Industrial Engineering and Operations Management Rome, Italy (2021)
53. Furstenau, L.B., Sott, M.K., Kipper, L.M., Machado, E.L., et al.: Link between sustainability and industry 4.0: trends, challenges and new perspectives. IEEE Access **8**, 140079–140096 (2020)
54. Morandín-Ahuerma, I., et al.: Socio–ecosystemic sustainability. Sustainability **11**(12), 3354 (2019)
55. Rupprecht, C.D.D., et al.: Multispecies sustainability. Global Sustain. **3**, E34 (2020)
56. Mattson, C.A., Pack, A.T., Lofthouse, V., Bhamra, T.: Using a product's sustainability space as a design exploration tool. Des. Sci. **5**, E1 (2019)
57. https://figes.com.tr/en/makaleler/matlab-and-simulink-for-model-based-systems-engineering
58. https://www.zuken.com/en/resource/vitech-improves-mbse-performance-and-usability-in-genesys-2022/
59. INCOSE: Systems Engineering Vision 2035, engineering solutions for a better world (2021)
60. Henderson, K., Salado, A.: Value and benefits of model-based systems engineering (MBSE): evidence from the literature. Syst. Eng. **24**(1), 51–66 (2021)
61. Cochran, D.S., Arinez, J.F., Duda, J.W., Linck, J.: A decomposition approach for manufacturing system design. J. Manuf. Syst. **20**(6), 371–389 (2001)

Author Index

E. Puik et al. (Eds.): ICAD 2023, LNNS 849, pp. 313–314, 2024.
https://doi.org/10.1007/978-3-031-49920-3